FOREWORD

W9-AAI-651

Aviation Weather Services, Advisory Circular 00-45F, is published jointly by the National Weather Service (NWS) and the Federal Aviation Administration (FAA). This publication supplements its companion manual Aviation Weather, Advisory Circular 00-6A, which documents weather theory and its application to the aviation community.

This advisory circular, AC 00-45F, explains U.S. aviation weather products and services. It details the interpretation and application of advisories, coded weather reports, forecasts, observed and prognostic weather charts, and radar and satellite imagery. Product examples and explanations are taken primarily from the Aviation Weather Center's Aviation Digital Data Service website (http://adds.aviationweather.noaa.gov/).

The AC 00-45F was written primarily by Robert A. Prentice and Douglas D. Streu, and edited by Cynthia Abelman. Additional contributors include Raymond Tanabe, Larry Burch, Duane Carpenter, Kathleen Schlachter, Lisa Glikbarg, Celia Miner, Jon Osterberg, Ronald Olson and Richard Akers.

An on-line version of this document, which includes links to additional information, can be found at:

http://www.srh.noaa.gov/faa/

Comments and suggestions for improving this publication are encouraged and should be directed to:

National Weather Service Coordinator, W/SR64
Federal Aviation Administration, AMA-514
Mike Monroney Aeronautical Center
P.O. Box 25082
Oklahoma City, OK 73125-0082

Or via e-mail at:

m9-amc-ama-pwb@faa.gov

Advisory Circular, AC 00-45F, supersedes AC 00-45E, Aviation Weather Services, revised 1999.

Visit **www.asa2fly.com/ac0045** to download an electronic copy
of this book. The PDF includes hyperlinks throughout for clickable
access to definitions, weather resources, and other
FAA publications for additional information on a given topic.

1 AVIATION WEATHER SERVICE PROGRAM

The aviation weather service program is a joint effort of the National Weather Service (NWS), the Federal Aviation Administration (FAA), the Department of Defense (DOD), and other aviation-oriented groups and individuals. This section discusses the civilian agencies of the U.S. Government and their observation, communication and forecast services to the aviation community.

1.1 National Oceanic and Atmospheric Administration (NOAA)

The National Oceanic and Atmospheric Administration (NOAA) is an agency of the Department of Commerce (DOC). NOAA conducts research and gathers data about the global oceans, atmosphere, space, and sun, and applies this knowledge to science and service which touches the lives of all Americans. Among its six major divisions are the National Environmental Satellite Data and Information Service (NESDIS) and the NWS.

1.1.1 National Environmental Satellite Data and Information Service (NESDIS)

The National Environmental Satellite Data and Information Service (NESDIS) manages the U.S. civil operational remote-sensing satellite systems, as well as other global information for meteorology, oceanography, solid-earth geophysics, and solar-terrestrial sciences. NESDIS provides this data to NWS meteorologists and a wide range of other users for operational weather forecasting.

1.1.1.1 Satellite Analysis Branch (SAB)

NESDIS' Satellite Analysis Branch (SAB) serves as the operational focal point for real-time imagery products and multi-disciplinary environmental analyses. The SAB's primary mission is to support disaster mitigation and warning services for U.S. Federal agencies and the international community. Routine environmental analyses are provided to forecasters and other environmental users and used in the numerical models of the NWS. The SAB schedules and distributes real-time satellite imagery products from global geostationary and polar-orbiting satellites to environmental users. The SAB coordinates the satellite and other information for the NOAA Volcanic Hazards Alert program under an agreement with the FAA and works with the NWS as part of the Washington, D.C. Volcanic Ash Advisory Center (VAAC).

1.1.2 National Weather Service (NWS)

The National Weather Service (NWS) provides weather data, forecasts and warnings for the United States, its territories, adjacent waters and ocean areas for the protection of life and property and the enhancement of the national economy. NWS data and products form a national information database and infrastructure that can be used by other government agencies, the private sector, the public and the global community. The following is a description of NWS offices associated with aviation weather.

1.1.2.1 National Centers for Environmental Prediction (NCEP)

The National Centers for Environmental Prediction (NCEP) is where virtually all global meteorological data is collected and analyzed. NCEP then provides a wide variety of national and international weather guidance products to NWS field offices, government agencies, emergency managers, private sector meteorologists, and meteorological organizations and societies throughout the world. NCEP is a critical resource in national and global weather prediction and is the starting point for nearly all weather forecasts in the U. S.

NCEP is comprised of nine distinct centers and the Office of the Director. Each center has its own specific mission. The following NCEP centers provide aviation weather products and services:

1.1.2.1.1 NCEP Central Operations (NCO)
NCEP's Central Operations (NCO) in Camp Springs, Maryland, sustains and executes the operational suite of the numerical analysis and forecast models and prepares NCEP products for dissemination. It also links all nine of the national centers together via computer and communications-related services.

1.1.2.1.2 Aviation Weather Center (AWC)
The Aviation Weather Center (AWC), a Meteorological Watch Office (MWO) for the International Civil Aviation Organization (ICAO), is located in Kansas City, Missouri. The AWC issues the following products in support of FAA air traffic controllers and the National Airspace System (NAS): Airman's Meteorological Information (AIRMETs), Significant Meteorological Information (SIGMETs), Convective SIGMETs, Area Forecasts (FAs), Significant Weather Prognostic Charts (low, middle, and high), Collaborative Convective Forecast Product (CCFP), National Convective Weather Forecast (NCWF), Current Icing Product (CIP), and Forecast Icing Potential (FIP).

1.1.2.1.3 Hydrometeorological Prediction Center (HPC)
The Hydrometeorological Prediction Center (HPC) in Camp Springs, Maryland, provides analysis and forecast products specializing in quantitative precipitation forecasts to five days, weather forecast guidance to seven days, real-time weather model diagnostics discussions and surface pressure and frontal analyses.

1.1.2.1.4 Storm Prediction Center (SPC)
The Storm Prediction Center (SPC) in Norman, Oklahoma, provides tornado and severe weather watches for the contiguous U. S. along with a suite of hazardous weather forecasts including the Alert Severe Weather Watch Bulletins and mesoscale guidance products

1.1.2.1.5 Tropical Prediction Center (TPC)
The Tropical Prediction Center (TPC) in Miami, Florida, provides official NWS forecasts of the movement and strength of tropical weather systems and issues the appropriate watches and warnings for the contiguous U.S. and surrounding areas. It also issues a suite of marine products covering the tropical Atlantic, Caribbean, Gulf of Mexico, and tropical eastern Pacific.

1.1.2.2 Alaskan Aviation Weather Unit (AAWU)
The Alaskan Aviation Weather Unit (AAWU), located in Anchorage, Alaska, is a MWO for the ICAO. The AAWU is responsible for the entire Anchorage Flight Information Region (FIR). They issue the following products for the airspace over Alaska and adjacent coastal waters: AIRMETs, SIGMETs, FAs, Graphic Area Forecasts, and Significant Weather Prognostic Charts (Low- and Mid-level – below flight level (FL) 250).

The AAWU is also designated as the Anchorage Volcanic Ash Advisory Center (VAAC). The VAAC area of responsibility includes the Anchorage FIR and Far Eastern Russia and is responsible for the issuance of Volcanic Ash Advisories (FVs).

1.1.2.3 Center Weather Service Unit (CWSU)

Center Weather Service Units (CWSUs) are units of NWS meteorologists under contract with the FAA that are stationed at and support the FAA's Air Route Traffic Control Centers (ARTCC).

CWSUs provide timely weather consultation, forecasts, and advice to managers within ARTCCs and to other supported FAA facilities. This information is based on monitoring, analysis, and interpretation of real-time weather data at the ARTCC through the use of all available data sources including radar, satellite, Pilot Weather Reports (PIREPs), and various NWS products such as Terminal Aerodrome Forecasts (TAFs), FAs, and inflight advisories.

Special emphasis is given to those weather conditions hazardous to aviation or which would impede the flow of air traffic within the NAS. Rerouting of aircraft around hazardous weather is based largely on forecasts provided by the CWSU meteorologist. They issue the following products in support of their respective ARTCC: Center Weather Advisories (CWA) and Meteorological Impact Statements (MIS).

1.1.2.4 Weather Forecast Office (WFO)

A NWS Weather Forecast Office (WFO) is a multi-purpose, local weather forecast center that produces, among its suite of services, aviation-related products. In support of aviation, WFOs issue Terminal Aerodrome Forecasts (TAFs), Transcribed Weather En Route Broadcasts (TWEBs) forecasts, with some offices issuing Airport Weather Warnings, and Soaring Forecasts.

WFO Honolulu is also designated as a Meteorological Watch Office (MWO) for ICAO. As a result of this unique designation, WFO Honolulu is the only WFO to issue the following text products: AIRMETs, SIGMETs, FAs, and Route Forecasts (ROFOR). WFO Honolulu serves as the Central Pacific Hurricane Center (CPHC). CPHC provides official NWS forecast of the movement and strength of tropical weather systems and issues the appropriate watches and warnings for the central Pacific including the State of Hawaii. WFO Honolulu also issues a suite of marine products covering a large portion of the Pacific Ocean.

1.1.2.5 NWS Office at the FAA Academy

The mission of the National Weather Service (NWS) Office at the FAA Academy is to provide weather training for Federal Aviation Administration (FAA) Air Traffic Controllers, write reference materials, and administer NWS certification examinations for FAA Pilot Weather Briefers and Tower Visibility Observers.

1.2 Federal Aviation Administration (FAA)

The FAA, a part of the Department of Transportation (DOT), provides a safe, secure, and efficient aerospace system that contributes to national security and the promotion of U.S. aerospace safety. As the leading authority in the international aerospace community, the FAA is responsive to the dynamic nature of user needs, economic conditions, and environmental concerns.

The FAA provides a wide range of services to the aviation community. The following is a description of those FAA facilities which are involved with aviation weather and pilot services.

1.2.1 Air Traffic Control System Command Center (ATCSCC)

The Air Traffic Control System Command Center (ATCSCC) is located in Herndon, Virginia. ATCSCC has the mission of balancing air traffic demand with system capacity. This ensures maximum safety and efficiency for the NAS while minimizing delays. The ATCSCC utilizes the Traffic Management System, aircraft situation display, monitor alert, the follow-on functions, and direct contact with ARTCC and terminal radar approach control facility (TRACON) traffic management units to manage flow on a national level.

Because weather is the most common reason for air traffic delays and re-routings, the ATCSCC is supported by Air Traffic Control System Command Center Weather Unit Specialists (ATCSCCWUS). These flight service specialists are responsible for the dissemination of meteorological information as it pertains to national air traffic flow management.

1.2.2 Air Route Traffic Control Center (ARTCC)

An ARTCC is a facility established to provide air traffic control service to aircraft operating on Instrument Flight Rules (IFR) flight plans within controlled airspace and principally during the en route phase of flight. When equipment capabilities and controller workload permit, certain advisory/assistance services may be provided to Visual Flight Rules (VFR) aircraft.

En route controllers become familiar with pertinent weather information and stay aware of current weather information needed to perform air traffic control duties. En route controllers advise pilots of hazardous weather that may impact operations within 150 NM of the controller's assigned sector(s).

1.2.3 Air Traffic Control Tower (ATCT) and Terminal Radar Approach Control (TRACON)

An ATCT is a terminal facility that uses air/ground communications, visual signaling, and other devices to provide ATC services to aircraft operating in the vicinity of an airport or on the movement area. It authorizes aircraft to land or take off at the airport controlled by the tower or to transit the Class D airspace area regardless of flight plan or weather conditions (IFR or VFR). A tower may also provide approach control services.

TRACONs manage the airspace from 10 to 40 miles outside of selected airports and below 13,000 feet. They also coordinate aircraft spacing as they approach and depart these airports.

Terminal controllers become familiar with pertinent weather information and stay aware of current weather information needed to perform air traffic control duties. Terminal controllers advise pilots of hazardous weather that may impact operations within 150 NM of the controller's assigned sector or area of jurisdiction. ATCTs and TRACONs may opt to broadcast hazardous weather information alerts only when any part of the area described is within 50 NM of the airspace under the ATCT's jurisdiction.

The tower controllers are also properly certified and act as official weather observers as required.

An Automatic Terminal Information Service (ATIS) is a continuous broadcast of recorded information in selected terminal areas. Its purpose is to improve controller effectiveness and to relieve frequency congestion by automating the repetitive transmission of non-controlled airport/terminal area and meteorological information.

1.2.4 Flight Service Station (FSS) / Automated Flight Service Station (AFSS)

Flight Service Stations (FSSs) and Automated Flight Service Stations (AFSSs) provide pilot weather briefings, en route weather, receive and process IFR and VFR flight plans, relay ATC clearances, and issue Notices to Airmen (NOTAMs). They also provide assistance to lost aircraft and aircraft in emergency situations, and conduct VFR search and rescue services.

1.3 Dissemination of Aviation Weather Products

The ultimate users of aviation weather services are pilots and aircraft dispatchers. Maintenance personnel may use the service to keep informed of weather that could cause possible damage to unprotected aircraft.

Pilots contribute to aviation weather services as well as use them. PIREPs help other pilots, dispatchers, briefers and forecasters as an observation of current conditions.

In the interest of safety and in compliance with Title 14, Code of Federal Regulations, all pilots should get a complete weather briefing before each flight. The pilot is responsible for ensuring he/she has all the information needed to make a safe flight.

1.3.1 Weather Briefings

Prior to every flight, pilots should gather all information vital to the nature of the flight. This includes an appropriate weather briefing obtained from a specialist at an FSS, AFSS, or via Direct User Access Terminal Service (DUATS).

To provide an appropriate weather briefing, specialists need to know which of the three types of briefings is needed - a standard, abbreviated or outlook. Other helpful information is whether the flight will be conducted VFR or IFR, aircraft identification and type, departure point, estimated time of departure (ETD), flight altitude, route of flight, destination, and estimated time en route (ETE).

This information is recorded in the flight plan system and a note is made regarding the type of weather briefing provided. If necessary, it can be referenced later to file or amend a flight plan. It is also used when an aircraft is overdue or is reported missing.

1.3.1.1 Standard Briefing

A standard briefing provides a complete weather picture and is the most detailed of all briefings. This type of briefing should be obtained prior to the departure of any flight and should be used during flight planning. A standard briefing provides the following information in sequential order if it is applicable to the route of flight.

1. Adverse Conditions - This includes information about adverse conditions that may influence a decision to cancel or alter the route of flight. Adverse conditions include significant weather such as thunderstorms, aircraft icing, turbulence, wind shear, reduced visibilities and other important items such as airport closings.

2. VFR Flight NOT RECOMMENDED (VNR) - If the weather for the route of flight is below VFR minimums, or if it is doubtful the flight could be made under VFR conditions due to the forecast weather, the briefer may state that VFR is not recommended. The pilot can then decide whether or not to continue the flight under VFR, but this advisory should be weighed carefully.

3. Synopsis - The synopsis is an overview of the larger weather picture. Fronts and major weather systems along or near the route of flight and weather which may affect the flight are provided.

4. Current Conditions - This portion of the briefing contains the current surface weather observations, pilot weather reports (PIREPs), satellite and radar data along the route of flight. If the departure time is more than 2 hours away, current conditions will not be included in the briefing.

5. En Route Forecast - The en route forecast is a summary of the weather forecast for the proposed route of flight.

6. Destination Forecast - The destination forecast is a summary of the expected weather for the destination airport at the estimated time of arrival (ETA).

7. Winds and Temperatures Aloft - Winds and temperatures aloft is a forecast of the winds at specific altitudes along the route of flight. However, the temperature information is provided only on request.

8. NOTAMs - This portion supplies Notice to Airmen (NOTAM) information pertinent to the route of flight which has not been published in the Notice to Airmen publication. Published NOTAM information is provided during the briefing only when requested.

9. ATC Delays - This is an advisory of any known air traffic control (ATC) delays that may affect the flight.

10. Other Information - At the end of the standard briefing, the specialist will provide the radio frequencies needed to open a flight plan and to contact En Route Flight Advisory Service (EFAS). Any additional information requested is also provided at this time.

1.3.1.2 Abbreviated Briefing

An abbreviated briefing is a shortened version of the standard briefing. It should be requested when a departure has been delayed or when specific weather information is needed to update a previous standard briefing. When this is the case, the weather specialist needs to know the time and source of the previous briefing so the necessary weather information will not be omitted inadvertently.

1.3.1.3 Outlook Briefing

An outlook briefing should be requested when a planned departure is 6 or more hours away. It provides initial forecast information that is limited in scope due to the timeframe of the planned flight. This type of briefing is a good source of flight planning information that can influence decisions regarding route of flight, altitude, and ultimately the "go, no-go" decision. A follow-up standard briefing prior to departure is advisable since an outlook briefing generally only contains information based on weather trends and existing weather in geographical areas at or near the departure airport.

The FSS/AFSS's purpose is to serve the aviation community. Pilots should not hesitate to ask questions and discuss factors they do not fully understand. The briefing should be considered complete only when the pilot has a clear picture of what weather to expect. Pilots should also make a final weather check immediately before departure if at all possible.

1.3.2 Direct Use Access Terminal Service (DUATS/DUAT)

The Direct User Access Terminal Service, which is funded by the FAA, allows any pilot with a current medical certificate to access weather information and file a flight plan via computer. Two methods of access are available to connect with DUATS. The first is on the Internet through Computer Sciences Corporation (CSC) at http://www.duats.com or Data Transformation Corporation at http://www.duat.com. The second method requires a modem and a communications program supplied by a DUATS provider. To access the weather information and file a flight plan by this method, pilots use a toll free telephone number to connect the user's computer directly to the DUATS computer. The current vendors of DUATS service and the associated phone numbers are listed in Chapter 7 of the *Aeronautical Information Manual (AIM)*.

1.3.3 Aviation Digital Data Service (ADDS)

The Aviation Digital Data Service (ADDS) provides the aviation community with text, digital and graphical forecasts, analyses, and observations of aviation-related weather variables. ADDS is a joint effort of NOAA Forecast Systems Laboratory (FSL), NCAR Research Applications Laboratory (RAL), and the AWC.

1.3.4 Telephone Information Briefing Service (TIBS)

The Telephone Information Briefing Service (TIBS) is a service prepared and disseminated by selected Automated Flight Service Stations. It provides continuous telephone recordings of meteorological and aeronautical information. Specifically, TIBS provides area and route briefings, as well as airspace procedures and special announcements, if applicable. It is designed to be a preliminary briefing tool and is not intended to replace a standard briefing from a flight service specialist. The TIBS service is available 24 hours a day and is updated when conditions change, but it can only be accessed by a TOUCH-TONE phone. The phone numbers for the TIBS service are listed in the Airport/Facility Directory (A/FD).

TIBS should also contain, but is not limited to: surface observations, TAFs, and winds/temperatures aloft forecasts.

Each AFSS provides at least four route and/or area briefings. As a minimum, area briefings encompass a 50 NM radius. Pilots have access to NOTAM data through: Area or route briefings, on separate channels that are designated specifically for NOTAMs, or by access to a briefer.

Separate channels are designated for each route, area, local meteorological/aeronautical information, special event, airspace procedures, etc.

The order and content of the TIBS recording is as follows:

1. Introduction. Includes the preparation time and the route and/or the area of coverage. The service area may be configured to meet the individual facility's needs.

2. Adverse Conditions. A summary of Convective SIGMETs, SIGMETs, AIRMETs, Center Weather Advisories, Alert Severe Weather Watch Bulletins, and any other available information that may adversely affect flight in the route/area.

3. VNR Statement. Included when current or forecast conditions, surface or aloft, would make the flight under visual flight rules doubtful.

4. Synopsis. A brief statement describing the type, location, and movement of weather systems and/or air masses that might affect the route or the area. This element may be combined with adverse conditions and/or the VNR element, in any order, when it will help to more clearly describe conditions.

5. Current Conditions. A summary of current weather conditions over the route/area. PIREPs are included on conditions reported aloft and a summary of observed radar echoes. Specific departure/destination observation may also be included.

6. Density Altitude. The statement "check density altitude" will be included for any weather reporting point with a field elevation of 2,000 feet MSL or above that meets certain temperature criteria.

7. En Route Forecast. A summary of appropriate forecast data provided in logical order, i.e., climb out, en route, and descent.

8. Winds Aloft. A summary of winds aloft forecast for the route/area as interpolated from forecast data for the local and/or the adjacent reporting locations for levels through 12,000 feet. The broadcast should include the levels from 3,000 to 12,000 feet, but usually includes at least two forecast levels above the surface.

9. Request for PIREPs. When weather conditions within the area or along the route meet requirements for soliciting PIREPs, a request will be included in the recording.

10. NOTAM information that affects the route/area may be included as part of the briefing, on a separate channel, or obtained by direct contact with a pilot weather briefer.

11. Military Training Activity. A statement is included in the closing announcement to contact a briefer for information on military training activity.

12. Closing Announcement.

TIBS services may be reduced during the hours of 1800-0600 local time only. Resumption of full broadcast service is adjusted seasonally to coincide with daylight hours. During the period of reduced broadcast, a recorded statement may indicate when the broadcast will be resumed and to contact Flight Service for weather briefing and other services.

For those pilots already in flight and needing weather information and assistance, the following services are provided by flight service stations. They can be accessed over the proper radio frequencies printed in flight information publications.

1.3.5 Hazardous Inflight Weather Advisory Service (HIWAS)
HIWAS is a national program for broadcasting hazardous weather information continuously over selected navigational aids (NAVAIDs). The broadcasts include advisories such as AIRMETs, SIGMETS, convective SIGMETs, and urgent PIREPs. These broadcasts are only a summary of the information, and pilots should contact an FSS/AFSS or En Route Flight Advisory Service (EFAS) for detailed information.

The HIWAS broadcast area is defined as the area within 150 NM of HIWAS outlets.

HIWAS broadcasts are not interrupted or delayed except for emergency situations, when an aircraft requires immediate attention, or for reasonable use of the voice override capability on specific HIWAS outlets in order to use the limited Remote Communications Outlet (RCO) to maintain en route communications. The service is provided 24-hours a day. An announcement is made for no hazardous weather advisories.

Hazardous weather information is recorded if it is occurring within the HIWAS broadcast area. The broadcast includes the following elements:

1. A statement of introduction including the appropriate area(s) and a recording time.

2. A summary of Convective SIGMETs, SIGMETs, AIRMETs, Urgent PIREPs, Aviation Watch Notification Messages, Center Weather Advisories, and any other weather such as isolated thunderstorms that are rapidly developing and increasing in intensity, or low ceilings and visibilities that are becoming widespread which are considered significant and are not included in a current hazardous weather advisory.

3. A request for PIREPs, if applicable.

4. A recommendation to contact AFSS/FSS/FLIGHT WATCH for additional details concerning hazardous weather.

Once the HIWAS broadcast is updated, an announcement will be made once on all communications/NAVAID frequencies except emergency, EFAS, and navigational frequencies already dedicated to continuous broadcast services. In the event a HIWAS broadcast area is out of service, an announcement is made on all communications/NAVAID frequencies except on emergency, EFAS, and navigational frequencies already dedicated to continuous broadcast services.

1.3.6 En Route Flight Advisory Service (EFAS)

The purpose of EFAS, radio call "FLIGHT WATCH" (FW), is to provide en route aircraft with timely and pertinent weather data tailored to a specific altitude and route using the most current available sources of aviation meteorological information.

EFAS specialists tailor en route flight advisories to the phase of flight that begins after climb out and ends with descent to land. Current weather and terminal forecast at the airport of first intended landing and/or the alternate airport is provided on request. When conditions dictate, EFAS specialists provide information on weather for alternate routes and/or altitudes to assist the pilot in the avoidance of hazardous flight conditions. The pilot is advised to contact the adjacent flight watch facility when adverse weather conditions along the intended route extend beyond the Flight Watch Area (FWA).

EFAS is NOT used for routine in-flight services; e.g., flight plan filing, position reporting, or full route (pre-flight) briefings. If a request for information is received that is not within the scope of EFAS, the pilot is advised of the appropriate AFSS/FSS to contact.

EFAS specialists suggest route or destination changes to avoid areas of weather that in the judgment of the specialists constitutes a threat to safe flight.

EFAS is provided on 122.0 MHz to aircraft below FL180. An assigned discrete frequency is used to provide EFAS to aircraft at FL180 and above. This frequency can also be used for communications with aircraft below FL180 when communication coverage permits. Aircraft operating at FL 180 or above that contact FW on frequency 122.0 MHz are advised to change to the discrete frequency for EFAS.

2 AVIATION WEATHER PRODUCT CLASSIFICATION AND POLICY

The demand for new and improved aviation weather products continues to grow and, with new products introduced to meet the demand, some confusion has resulted in the aviation community regarding the relationship between regulatory requirements and the new weather products.

This section will clarify that relationship by providing:

- classification of the weather products and policy guidance in their use,
- descriptions of the types of aviation weather information, and
- categorization of the sources of aviation weather information.

2.1 Classification of Aviation Weather Products

The FAA has developed two classifications of aviation weather products: *primary* weather products, and *supplementary* weather products. The classifications are meant to eliminate confusion by differentiating between weather products that may be used to meet regulatory requirements and other weather products that may only be used to improve situational awareness.

All flight-related, aviation weather decisions must be based on the primary weather products. Supplementary weather products augment the primary products by providing additional weather information, but may not be used as stand-alone products to meet aviation weather regulatory requirements or without the relevant primary products. When discrepancies exist between primary and supplementary products pertaining to the same weather phenomena, pilots must base flight-related decisions on the primary weather product. Furthermore, multiple primary products may be necessary to meet all aviation weather regulatory requirements.

Aviation weather products produced by the federal government (NWS) are primary products unless designated as a supplementary product by the FAA. In addition, the FAA may choose to restrict certain weather products to specific types of usage or classes of user. Any limitations imposed by the FAA on the use of a product will appear in the product label.

2.1.1 Primary Weather Product Classification
A primary weather product is an aviation weather product that meets all of the regulatory requirements and safety needs for use in making weather-related flight decisions.

Note: Sections 3 through 8 of this Advisory Circular are considered Primary Weather Products.

2.1.2 Supplementary Weather Product Classification
A supplementary weather product is an aviation weather product that may be used for enhanced situational awareness. A supplementary weather product must only be used in conjunction with one or more primary weather products. In addition, the FAA may further restrict the use of the supplementary weather products through limitations described in the product label.

Note: Section 9 of this Advisory Circular contains information on Supplementary Weather Products.

2.2 Types of Aviation Weather Information

The FAA has identified the following three distinct types of weather information that may be needed to conduct aircraft operations: observations, analyses, and forecasts.

2.2.1 Observations
Observations are raw weather data collected by some type of sensor(s). The observations can either be in situ (e.g. surface or airborne) or remote (e.g. weather radar, satellite, profiler, and lightning).

2.2.2 Analysis
Analyses of weather information are an enhanced depiction and/or interpretation of observed weather data.

2.2.3 Forecasts
Forecasts are the predictions of the development and/or movement of weather phenomena based on meteorological observations and various mathematical models.

In-flight weather advisories, including Significant Meteorological Information (SIGMET), Convective SIGMETs, Airman's Meteorological Information (AIRMET), Center Weather Advisories (CWA), and Meteorological Impact Statements (MIS), are considered forecast weather information products.

2.3 Categorizing Aviation Weather Sources

The regulations pertaining to aviation weather reflect that, historically, the federal government was the only source of aviation weather information. That is, the FAA and NWS, or its predecessor organizations, were solely responsible for the collection and dissemination of weather data, including forecasts. Thus, the term "approved source(s)" referred exclusively to the federal government. The federal government is no longer the only source of weather information, due to the growing sophistication of aviation operations and scientific and technological advances.

Since all three types of weather information defined in paragraph 2.3 are not available from all sources of aviation weather information, the FAA has categorized the sources as follows: federal government, Enhanced Weather Information System (EWINS), and commercial weather information providers.

2.3.1 Federal Government
The FAA and NWS collect weather observations. The NWS analyzes the observations, and produces forecasts, including in-flight aviation weather advisories (e.g., SIGMETs). The FAA and NWS disseminate meteorological observations, analyses, and forecast products through a variety of systems. The federal government is the only approval authority for sources of weather observations (e.g., contract towers and airport operators).

Commercial weather information providers contracted by the FAA to provide weather observations (e.g., contract towers) are included in the federal government category of approved sources by virtue of maintaining required technical and quality assurance standards under FAA and NWS oversight.

2.3.2 Enhanced Weather Information System (EWINS)

EWINS is an FAA-approved proprietary system for tracking, evaluating, reporting, and forecasting the presence or absence of adverse weather phenomena. EWINS is authorized to produce flight movement forecasts, adverse weather phenomena forecasts, and other meteorological advisories.

To receive FAA approval, EWINS-approved source must have sufficient procedures, personnel, and communications and data processing equipment to effectively obtain, analyze, and disseminate aeronautical weather data. For a full explanation of the requirements for EWINS approval, see the *Air Transportation Operations Inspector's Handbook*, Order 8400.10, volume 3, chapter 7, section 5. An EWINS-approved source may produce weather analyses and forecasts based on meteorological observations provided by the federal government. Approval to use EWINS weather products is issued on a case by case basis and is currently only applicable to FAR part 121 and 135 certificate holders, who may either act as their own EWINS or contract those services from a separate entity. For these approved users, the weather analyses and forecasts produced by their approved EWINS are considered primary weather products as defined in paragraph 2.2.1, Primary Weather Products.

2.3.3 Commercial Weather Information Providers

Commercial weather providers are a major source of weather products for the aviation community. In general, they produce proprietary weather products based on NWS products with formatting and layout modifications but no material changes to the weather information itself. This is also referred to as "repackaging."

Commercial providers may also produce forecasts, analyses, and other proprietary weather products and substantially alter the information contained in NWS-produced products. Hence, operators and pilots contemplating using such services should request and/or review an appropriate description of services and provider disclosure. This should include, but is not limited to,

- the type of weather product (e.g., current weather or forecast weather),
- the currency of the product (i.e., product issue and valid times), and
- the relevance of the product.

Pilots and operators should be cautious when using unfamiliar products, or products not supported by FAA/NWS technical specifications. Commercially-available proprietary weather products that substantially alter NWS-produced weather products, or information, may only be approved for use by part 121 or part 135 operators or fractional ownership programs if the commercial provider is EWINS-qualified (see paragraph 2.4.2, above). Government products that are only repackaged and not altered, or products produced by EWINS-approved source, are considered primary weather products as defined in paragraph 2.2.1, Primary Weather Products.

3 OBSERVED TEXT PRODUCTS

3.1 Aviation Routine Weather Reports (METAR) and Selected Special Weather Reports (SPECI)

Surface weather observations are fundamental to all meteorological services. Observations are the basic information upon which forecasts and warnings are made in support of a wide range of weather sensitive activities within the public and private sectors, including aviation.

Although the METAR/SPECI code is used worldwide, each country is allowed to make modifications or exceptions to the code for use in their particular country. This section will focus on the U.S. modifications and exceptions. METAR/SPECIs are available online at: http://adds.aviationweather.gov/metars/

3.1.1 Aviation Routine Weather Report (METAR)

Aviation Routine Weather Report (METAR) is the primary observation code used in the U. S. to satisfy World Meteorological Organization (WMO) and International Civil Aviation Organization (ICAO) requirements for reporting surface meteorological data. A METAR report includes the airport identifier, time of observation, wind, visibility, runway visual range, present weather phenomena, sky conditions, temperature, dew point, and altimeter setting. Excluding the airport identifier and the time of observation, this information is collectively referred to as "the body of the report." As an addition, coded and/or plain language information elaborating on data in "the body of the report" may be appended to the end of the METAR in a section coded as "Remarks." The contents of the "Remarks" section vary with the type of reporting station. The METAR may be abridged at some designated stations only including a few of the mentioned elements.

3.1.2 Selected Special Weather Report (SPECI)

A Selected Special Weather Report (SPECI) is an unscheduled report taken when any of the criteria given in Table 3-1 are observed during the interim period between the hourly reports. SPECI contains all data elements found in a METAR plus additional plain language information which elaborates on data in the body of the report. All SPECIs are made as soon as possible after the relevant criteria are observed.

Whenever SPECI criteria are met at the time of the routine METAR, a METAR is issued.

Table 3-1. SPECI Criteria

1	Wind Shift	Wind direction changes by 45 degrees or more in less than 15 minutes and the wind speed is 10 knots or more throughout the wind shift.
2	Visibility	Surface visibility as reported in the body of the report decreases to less than, or if below, increases to equal or exceed: a. 3 miles b. 2 miles c. 1 mile d. The lowest standard instrument approach procedure minimum as published in the National Ocean Service (NOS) *U.S Instrument Procedures*. If none published use ½ mile.
3	Runway Visual Range (RVR)	The highest value from the designated RVR runway decreases to less than, or if below, increases to equal or exceed 2,400 feet during the preceding 10 minutes. U.S. military stations may not report a SPECI based on RVR.
4	Tornado, Funnel Cloud, or Waterspout	a. is observed. b. disappears from sight, or ends.
5	Thunderstorm	a. begins (a SPECI is not required to report the beginning of a new thunderstorm if one is currently reported). b. ends.
6	Precipitation	a. hail begins or ends. b. freezing precipitation begins, ends, or changes intensity. c. ice pellets begin, end, or change intensity
7	Squalls	When they occur
8	Ceiling	The ceiling (rounded off to reportable values) forms or dissipates below, decreases to less than, or if below, increases to equal or exceed: a. 3,000 feet. b. 1,500 feet c. 1,000 feet d. 500 feet e. The lowest standard instrument approach procedure minimum as published in the National Ocean Service (NOS) *U.S Instrument Procedures*. If none published, use 200 feet.
9	Sky Condition	A layer of clouds or obscurations aloft is present below 1,000 feet and no layer aloft was reported below 1,000 feet in the preceding METAR or SPECI.
10	Volcanic Eruption	When an eruption is first noted
11	Aircraft Mishap	Upon notification of an aircraft mishap, unless there has been an intervening observation
12	Miscellaneous	Any other meteorological situation designated by the responsible agency of which , in the opinion of the observer, is critical.

3.1.3 Format

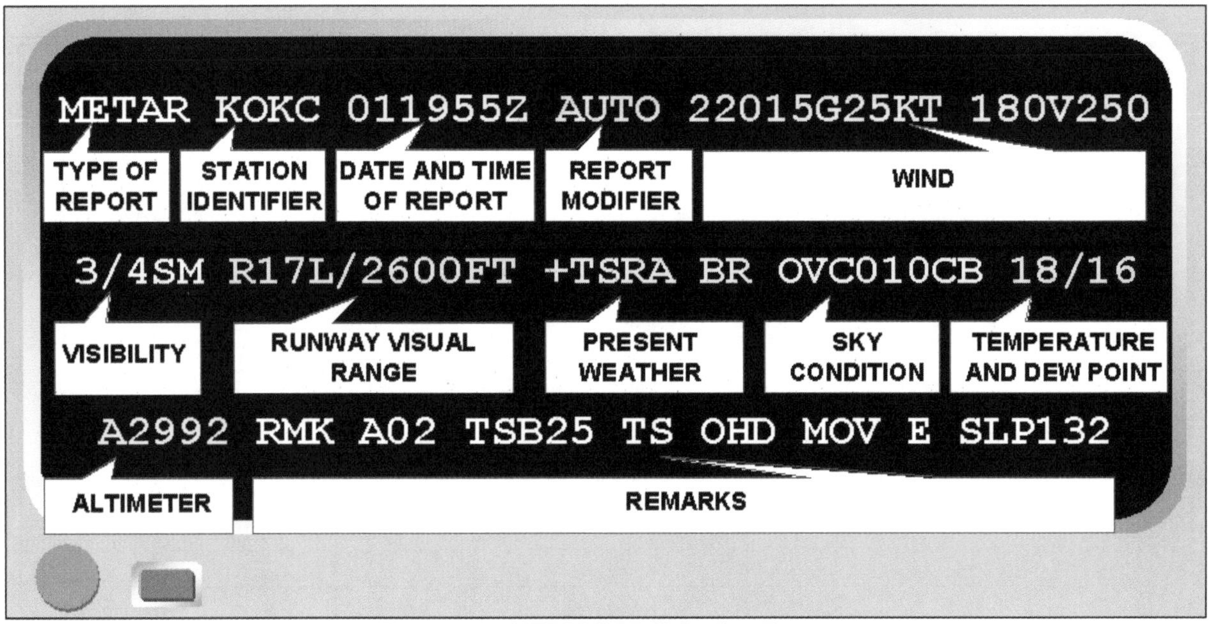

METAR KOKC 011955Z AUTO 22015G25KT 180V250

| TYPE OF REPORT | STATION IDENTIFIER | DATE AND TIME OF REPORT | REPORT MODIFIER | WIND |

3/4SM R17L/2600FT +TSRA BR OVC010CB 18/16

| VISIBILITY | RUNWAY VISUAL RANGE | PRESENT WEATHER | SKY CONDITION | TEMPERATURE AND DEW POINT |

A2992 RMK A02 TSB25 TS OHD MOV E SLP132

| ALTIMETER | REMARKS |

Figure 3-1. METAR/SPECI Coding Format

A METAR/SPECI (Figure 3-1) has two major sections: the Body (consisting of a maximum of 11 groups) and the Remarks (consisting of 2 categories). Together, the body and remarks make up the complete METAR/SPECI. When an element does not occur, or cannot be observed, the corresponding group is omitted from that particular report.

3.1.3.1 Type of Report

METAR KOKC 011955Z AUTO 22015G25KT 180V250 3/4SM R17L/2600FT +TSRA BR OVC010CB 18/16 A2992 RMK AO2 TSB25 TS OHD MOV E SLP132

The type of report, **METAR** or **SPECI** precedes the body of all reports.

3.1.3.2 Station Identifier

METAR **KOKC** 011955Z AUTO 22015G25KT 180V250 3/4SM R17L/2600FT +TSRA BR OVC010CB 18/16 A2992 RMK AO2 TSB25 TS OHD MOV E SLP132

The station identifier, in ICAO format, is included in all reports to identify the station to which the coded report applies.

The ICAO airport code is a four-letter alphanumeric code designating each airport around the world. The ICAO codes are used for flight planning by air traffic controllers and airline operation departments. These codes are not the same as the International Air Transport Association (IATA) codes encountered by the general public used for reservations, baggage handling and in airline timetables. ICAO codes are also used to identify weather stations located on- or off-airport.

Unlike the IATA codes, the ICAO codes have a regional structure. For example, the first letter is allocated by continent (Figure 3-2), the second is a country within the continent; the remaining two are used to identify each airport.

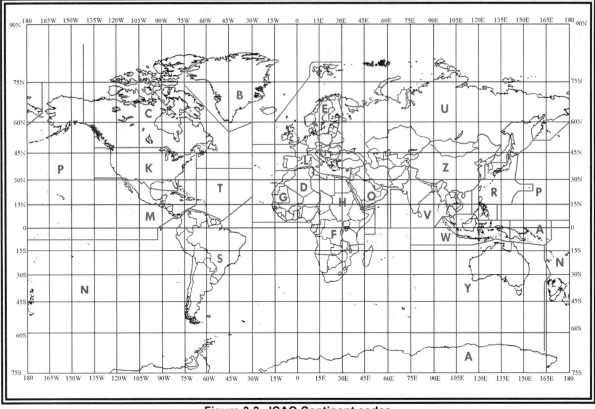

Figure 3-2. ICAO Continent codes

In the contiguous U. S., ICAO station identifiers are coded **K** followed by the three-letter IATA identifier. For example, the Seattle, Washington (IATA identifier SEA) becomes the ICAO identifier KSEA.

ICAO station identifiers in Alaska, Hawaii, and Guam begin with the continent code P, followed by the proper country code (A, H, and G respectively), and the two-letter airport identifier.

Examples:

PANC	Anchorage, AK
PAOM	Nome, AK
PHNL	Honolulu, HI
PHKO	Keahole Point, HI
PGUM	Agana, Guam
PGUA	Anderson AFB, Guam

Canadian station identifiers begin with C, followed by the country code, and the two-letter airport identifier.

Examples:

CYYZ	Toronto, Canada
CYYC	Calgary Canada
CYQB	Quebec, Canada
CYXU	London, Canada
CZUM	Churchill Falls, Canada

Mexican and western Caribbean station identifiers begin with M, followed by the proper country code and two-letter airport identifier.

Examples:

MMMX	Mexico City, Mexico
MUGM	Guantanamo Bay, Cuba
MDSD	Santo Domingo, Dominican Republic
MYNN	Nassau, Bahamas

Eastern Caribbean station identifiers begin with T, followed by the proper country code, and airport identifier.

Examples:

TJSJ	San Juan, Puerto Rico
TIST	Saint Thomas, Virgin Islands

For a list of Alaskan, Hawaiian, Canadian, Mexican, Pacific, and Caribbean ICAO identifiers see FAA Order 7350.7. For a complete worldwide listing, see ICAO Document 7910, "Location Indicators." Both are available on-line.

3.1.3.3 Date and Time of Report

```
METAR KOKC 011955Z AUTO 22015G25KT 180V250 3/4SM R17L/2600FT +TSRA BR
OVC010CB 18/16 A2992 RMK AO2 TSB25 TS OHD MOV E SLP132
```

The date and time is coded in all reports as follows: the day of the month is the first two digits (**01**) followed by the hour (**19**), and the minutes (**55**). The coded time of observations is the actual time of the report or when the criteria for a SPECI is met or noted. If the report is a correction to a previously disseminated report, the time of the corrected report is the same time used in the report being corrected. The date and time group always ends with a **Z** indicating Zulu time (or UTC). For example, METAR KOKC 011955Z would be disseminated as the 2000 hour scheduled report for station KOKC taken on the 1st of the month at 1955 UTC.

3.1.3.4 Report Modifier (As Required)

```
METAR KOKC 011955Z AUTO 22015G25KT 180V250 3/4SM R17L/2600FT +TSRA BR
OVC010CB 18/16 A2992 RMK AO2 TSB25 TS OHD MOV E SLP132
```

The report modifier, **AUTO**, identifies the METAR/SPECI as a fully automated report with no human intervention or oversight. In the event of a corrected METAR or SPECI, the report modifier, **COR**, is substituted for AUTO.

3.1.3.5 Wind Group

```
METAR KOKC 011955Z AUTO 22015G25KT 180V250 3/4SM R17L/2600FT +TSRA BR
OVC010CB 18/16 A2992 RMK AO2 TSB25 TS OHD MOV E SLP132
```

Wind is the horizontal motion of air past a given point. It is measured in terms of velocity, which is a vector that includes direction and speed. It indicates the direction the wind is coming FROM.

In the wind group, the wind direction is coded as the first three digits (**220**) and is determined by averaging the recorded wind direction over a 2-minute period. It is coded in tens of degrees relative to true north using three figures. Directions less than 100 degrees are preceded with a **0**. For example, a wind direction of 90^0 is coded as **090**.

Immediately following the wind direction is the wind speed coded in two or three digits (**15**). Wind speed is determined by averaging the speed over a 2-minute period and is coded in whole knots using the units, tens digits and, when required, the hundreds digit. When wind speeds are less than 10 knots, a leading zero is used to maintain at least a two digit wind code. For example, a wind speed of 8 knots will be coded **08KT**. The wind group is always coded with a **KT** to indicate wind speeds are reported in knots. Other countries may use kilometers per hour (KPH) or meters per second (MPS) instead of knots.

Examples:

> **05008KT** ┈┈┈┈→ Wind 50 degrees at 8 knots
> **15014KT** ┈┈┈┈→ Wind 150 degrees at 14 knots
> **340112KT** ┈┈┈┈→ Wind 340 degrees at 112 knots

3.1.3.5.1 Wind Gust
Wind speed data for the most recent 10 minutes is examined to evaluate the occurrence of gusts. Gusts are defined as rapid fluctuations in wind speed with a variation of 10 knots or more between peaks and lulls. The coded speed of the gust is the maximum instantaneous wind speed.

Wind gusts are coded in two or three digits immediately following the wind speed. Wind gusts are coded in whole knots using the units, tens, and, if required, the hundreds digit. For example, a wind out of the west at 20 knots with gusts to 35 knots would be coded **27020G35KT**.

3.1.3.5.2 Variable Wind Direction (speed 6 knots or less)
Wind direction may be considered variable when, during the previous 2-minute evaluation period, the wind speed was 6 knots or less. In this case, the wind may be coded as **VRB** in place of the 3-digit wind direction. For example, if the wind speed was recorded as 3 knots, it would be coded **VRB03KT**.

3.1.3.5.3 Variable Wind Direction (speed greater than 6 knots)
Wind direction may also be considered variable when, during the 2-minute evaluation period, it varies by 60 degrees or more and the speed is greater than 6 knots. In this case a variable wind direction group immediately follows the wind group. The directional variability is coded in a clockwise direction and consists of the extremes of the wind directions separated by a **V**. For

example, if the wind is variable from 180° to 240° at 10 knots, it would be coded **21010KT 180V240**.

3.1.3.5.4 Calm Wind

When no motion of air is detected, the wind is reported as calm. A calm wind is coded as **00000KT**.

3.1.3.6 Visibility Group

```
METAR KOKC 011955Z AUTO 22015G25KT 180V250 3/4SM R17L/2600FT +TSRA BR
OVC010CB 18/16 A2992 RMK AO2 TSB25 TS OHD MOV E SLP132
```

Visibility is a measure of the opacity of the atmosphere.

Prevailing visibility is the reported visibility considered representative of recorded visibility conditions at the station during the time of observation. It is the greatest distance that can be seen throughout at least half of the horizon circle, not necessarily continuous.

Surface visibility is the prevailing visibility from the surface at manual stations or the visibility derived from sensors at automated stations.

The visibility group is coded as the surface visibility in statute miles. A space is coded between whole numbers and fractions of reportable visibility values. The visibility group ends with **SM** to indicate that the visibility is in statute miles. For example, a visibility of one and a half statute miles is coded **1 1/2SM**. Other countries may use meters (no code).

Automated stations use an **M** to indicate "less than." For example, **M1/4SM** means a visibility of less than one-quarter statute mile.

3.1.3.7 Runway Visual Range (RVR) Group

```
METAR KOKC 011955Z AUTO 22015G25KT 180V250 3/4SM R17L/2600FT +TSRA BR
OVC010CB 18/16 A2992 RMK AO2 TSB25 TS OHD MOV E SLP132
```

The runway visual range (RVR) is an instrument-derived value representing the horizontal distance a pilot may see down the runway.

RVR is reported whenever the station has RVR equipment and prevailing visibility is 1 statute mile or less and/or the RVR for the designated instrument runway is 6,000 feet or less. Otherwise the RVR group is omitted.

Runway visual range is coded in the following format: the initial **R** is code for runway and is followed by the runway number. When more than one runway is defined with the same runway number a directional letter is coded on the end of the runway number. Next is a solidus /; followed by the visual range in feet and then **FT** completes the RVR report. For example, an RVR value for Runway 01L of 800 feet would be coded **R01L/0800FT**. Other countries may use meters.

RVR values are coded in increments of 100 feet up to 1,000 feet, increments of 200 feet from 1,000 feet to 3,000 feet, and increments of 500 feet from 3,000 feet to 6,000 feet. Manual RVR is not reported below 600 feet. At automated stations, RVR may be reported for up to four designated runways.

When the RVR varies by more than one reportable value, the lowest and highest values will be shown with **V** between them indicating variable conditions. For example, the 10-minute RVR for runway 01L varying between 600 and 1,000 feet would be coded **R01L/0600V1000FT**.

If RVR is less than its lowest reportable value, the visual range group is preceded by **M**. For example, an RVR for runway 01L of less than 600 feet is coded **R01L/M0600FT**.

If RVR is greater than its highest reportable value, the visual range group is preceded by a **P**. For example, an RVR for runway 27 of greater than 6,000 feet will be coded **R27/P6000FT**.

3.1.3.8 Present Weather Group

```
METAR KOKC 011955Z AUTO 22015G25KT 180V250 3/4SM R17L/2600FT +TSRA BR
OVC010CB 18/16 A2992 RMK AO2 TSB25 TS OHD MOV E SLP132
```

Present weather includes precipitation, obscurations, and other weather phenomena. The appropriate notations found in Table 3-2 are used to code present weather.

Table 3-2. METAR/SPECI Notations for Reporting Present Weather[1]

QUALIFIER		WEATHER PHENOMENA		
INTENSITY OR PROXIMITY	DESCRIPTOR	PRECIPITATION	OBSCURATION	OTHER
1	2	3	4	5
- Light	**MI** Shallow	**DZ** Drizzle	**BR** Mist	**PO** Dust/Sand whirls
Moderate[2]	**PR** Partial	**RA** Rain	**FG** Fog	**SQ** Squalls
+ Heavy	**BC** Patches	**SN** Snow	**FU** Smoke	**FC** Funnel Cloud, Tornado, or Waterspout[4]
VC In the Vicinity[3]	**DR** Low Drifting	**SG** Snow Grains	**VA** Volcanic Ash	**SS** Sandstorm
	BL Blowing	**IC** Ice Crystals (Diamond Dust)	**DU** Widespread Dust	**DS** Duststorm
	SH Shower(s)	**PL** Ice Pellets	**SA** Sand	
	TS Thunderstorms	**GR** Hail	**HZ** Haze	
	FZ Freezing	**GS** Small Hail and/or Snow Pellets	**PY** Spray	
		UP Unknown Precipitation		

1. The weather groups are constructed by considering columns 1 to 5 in the table above in sequence, i.e., intensity followed by description, followed by weather phenomena, e.g., heavy rain shower(s) is coded as +SHRA.
2. To denote moderate intensity no entry or symbol is used.
3. See text for vicinity definitions.
4. Tornadoes and waterspouts are coded as +FC.

Separate groups are used for each type of present weather. Each group is separated from the other by a space. METAR/SPECI reports contain no more than three present weather groups.

When more than one type of present weather is reported at the same time, present weather is reported in the following order:

- Tornadic activity – Tornado, Funnel Cloud, or Waterspout.
- Thunderstorm(s) with and without associated precipitation.
- Present weather in order of decreasing dominance, i.e., the most dominant type is reported first.
- Left-to-right in Table 3-2 (Columns 1 through 5).

Qualifiers may be used in various combinations to describe weather phenomena. Present weather qualifiers fall into two categories: intensity (Section 3.1.3.8.1) or proximity (Section 3.1.3.8.2) and descriptors (Section 3.1.3.8.3).

3.1.3.8.1 Intensity Qualifier
The intensity qualifiers are light, moderate, and heavy. They are coded with precipitation types except ice crystals (**IC**) and hail (**GR** or **GS**) including those associated with a thunderstorm (**TS**) and those of a showery nature (**SH**). Tornadoes and waterspouts are coded as heavy (**+FC**). No intensity is ascribed to the obscurations of blowing dust (**BLDU**), blowing sand (**BLSA**), and blowing snow (**BLSN**). Only moderate or heavy intensity is ascribed to sandstorm (**SS**) and duststorm (**DS**).

When more than one form of precipitation is occurring at a time or precipitation is occurring with an obscuration, the reported intensities are not cumulative. The reported intensity will not be greater than the intensity for each form of precipitation.

3.1.3.8.2 Proximity Qualifier
Weather phenomena occurring beyond the point of observation (between 5 and 10 statute miles) are coded as in the vicinity (**VC**). VC can be coded in combination with thunderstorm (**TS**), fog (**FG**), shower(s) (**SH**), well-developed dust/sand whirls (**PO**), blowing dust (**BLDU**), blowing sand (**BLSA**), blowing snow (**BLSN**), sandstorm (**SS**), and duststorm (**DS**). Intensity qualifiers are not coded in conjunction with **VC**.

For example, **VCFG** can be decoded as meaning some form of fog is between 5 and 10 statute miles of the point of observation. If **VCSH** is coded, showers are occurring between 5 and 10 statute miles of the point of observation.

Weather phenomena occurring at the point of observation (at the station) or in the vicinity of the point of observation are coded in the body of the report. Weather phenomena observed beyond 10SM from the point of observation (at the station) is not coded in the body but may be coded in the remarks section (Section 3.1.3.12).

3.1.3.8.3 Descriptor Qualifier
Descriptors are qualifiers which further amplify weather phenomena and are used in conjunction with some types of precipitation and obscurations. The descriptor qualifiers are: shallow (**MI**), partial (**PR**), patches (**BC**), low drifting (**DR**), blowing (**BL**), shower(s) (**SH**), thunderstorm (**TS**), and freezing (**FZ**).

Only one descriptor is coded for each weather phenomena group, e.g., **FZDZ**.

The descriptors shallow (**MI**), partial (**PR**), and patches (**BC**) are only coded with **FG**, e.g., **MIFG**. Mist (**BR**) is not coded with any descriptor.

The descriptors low drifting (**DR**) and blowing (**BL**) will only be coded with dust (**DU**), sand (**SA**), and snow (**SN**), e.g., **BLSN** or **DRSN**. **DR** is coded with **DU**, **SA**, or **SN** for raised particles drifting less than six feet above the ground.

When blowing snow is observed with snow falling from clouds, both phenomena are reported, e.g., **SN BLSN**. If blowing snow is occurring and the observer cannot determine whether or not snow is also falling, then **BLSN** is reported. Spray (**PY**) is coded only with blowing (**BL**).

The descriptor for showery-type precipitation (**SH**) is coded only with one or more of the precipitation qualifiers for rain (**RA**), snow (**SN**), ice pellets (**PL**), small hail (**GS**), or large hail (**GR**). The **SH** descriptor indicates showery-type precipitation. When any type of precipitation is coded with **VC**, the intensity and type of precipitation is not coded.

The descriptor for thunderstorm (**TS**) may be coded by itself when the thunderstorm is without associated precipitation. A thunderstorm may also be coded with the precipitation types of rain (**RA**), snow (**SN**), ice pellets (**PL**), small hail and/or snow pellets (**GS**), or hail (**GR**). For example, a thunderstorm with snow and small hail and/or snow pellets would be coded as **TSSNGS**. **TS** are not coded with **SH**.

The descriptor freezing (**FZ**) is only coded in combination with fog (**FG**), drizzle (**DZ**), or rain (**RA**), e.g., **FZRA**. **FZ** is not coded with **SH**.

3.1.3.8.4 Precipitation
Precipitation is any of the forms of water particles, whether liquid or solid, that falls from the atmosphere and reaches the ground. The precipitation types are: drizzle (**DZ**), rain (**RA**), snow (**SN**), snow grains (**SG**), ice crystals (**IC**), ice pellets (**IP**), hail (**GR**), small hail and/or snow pellets (**GS**), and unknown precipitation (**UP**). **UP** is reported if an automated station detects the occurrence of precipitation but the precipitation sensor cannot recognize the type.

Up to three types of precipitation may be coded in a single present weather group. They are coded in order of decreasing dominance based on intensity.

3.1.3.8.5 Obscuration
Obscurations are any phenomenon in the atmosphere, other than precipitation, reducing the horizontal visibility. The obscuration types are: mist (**BR**), fog (**FG**), smoke (**FU**), volcanic ash (**VC**), widespread dust (**DU**), sand (**SA**), haze (**HZ**), and spray (**PY**). Spray (**PY**) is coded only as **BLPY**.

With the exception of volcanic ash, low drifting dust, low drifting sand, low drifting snow, shallow fog, partial fog, and patches (of) fog, an obscuration is coded in the body of the report if the surface visibility is less than 7 miles or considered operationally significant. Volcanic ash is always reported when observed.

3.1.3.8.6 Other Weather Phenomena
Other weather phenomena types include: well-developed dust/sand whirls (**PO**), sand storms (**SS**), dust storms (**DS**), squalls (**SQ**), funnel clouds (**FC**), and tornados and waterspouts (**+FC**).

Examples:

–DZ	Light drizzle
–RASN	Light rain and snow
SN BR	(Moderate) snow, mist
–FZRA FG	Light freezing rain, fog
SHRA	(Moderate) rain shower
VCBLSA	Blowing sand in the vicinity
–RASN FG HZ	Light rain and snow, fog, haze
TS	Thunderstorm (without precipitation)

+TSRA ----------→ Thunderstorm, heavy rain

+FC TSRAGR BR ·········→ Tornado, thunderstorm, (moderate) rain, hail, mist

3.1.3.9 Sky Condition Group

METAR KOKC 011955Z AUTO 22015G25KT 180V250 3/4SM R17L/2600FT +TSRA BR
OVC010CB 18/16 A2992 RMK AO2 TSB25 TS OHD MOV E SLP132

Sky condition is a description of the appearance of the sky. It is coded as: sky condition, vertical visibility, or clear skies.

The sky condition group is based on the amount of sky cover (the first three letters) followed by the height of the base of the sky cover (final three digits). No space is between the amount of sky cover and the height of the layer. The height of the layer is recorded in feet Above Ground Level (AGL).

Sky condition is coded in ascending order and ends at the first overcast layer. At mountain stations, if the layer is below station level, the height of the layer will be coded as *///*.

Vertical visibility is coded as **VV** followed by the vertical visibility into the indefinite ceiling. No space is between the group identifier and the vertical visibility. Figure 3-3 illustrates the effect of an obscuration on the vision from a descending aircraft.

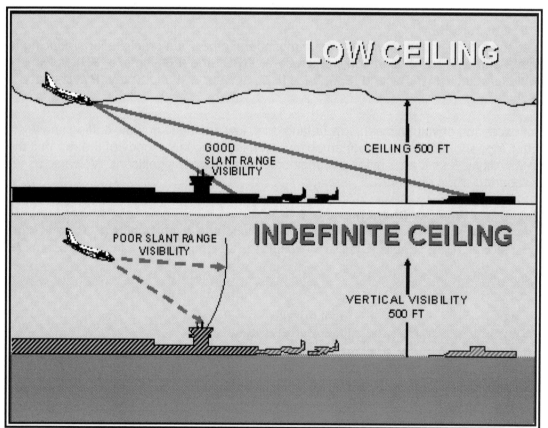

Figure 3-3. Obscuration Effects on Slant Range Visibility
The ceiling is 500 feet in both examples, but the indefinite ceiling example (bottom) produces a more adverse
impact to landing aircraft. This is because an obscuration (e.g., fog, blowing dust, snow, etc.) limits runway

acquisition due to reduced slant range visibility. **This pilot would be able to see the ground but not the runway. If the pilot was at approach minimums, the approach could not be continued and a missed approach must be executed.**

Clear skies are coded in the format, **SKC** or **CLR**. When **SKC** is used, an observer indicates no layers are present; and **CLR** is used by automated stations to indicate no layers are detected at or below 12,000 feet.

Each coded layer is separated from the others by a space. Each layer reported is coded by using the appropriate reportable contraction seen in Table 3-3. A report of clear skies (**SKC** or **CLR**) is a complete layer report within itself. The abbreviations **FEW**, **SCT**, **BKN**, and **OVC** will be followed, without a space, by the height of the layer.

Table 3-3. METAR/SPECI Contractions for Sky Cover

Reportable Contraction	Meaning	Summation Amount of Layer
VV	Vertical Visibility	8/8
SKC or CLR[1]	Clear	0
FEW[2]	Few	1/8 – 2/8
SCT	Scattered	3/8 – 4/8
BKN	Broken	5/8 – 7/8
OVC	Overcast	8/8

1. The abbreviation **CLR** will be used at automated stations when no layers at or below 12,000 feet are reported; the abbreviation **SKC** will be used at manual stations when no layers are reported.
2. Any layer amount less than 1/8 is reported as FEW.

The height is coded in hundreds of feet above the surface using three digits in accordance with Table 3-4.

Table 3-4. METAR/SPECI Increments of Reportable Values of Sky Cover Height

Range of Height Values (feet)	Reportable Increment (feet)
Less than or equal to 5,000	To nearest 100
5,001 to 10,000	To nearest 500
Greater than 10,000	To nearest 1,000

The ceiling is the lowest layer aloft reported as broken or overcast. If the sky is totally obscured with ground based clouds, the vertical visibility is the ceiling.

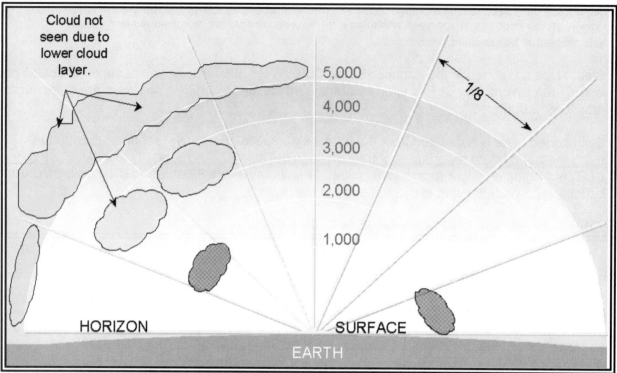

Figure 3-4. METAR/SPECI Sky Condition Coding

Clouds at 1,200 feet obscure 2/8ths of the sky (FEW). Higher clouds at 3,000 feet obscure an additional 1/8th of the sky, and because the observer cannot see above the 1,200-foot layer, he is to assume that the higher 3,000-foot layer also exists above the lower layer (SCT). The highest clouds at 5,000 feet obscure 2/8ths of the sky, and again since the observer cannot see past the 1,200 and 3,000-foot layers, he is to assume the higher 5,000-foot layer also exists above the lower layers (BKN). The sky condition group would be coded as: FEW012 SCT030 BKN050.

At manual stations, cumulonimbus (**CB**) or towering cumulus (**TCU**) is appended to the associated layer. For example, a scattered layer of towering cumulus at 1,500 feet would be coded **SCT015TCU** and would be followed by a space if there were additional higher layers to code.

Examples:

SKC	No layers are present
CLR	No layers are detected at or below 12,000 feet AGL
FEW004	Few at 400 feet AGL
SCT023TCU	Scattered layer of towering cumulus at 2,300 feet
BKN105	Broken layer (ceiling) at 10,500 feet
OVC250	Overcast layer (ceiling) at 25,000 feet
VV001	Indefinite ceiling with a vertical visibility of 100 feet
FEW012 SCT046	Few clouds at 1,200 feet, scattered layer at 4,600 feet
SCT033 BKN085	Scattered layer at 3,300 feet, broken layer (ceiling) at 8,500 feet
SCT018 OVC032CB	Scattered layer at 1,800 feet, overcast layer (ceiling) of cumulonimbus at 7,500 feet
SCT009 SCT024 BKN048	Scattered layer at 900 feet, scattered layer at 2,400 feet, broken layer (ceiling) at 4,800 feet

3.1.3.10 Temperature/Dew Point Group

```
METAR KOKC 011955Z AUTO 22015G25KT 180V250 3/4SM R17L/2600FT +TSRA BR
OVC010CB 18/16 A2992 RMK AO2 TSB25 TS OHD MOV E SLP132
```

Temperature is the degree of hotness or coldness of the ambient air seems as measured by a suitable instrument. Dew point is the temperature to which a given parcel of air must be cooled at constant pressure and constant water vapor content for the air to become fully saturated.

Temperature and dew point are coded as two digits rounded to the nearest whole degree Celsius. For example, a temperature of 0.3°C would be coded at **00**. Sub-zero temperatures and dew points are prefixed with an **M**. For example, a temperature of 4°C with a dew point of – 2°C would be coded as **04/M02**; a temperature of –2°C would be coded as **M02**.

If temperature is not available, the entire temperature/dew point group is not coded. If dew point is not available, temperature is coded followed by a solidus, **/,** and no entry made for dew point. For example, a temperature of 1.5°C and a missing dew point would be coded as **02/.**

3.1.3.11 Altimeter

```
METAR KOKC 011955Z AUTO 22015G25KT 180V250 3/4SM R17L/2600FT +TSRA BR
OVC010CB 18/16 A2992 RMK AO2 TSB25 TS OHD MOV E SLP132
```

The altimeter setting group codes the current pressure at elevation. This setting is then used by aircraft altimeters to determine the true altitude above a fixed plane of mean sea level.

The altimeter group always starts with an **A** (the international indicator for altimeter in inches of mercury) and is followed by the four digit group representing the pressure in tens, units, tenths, and hundredths of inches of mercury. The decimal point is not coded. For example, an altimeter setting of 29.92 inches of Mercury would be coded as **A2992**.

3.1.3.12 Remarks (RMK)

```
METAR KOKC 011955Z AUTO 22015G25KT 180V250 3/4SM R17L/2600FT +TSRA BR
OVC010CB 18/16 A2992 RMK AO2 TSB25 TS OHD MOV E SLP132
```

Remarks are included in all METAR and SPECI, when appropriate.

Remarks are separated from the body of the report by the contraction **RMK**. When no remarks are necessary, the contraction **RMK** is not required.

METAR/SPECI remarks fall into two categories: (1) Automated, Manual, and Plain Language, and (2) Additive Maintenance Data.

Table 3-5. METAR/SPECI Order of Remarks

Automated, Manual, and Plain Language				Additive and Automated Maintenance Data	
1.	Volcanic Eruptions	14.	Hailstone Size	27.	Precipitation*
2.	Funnel Cloud	15.	Virga	28.	Cloud Types*
3.	Type of Automated Station	16.	Variable Ceiling Height	29.	Duration of Sunshine*
4.	Peak Wind	17.	Obscurations	30.	Hourly Temperature and Dew Point
5.	Wind Shift	18.	Variable Sky Condition	31.	6-Hourly Maximum Temperature*
6.	Tower or Surface Visibility	19.	Significant Cloud Types	32.	6-Hourly Minimum Temperature*
7.	Variable Prevailing Visibility	20.	Ceiling Height at Second Location	33.	24-Hour Maximum and Minimum Temperature*
8.	Sector Visibility	21.	Pressure Rising or Falling Rapidly	34.	3-Hourly Pressure Tendency*
9.	Visibility at Second Location	22.	Sea-Level Pressure	35.	Sensor Status Indicators
10.	Lightning	23.	Aircraft Mishap	36.	Maintenance Indicator
11.	Beginning and Ending of Precipitation	24.	No SPECI Reports Taken		Note: Additive data is primarily used by the National Weather Service for climatological purposes.
12.	Beginning and Ending of Thunderstorms	25.	Snow Increasing Rapidly		* These groups should have no direct impact on the aviation community and will not be discussed in this document.
13.	Thunderstorm Location	26.	Other Significant Information		

Remarks are made in accordance with the following:

- Time entries are made in minutes past the hour if the time reported occurs during the same hour the observation is taken. Hours and minutes are used if the hour is different;

- Present weather coded in the body of the report as **VC** may be further described, i.e., direction from the station, if known. Weather phenomena beyond 10 statute miles of the point(s) of observation are coded as distant (**DSNT**) followed by the direction from the station. For example, precipitation of unknown intensity within 10 statute miles east of the station would be coded as **VCSH E**; lightning 25 statute miles west of the station would be coded as **LTG DSNT W**;

- Distance remarks are in statute miles except for automated lightning remarks which are in nautical miles;

- Movement of clouds or weather, when known, is coded with respect to the direction toward which the phenomena are moving. For example, a thunderstorm moving toward the northeast would be coded as **TS MOV NE**;

- Directions use the eight points of the compass coded in a clockwise order; and

- Insofar as possible, remarks are entered in the order they are presented in the following paragraphs (and Table 3-5).

3.1.3.13 Automated, Manual, and Plain Language Remarks
These remarks generally elaborate on parameters reported in the body of the report. Automated and manual remarks may be generated either by an automated station or observer. Plain language remarks are only provided from an observer.

3.1.3.13.1 Volcanic Eruptions (Plain Language)
Volcanic eruptions are coded in plain language and contain the following, when known:

- **Name** of volcano
- **Latitude and longitude** or the direction and approximate distance from the station
- **Date/Time** (UTC) of the eruption
- Size **description**, approximate height, and direction of movement **of the ash cloud**
- Any **other pertinent data** about the eruption

For example, a remark on a volcanic eruption would look like the following:

```
MT. AUGUSTINE VOLCANO 70 MILES SW ERUPTED AT 231505 LARGE ASH CLOUD
EXTENDING TO APRX 30000 FEET MOVING NE.
```

Pre-eruption volcanic activity is not coded. Pre-eruption refers to unusual and/or increasing volcanic activity which could presage a volcanic eruption.

3.1.3.13.2 Funnel Cloud
At manual stations, tornadoes, funnel clouds, and waterspouts are coded in the following format: Tornadic activity, **TORNADO**, **FUNNEL CLOUD**, or **WATERSPOUT,** followed by the beginning and/or ending time, followed by the location and/or direction of the phenomena from the station, and/or movement, when known. For example, **TORNADO B13 6 NE** would indicate that a tornado began at 13 minutes past the hour and was 6 statute miles northeast of the station.

3.1.3.13.3 Type of Automated Station
AO1 or **AO2** are coded in all METAR/SPECI from automated stations. Automated stations without a precipitation discriminator are identified as **AO1**; automated stations with a precipitation discriminator are identified as **AO2**.

3.1.3.13.4 Peak Wind
Peak wind is coded in the following format: the remark identifier **PK WND**, followed by the direction of the wind (first three digits), peak wind speed (next two or three digits) since the last METAR, and the time of occurrence. A space is between the two elements of the remark identifier and the wind direction/speed group; a solidus, /, (without spaces) separates the wind

direction/speed group and the time. For example, a peak wind of 45 knots from 280 degrees which occurred at 15 minutes past the hour is coded **PK WND 28045/15**.

3.1.3.13.5 Wind Shift

Wind shift is coded in the format: the remark identifier **WSHFT**, followed by the time the wind shift began. The contraction **FROPA** is entered following the time if there is reasonable data to consider the wind shift was the result of a frontal passage. A space is between the remark identifier and the time and, if applicable, between the time and the frontal passage contraction. For example, a remark reporting a wind shift accompanied by a frontal passage that began at 30 minutes after the hour would be coded as **WSHFT 30 FROPA**.

3.1.3.13.6 Tower or Surface Visibility

Tower or surface visibility is coded in the following format: tower **TWR VIS** or surface **SFC**, followed by the observed tower/surface visibility value. A space is coded between each of the remark elements. For example, the control tower visibility of 1 ½ statute miles would be coded **TWR VIS 1 1/2**.

3.1.3.13.7 Variable Prevailing Visibility

Variable prevailing visibility is coded in the following format: the remark identifier **VIS**, followed the lowest and highest visibilities evaluated separated by the letter **V**. A space follows the remark identifier and no spaces are between the letter **V** and the lowest/highest values. For example, a visibility that was varying between 1/2 and 2 statute miles would be coded **VIS 1/2V2**.

3.1.3.13.8 Sector Visibility (Plain Language)

Sector visibility is coded in the following format: the remark identifier **VIS**, followed by the sector referenced to 8 points of the compass, and the sector visibility in statute miles. For example, a visibility of 2 1/2 statute miles in the northeastern octant is coded **VIS NE 2 1/2**.

3.1.3.13.9 Visibility at Second Location

At designated automated stations, the visibility at a second location is coded in the following format: the remark identifier **VIS**, followed by the measured visibility value and the specific location of the visibility sensor(s) at the station. This remark will only be generated when the condition is lower than that contained in the body of the report. For example, a visibility of 2 1/2 statute miles measured by a second sensor located at runway 11 is coded **VIS 2 1/2 RWY11**.

3.1.3.13.10   Lightning

When lightning is observed at a manual station, the frequency, type of lightning and location is reported. The contractions for the type and frequency of lightning are based on Table 3-6, for example, **OCNL LTGICCG NW**, **FRQ LTG VC**, or **LTG DSNT W**.

When lightning is detected by an <u>automated</u> system:

- Within 5 nautical miles of the Airport Location Point (ALP), it is reported as **TS** in the body of the report with no remark;

- Between 5 and 10 nautical miles of the ALP, it is reported as **VCTS** in the body of the report with no remark; and

- Beyond 10 but less than 30 nautical miles of the ALP, it is reported in remarks only as **LTG DSNT** followed by the direction from the ALP.

Table 3-6. METAR/SPECI Type and Frequency of Lightning

Type of Lightning		
Type	**Contraction**	**Definition**
Cloud-ground	CG	Lightning occurring between cloud and ground.
In-cloud	IC	Lightning which takes place within the cloud.
Cloud-cloud	CC	Streaks of lightning reaching from one cloud to another.
Cloud-air	CA	Streaks of lightning which pass from a cloud to the air, but do not strike the ground.
Frequency of Lightning		
Frequency	**Contraction**	**Definition**
Occasional	OCNL	Less than 1 flash/minute.
Frequent	FRQ	About 1 to 6 flashes/minute.
Continuous	CONS	More than 6 flashes/minute.

3.1.3.13.11 Beginning and Ending of Precipitation

At designated stations, the beginning and ending time of precipitation is coded in the following format: the type of precipitation, followed by either a **B** for beginning or an **E** for ending, and the time of occurrence. No spaces are coded between the elements. The coded times of the precipitation start and stop times are found in the remarks section of the next METAR. The times are not required to be in the SPECI. The intensity qualifiers are coded. For example, if rain began at 0005 and ended at 0030 and then snow began at 0020 and ended at 0055, the remarks would be coded as **RAB05E30SNB20E55**. If the precipitation were showery, the remark is coded **SHRAB05E30SHSNB20E55**. If rain ended and snow began at 0042, the remark would be coded as **RAESNB42**.

3.1.3.13.12 Beginning and Ending of Thunderstorms

The beginning and ending of thunderstorms are coded in the following format: **TS** for thunderstorms, followed by either a **B** for beginning or an **E** for ending and the time of occurrence. No spaces are between the elements. For example, if a thunderstorm began at 0159 and ended at 0230, the remark is coded **TSB0159E30**.

3.1.3.13.13 Thunderstorm Location (Plain Language)

Thunderstorm locations are coded in the following format: the thunderstorm identifier, **TS**, followed by location of the thunderstorm(s) from the station and the direction of movement when known. For example, a thunderstorm southeast of the station and moving toward the northeast is coded **TS SE MOV NE**.

3.1.3.13.14 Hailstone Size (Plain Language)

At designated stations the hailstone size is coded in the following format: the hail identifier **GR**, followed by the size of the largest hailstone. The hailstone size is coded in ¼ inch increments. For example, **GR 1 3/4** would indicate that the largest hailstone were 1 ¾ inches in diameter. If small hail or snow pellets, **GS,** is coded in the body of the report, no hailstone size remark is required.

3.1.3.13.15 Virga (Plain Language)

Virga is coded in the following format: the identifier **VIRGA**, followed by the direction from the station. The direction of the phenomena from the station is optional, e.g., **VIRGA** or **VIRGA SW**.

3.1.3.13.16 Variable Ceiling Height

The variable ceiling height is coded in the following format: the identifier **CIG**, followed by the lowest ceiling height recorded, **V** denoting variability between two values, and ending with the highest ceiling height. A single space follows the identifier with no other spaces between the letter **V** and the lowest/highest ceiling values. For example, **CIG 005V010** would indicate a ceiling is variable between 500 and 1,000 feet.

3.1.3.13.17 Obscurations (Plain Language)

Obscurations, surface-based or aloft, are coded in the following format: the weather identifier causing the obscuration at the surface or aloft followed by the sky cover of the obscuration aloft (FEW, SCT, BKN, OVC) or at the surface (FEW, SCT, BKN), and the height. Surface-based obscurations have a height of **000**. A space separates the weather causing the obscuration and the sky cover; no space is between the sky cover and the height. For example, fog hiding 3/8 to 4/8 of the sky is coded **FG SCT000**; a broken layer at 2,000 feet composed of smoke is coded **FU BKN020**.

3.1.3.13.18 Variable Sky Condition (Plain Language)

Variable sky condition remarks are coded in the following format: the two operationally significant sky conditions (FEW, SCT, BKN, OVC) separated by spaces and **V** denoting the variability between the two ranges. If several layers have the same condition amount, the layer height of the variable layer is coded. For example, a cloud layer at 1,400 feet varying between broken and overcast is coded **BKN014 V OVC**.

3.1.3.13.19 Significant Cloud Types (Plain Language)

Significant cloud type remarks are coded in all reports.

3.1.3.13.19.1 Cumulonimbus or Cumulonimbus Mammatus

Cumulonimbus or Cumulonimbus Mammatus not associated with thunderstorms are coded in the following format: the cloud type (**CB** or **CBMAM**) followed by the direction from the station and the direction of movement when known. The cloud type, location, direction, and direction of movement entries are separated from each other by a space. For example, a CB up to 10 statute miles west of the station moving toward the east would be coded **CB W MOV E**. If the CB was more than 10 statute miles to the west, the remark is coded **CB DSNT W**.

Cumulonimbus (CB) always evolves from the further development of towering cumulus (TCU). The unusual occurrence of lightning and thunder within or from a CB leads to its popular titl, thunderstorm. A thunderstorm usually contains severe or greater turbulence, severe icing, low level wind shear (LLWS), and instrument flight rules (IFR) conditions.

Figure 3-5. Cumulonimbus (CB) Example
CB always evolves from the further development of towering cumulus (TCU). The usual occurrence of lightning and thunder within or from a CB leads to its popular title, thunderstorm. A thunderstorm usually contains severe or greater turbulence, severe icing, low level wind shear (LLWS), and instrument flight rules (IFR) conditions. (Copyright Robert A. Prentice, 1990)

Figure 3-6. Cumulonimbus Mammatus (CBMAM) Example
Cumulonimbus Mammatus (CBMAM) (also called mammatus) appears as hanging protuberances, like pouches, on the undersurface of a cloud. (Copyright Robert A. Prentice, 1993)

3.1.3.13.19.2 Towering Cumulus

Towering cumulus clouds are coded in the following format: the identifier **TCU** followed by the direction from the station. The cloud type and direction entries are separated by a space. For example, a towering cumulus cloud up to 10 statute miles west of the station is coded as **TCU W**.

Figure 3-7. Towering Cumulus (TCU) Example
Towering Cumulus (TCU). TCU is produced by strong convective updrafts and, thus, indicates turbulence. Icing is typically found above the freezing level. TCU often transforms into cumulonimbus (CB). (Copyright Charles A. Doswell, III, 1977)

3.1.3.13.19.3Altocumulus Castellanus

Altocumulus Castellanus is coded in the following format: the identifier **ACC** followed by direction from the station. The cloud type and direction entries are separated by a space. For example, an altocumulus cloud 5 to 10 statute miles northwest of the station is coded **ACC NW**.

Figure 3-8. Altocumulus Castellanus (ACC) Example
Altocumulus Castellanus (ACC). ACC indicates convective turbulence aloft from the top of the cloud to its base and usually an undetermined height below cloud base as well. (Photo courtesy of National Severe Storms Laboratory/University of Oklahoma)

3.1.3.13.19.4 Standing Lenticular or Rotor Clouds

Stratocumulus (**SCSL**), altocumulus (**ACSL**), or cirrocumulus (**CCSL**), or rotor clouds are coded in the following format: the cloud type followed by the direction from the station. The cloud type and direction entries are separated by a space. For example, altocumulus standing lenticular clouds observed southwest through west of the station are coded **ACSL SW-W**; an apparent rotor cloud 5 to 10 statute miles northeast of the station is coded **APRNT ROTOR CLD NE**; and cirrocumulus clouds south of the station are coded **CCSL S**.

Figure 3-9. Standing Lenticular and Rotor Clouds Example
From top to bottom: Cirrocumulus standing lenticular (CCSL), altocumulus standing lenticular (ACSL), and rotor cloud. These clouds are characteristic of mountain waves. Mountain waves can occasionally produce violent downslope windstorms. Intense mountain waves can present a significant hazard to aviation by producing severe or even extreme turbulence that extends upward into the lower stratosphere.

3.1.3.13.20 Ceiling Height at Second Location

At designated stations, the ceiling height at a second location is coded in the following format: the identifier **CIG** followed by the measured height of the ceiling and the specific location of the ceilometer(s) at the station. This remark is only generated when the ceiling is lower than that contained in the body of the report. For example, if the ceiling measured by a second sensor located at runway 11 is broken at 200 feet, the remark would be **CIG 002 RWY11**.

3.1.3.13.21   Pressure Rising or Falling Rapidly

At designated stations, the reported pressure is evaluated to determine if a pressure change is occurring. If the pressure is rising or falling at a rate of at least 0.06 inch per hour and the pressure change totals 0.02 inch or more at the time of the observation, a pressure change remark is reported. When the pressure is rising or falling rapidly at the time of observation, the remark **PRESRR** (pressure rising rapidly) or **PRESFR** (pressure falling rapidly) is included in the remarks.

3.1.3.13.22   Sea-Level Pressure

At designated stations, the sea-level pressure is coded in the following format: the identifier **SLP** immediately followed by the sea level pressure in hectopascals. The hundreds and thousands units are not coded and must be inferred. For example, a sea-level pressure of 998.2 hectopascals is coded as **SLP982**. A sea-level pressure of 1013.2 hectopascals would be coded as **SLP132**. For a METAR, if sea-level pressure is not available, it is coded as **SLPNO**.

3.1.3.13.23   Aircraft Mishap (Plain Language)

If a SPECI report is taken to document weather conditions when notified of an aircraft mishap, the remark **ACFT MSHP** is coded in the report but the SPECI not transmitted.

3.1.3.13.24   No SPECI Reports Taken (Plain Language)

At manual stations where SPECIs are not taken, the remark **NOSPECI** is coded to indicate no changes in weather conditions will be reported until the next METAR.

3.1.3.13.25   Snow Increasing Rapidly

At designated stations, the snow increasing rapidly remark is reported, in the NEXT METAR, whenever the snow depth increases by 1 inch or more in the past hour. The remark is coded in the following format: the remark indicator **SNINCR**, the depth increase in the past hour, and the total depth of snow on the ground at the time of the report. The depth of snow increase in the past hour and the total depth on the ground are separated from each other by a solidus, /. For example, a snow depth increase of 2 inches in the past hour with a total depth on the ground of 10 inches is coded **SNINCR 2/10**.

3.1.3.13.26   Other Significant Information (Plain Language)

Agencies may add to a report other information significant to their operations, such as information on fog dispersal operations, runway conditions, **FIRST** or **LAST** reports from station, etc.

3.1.3.14 Additive and Automated Maintenance Data

Additive data groups (Table 3-5) are only reported at designated stations and are primarily used by the NWS for climatological purposes. Most have no direct impact on the aviation community but a few are discussed below.

3.1.3.14.1 Hourly Temperature and Dew Point

At designated stations, the hourly temperature and dew point group are further coded to the tenth of a degree Celsius. For example, a recorded temperature of +2.6°C and dew point of -1.5°C would be coded as **T00261015**.

The format for the coding is as follows:

T   Group indicator

0   Indicates the following number is positive; a **1** would be used if the temperature was reported as negative at the time of observation

026   Temperature disseminated to the nearest 10th and read as 02.6

1   Indicates the following number is negative; a **0** would be used if the number was reported as positive at the time of observation

015   Dew Point disseminated to the nearest 10th and read as 01.5

No spaces are between the entries. For example, a temperature of 2.6°C and dew point of −1.5°C is reported in the body of the report as **03/M01** and the hourly temperature and dew point group as **T00261015**. If the dew point is missing only the temperature is reported; if the temperature is missing the hourly temperature and dew point group is not reported.

3.1.3.14.2 Maintenance Data Groups

The following maintenance data groups, Sensor Status Indicators and the Maintenance Indicator, are only reported from automated stations.

3.1.3.14.2.1 Sensor Status Indicators

Sensor status indicators are reported as indicated below:

- If the Runway Visual Range is missing and would normally be reported, **RVRNO** is coded

- When automated stations are equipped with a present weather identifier and the sensor is not operating, the remark **PWINO** is coded

- When automated stations are equipped with a tipping bucket rain gauge and the sensor is not operating, **PNO** is coded

- When automated stations are equipped with a freezing rain sensor and the sensor is not operating, the remark **FZRANO** is coded

- When automated stations are equipped with a lightning detection system and the sensor is not operating, the remark **TSNO** is coded

- When automated stations are equipped with a secondary visibility sensor and the sensor is not operating, the remark **VISNO LOC** is coded

- When automated stations are equipped with a secondary ceiling height indicator and the sensor is not operating, the remark **CHINO LOC** is coded

3.1.3.14.2.2 Maintenance Indicator

A maintenance indicator, **$**, is coded when an automated system detects maintenance is needed on the system.

3.1.4 Examples of METAR Reports, Explanations, and Phraseology

```
METAR KMKL 021250Z 33018KT 290V360 1/2SM R31/2600FT SN BLSN FG VV008
00/M03 A2991 RMK AO2 RAESNB42 SLPNO T00111032
```

METAR	▸	Aviation Routine Weather Report
KMKL	▸	United States Jackson McKellar-Sipes Regional Airport, Tennessee
021250Z	▸	The 2nd day of the month, 1300 hour scheduled report taken at 1250 UTC
33018KT	▸	Wind 330 degrees at 18 knots
290V360	▸	Wind direction variable between 290 and 360 degrees
1/2SM	▸	Visibility one-half statute mile
R31/2600FT	▸	Runway 31, runway visual range on runway 2,600 feet
SN	▸	Moderate snow
BLSN FG	▸	Blowing snow and fog
VV008	▸	Indefinite ceiling, vertical visibility 800 feet AGL
00/M03	▸	Temperature 0°C, dew point -3°C
A2991	▸	Altimeter, 29.91 inches of mercury
RMK	▸	Remarks
AO2	▸	Automated station with a precipitation discriminator
RAESNB42	▸	Rain ended at four two, snow began at four two past the hour
SLPNO	▸	Sea-level pressure not available
T00111032	▸	Temperature 1.1°C, dew point -3.2°C

Jackson McKellar-Sipes Regional Airport, wind three three zero at one eight, wind variable between two niner zero and three six zero, visibility one-half, runway three one R-V-R, two thousand six hundred, snow, blowing snow, fog, indefinite ceiling eight hundred, temperature zero, dew point minus three, altimeter two niner niner one, remarks rain ended and snow began at four two past the hour.

METAR KIPT 191254Z 00000KT 1 1/2SM -RA BR SCT034 BKN100 19/18 A2993 RMK AO2 RAB24 SLP133 P0001 T01890178

METAR	▸	Aviation Routine Weather Report
KIPT	▸	United States Williamsport Regional Airport, Pennsylvania
191254Z	▸	19th day of the month, the 1300 hour scheduled report taken 1254 UTC
00000KT	▸	Wind calm
1 1/2SM	▸	Visibility one and one-half statute mile
-RA BR	▸	Light rain, mist
SCT034 BKN100	▸	Scattered 3,400 feet AGL, ceiling broken 10,000 feet AGL
19/18	▸	Temperature 19 degrees Celsius, Dew Point 18 degrees Celsius
A2993	▸	Altimeter, 29.93 inches of mercury
RMK	▸	Remarks
AO2	▸	Automated station with a precipitation discriminator
RAB24	▸	Rain began at 1224 UTC
SLP133	▸	Sea level pressure 1013.3 hectopascals
P0001	▸	Precipitation over the past hour 00.01 inch
T01890178	▸	Temperature 18.9 degrees Celsius, dew point 17.8 degrees Celsius

Williamsport Regional Airport, wind calm, visibility one and one half, light rain, mist, three thousand four hundred scattered, ceiling one zero thousand broken, temperature one niner, dew point one eight, altimeter two niner niner three, remarks rain began at two four past the hour.

SPECI KCVG 312228Z 28024G36KT 3/4SM +TSRA SQ BKN008 OVC020CB 28/23
A3000 RMK TSB24 TS OHD MOV E

SPECI	→ Aviation Selected Special Weather Report
KCVG	→ United States Covington Cincinnati/Northern Kentucky International Airport, Kentucky
312228Z	→ The 31st of the month Special report taken at 2228 UTC
28024G36KT	→ Wind 280 degrees at 24 knots, gusts 36 knots
3/4SM	→ Visibility three-quarters statute mile
+TSRA SQ	→ Thunderstorm with heavy rain and squalls
BKN008 OVC020CB	→ Ceiling broken 800 feet AGL, overcast 2,000 feet AGL cumulonimbus
28/23	→ Temperature 28°C, dew point 23°C
A3000	→ Altimeter 30.00 inches of mercury
RMK	→ Remarks
TSB24	→ Thunderstorm began at two four minutes past the hour
TS OHD MOV E	→ Thunderstorm overhead moving east

Covington Cincinnati/Northern Kentucky International Airport, special report, two eight observation, wind two eight zero at two four, gusts three six, visibility three-quarters, thunderstorm, heavy rain, squall, ceiling eight hundred broken, two thousand overcast cumulonimbus, temperature two eight, dew point two three, altimeter three zero zero zero, thunderstorm began two four, thunderstorm overhead, moving east."

METAR KLAX 191350Z 08004KT 4SM HZ OVC009 18/16 A2997 RMK AO2 SLP147
T01830156

METAR	→ Aviation Routine Weather Report
KLAX	→ United States Los Angeles International Airport, California
191350Z	→ The 19th day of the month, the 1400 hour scheduled report at 1350 UTC
08004KT	→ Wind 80 degrees at 4 knots
4SM	→ Visibility 4 statute miles
HZ	→ Haze
OVC009	→ Ceiling overcast 900 feet AGL
18/16	→ Temperature 18°C, dew point 16°C
A2997	→ Altimeter 29.97 inches of mercury
RMK	→ Remarks
AO2	→ Automated observation with precipitation discriminator
SLP147	→ Sea level pressure 1014.7 hectopascals
T01830156	→ Temperature 18.3°C, dew point 15.6°C

Los Angeles International Airport, wind zero eight zero at four, visibility four, haze, ceiling niner hundred overcast, temperature one eight, dew point one six, altimeter two niner niner seven.

SPECI KDEN 241310Z 09014G35KT 1/4SM +SN FG VV002 01/01 A2975 RMK AO2
TWR VIS 1/2 RAESNB08

SPECI	→ Aviation Selected Special Weather Report
KDEN	→ United States Denver International Airport, Colorado

241310Z	........▸	The 24th of the month, Special report taken at 1310 UTC
09014G35KT	........▸	Wind 90 degrees at 14 knots, gusts to 35 knots
1/4SM	........▸	Visibility one-quarter statute mile
+SN FG	........▸	Heavy snow, fog
VV002	........▸	Indefinite ceiling, vertical visibility 200 feet AGL
01/01	........▸	Temperature 1°C, dew point 1°C
A2975	........▸	Altimeter 29.75 inches of mercury
RMK	........▸	Remarks
AO2	........▸	Automated observation with precipitation discriminator
TWR VIS 1/2	........▸	Tower visibility one-half statute mile
RAE08SNB08	........▸	Rain ended at 08 past the hour and snow began at 08 minutes past the hour

Denver International Airport, wind zero niner zero at one four, gusts three five, visibility one-quarter, heavy snow, fog, indefinite ceiling two hundred, temperature one, dew point one, altimeter two niner seven five, remarks tower visibility one half, ran ended and snow began at zero eight.

METAR KSPS 301656Z 06014KT 020V090 3SM -TSRA FEW040 BKN060CB 12/ A2982
RMK OCNL LTGICCG NE TSB17 TS E MOV NE PRESRR SLP093

METAR	........▸	Aviation Routine Weather Report
KSPS	........▸	United States Sheppard Air Force Base/Wichita Falls Municipal Airport, Texas
301656Z	........▸	The 30th day of the month, the 1700 scheduled report taken at 1656 UTC
06014KT 020V090	▸	Wind 60 degrees at 14 knots, wind variable between 020 and 090 degrees
3SM	........▸	Visibility 3 statute miles
-TSRA	........▸	Thunderstorm, light rain
FEW040 BKN060CB	▸	Few 4,000 feet AGL, ceiling broken 6,000 feet AGL cumulonimbus
12/	........▸	Temperature 12°C, dew point missing
A2982	........▸	Altimeter 29.82 inches of mercury
RMK	........▸	Remarks
OCNL LTGICCG NE	▸	Occasional lightning in cloud, cloud-to-ground northeast
TSB17	........▸	Thunderstorm began at 17 minutes past the hour
TS E MOV NE	........▸	Thunderstorm east moving northeast
PRESRR	........▸	Pressure rising rapidly
SLP093	........▸	Sea-level pressure 1009.3 hectopascals

Sheppard Air Force Base/Wichita Falls Municipal Airport, automated, wind zero six zero at one four, wind variable between zero two zero and zero niner zero, visibility three, thunderstorm, light rain, few clouds at four thousand, ceiling six thousand broken cumulonimbus, temperature one two, dew point missing, remarks occasional lightning in-cloud, cloud-to-ground northeast, thunderstorm began at one seven, thunderstorm east moving northeast, pressure rising rapidly.

SPECI KBOS 051237Z VRB02KT 3/4SM R15R/4000FT BR OVC004 05/05 A2998 RMK
AO2 CIG 002V006 T00520048

SPECI	........▸	Aviation Selected Special Weather Report

KBOS	United States Boston, Massachusetts
051237Z	The 5th of the month, Special report taken at 1237 UTC
VRB02KT	Wind variable at 2 knots
3/4SM	Visibility three-quarters statute mile
R15R/4000FT	Runway 15R, visual range on runway 4,000 feet
BR	Mist
OVC004	Ceiling overcast 400 feet AGL
05/05	Temperature 5°C, dew point 5°C
A2998	Altimeter 29.98 inches of mercury
RMK	Remarks
AO2	Automated observation with precipitation discriminator
CIG 002V006	Ceiling variable between 200 to 600 feet
T00520048	Temperature 5.2°C, dew point 4.8°C

Boston General Edward Lawrence Logan International Airport, special report, three seven observation, wind variable at two, visibility three-quarters, runway one five right R-V-R four thousand, mist, ceiling four hundred overcast, temperature five, dew point five, altimeter two niner niner eight, remarks, ceiling variable between two hundred and six hundred.

3.2 Pilot Weather Reports (PIREP)

No report is timelier than the one made from the flight deck of aircraft in flight. In fact, aircraft in flight are the only means of observing actual icing and turbulence conditions. Pilots welcome pilot weather reports (PIREPs) as well as pilot weather briefers and forecasters. Pilots should report any observation, good or bad, to assist other pilots with flight planning and preparation. If conditions were forecasted to occur but not encountered, a pilot should also report this inaccuracy. This will help the NWS verify forecast products and create more accurate products for the aviation community. Pilots should help themselves, the aviation public, and the aviation weather forecasters by providing PIREPs.

Pipe Up with a PIREP and help the aviation community operate more safely and effectively.

PIREPs are available in the internet at the Aviation Digital Data Service (ADDS) web page at: http://adds.aviationweather.gov/pireps/

3.2.1 Format
A PIREP is transmitted in a prescribed format (Figure 3-7). Required elements for all PIREPs are: message type, location, time, altitude/flight level, type aircraft, and at least one other element to describe the reported phenomena. The other elements will be omitted when no data is reported with them. All altitude references are mean sea level (MSL) unless otherwise noted. Distance for visibility is in statute miles and all other distances are in nautical miles. Time is reported in Universal Time Coordinated (UTC).

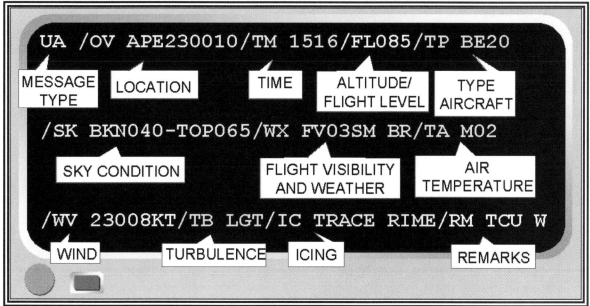

Figure 3-10. Pilot Weather Report (PIREP) Coding Format

3.2.1.1 Message Type (UUA/UA)
The two types of PIREPs are Urgent (**UUA**) and Routine (**UA**).

3.2.1.1.1 Urgent PIREPs
Urgent (**UUA**) PIREPs contain information about:

- Tornadoes, funnel clouds, or waterspouts

- Severe or extreme turbulence (including Clear Air Turbulence)

- Severe icing

- Hail

- Low Level Wind Shear (**LLWS**) within 2,000 feet of the surface. LLWS PIREPS are classified as **UUA** if the pilot reports air speed fluctuations of 10 knots or more or if air speed fluctuations are not reported but LLWS is reported, the PIREP is classified as **UUA**.

- Volcanic ash clouds

- Any other weather phenomena reported which are considered by the briefer as being hazardous, or potentially hazardous, to flight operations.

3.2.1.1.2 Routine PIREPs
Routine PIREPs are issued after receiving a report from a pilot that does not contain any urgent information as listed in Section 3.2.1.1.1.

3.2.1.2 Location (/OV)
The Location (/**OV**) can be referenced either by geographical position or by route segment.

3.2.1.2.1 Location
Location can be referenced to a VHF NAVAID or an airport, using either the three-letter International Air Transport Association (IATA) or four letter International Civil Aviation Organization (ICAO) identifier. If appropriate, the PIREP is encoded using the identifier, then three digits to define a radial and three digits to define the distance in nautical miles.

Examples:

/OV APE	Over Appleton VOR
/OV KJFK	Over John F. Kennedy International Airport, New York City, NY
/OV APE230010	230 degrees at 10 nautical miles from the Appleton VOR
/OV KJFK107080	107 degrees at 80 nautical miles from John F. Kennedy International Airport, New York City, New York

3.2.1.2.1.1 Route Segment
A PIREP can also be referenced using two or more fixes to describe a route.

Examples:

/OV KSTL-KMKC	From Lambert-Saint Louis International Airport, Missouri to Charles B. Wheeler Downtown Airport, Kansas City, Missouri
/OV KSTL090030-KMKC045015	From 90 degrees at 30 nautical miles from Lambert-Saint Louis International Airport, Missouri to 45 degrees at 15 nautical miles from Charles B. Wheeler Downtown Airport, Kansas City, Missouri

3.2.1.3 Time (/TM)

Time (/**TM**) is the time that the reported phenomenon occurred or was encountered. It is coded in four digits UTC.

Example:

`/TM 1315` ⟶ 1315 UTC

3.2.1.4 Altitude/Flight Level (/FL)

The Altitude/Flight Level (/**FL**) is the altitude in hundreds of feet MSL where the phenomenon was first encountered. If not known, **UNKN** is entered. If the aircraft was climbing or descending, the appropriate contraction (**DURC** or **DURD**) is entered in the remarks (/**RM**). If the condition was encountered within a layer, the altitude range is entered within the appropriate element that describes the condition.

Examples:

`/FL085` ⟶ 8,500 feet MSL
`/FL310` ⟶ Flight Level 310
`/FLUNKN /RM DURC` ⟶ Flight Level unknown, remarks, during climb

3.2.1.5 Aircraft Type (/TP)

Aircraft Type (/**TP**) is entered. If not known, **UNKN** is entered. Icing and turbulence reports always include aircraft type.

Examples:

`/TP BE20` ⟶ Super King Air 200
`/TP SR22` ⟶ Cirrus 22
`/TP P28R` ⟶ Piper Arrow
`/TP UNKN` ⟶ Type unknown

3.2.1.6 Sky Condition (/SK)

Sky Condition (/**SK**) group is used to report height of cloud bases, tops, and cloud cover. The height of the base of a layer of clouds is coded in hundreds of feet MSL. The top of a layer is entered in hundreds of feet MSL preceded by the word **-TOP**. If reported as clear above the highest cloud layer, **SKC** is coded following the reported level.

Examples:

`/BKN040-TOP065` ⟶ Base of broken layer 4,000 feet MSL, top 6,500 feet MSL
`/SK OVC100-TOP110/ SKC` ⟶ Base of an overcast layer 10,000 feet MSL, top 11,000 feet MSL, clear above
`/SK OVC015-TOP035/OVC230` ⟶ Base of an overcast layer 1,500 feet MSL, top 3,500 feet MSL, base of an overcast layer 23,000 feet MSL
`/SK OVC-TOP085` ⟶ Overcast layer, top 8,500 feet MSL

Cloud cover amount ranges are entered with a hyphen separating the amounts; i.e., **BKN-OVC**.

Examples:

`/SK SCT-BKN050-TOP100` ⟶ Base of a scattered to broken layer 5,000 feet MSL, top 10,000 feet MSL

`/SK BKN-OVCUNKN-TOP060/BKN120-TOP150/ SKC` ⟶ Base of a broken to overcast layer unknown, top 6,000 feet MSL, base of a broken layer 12,000 feet MSL, top 15,000 feet MSL, clear above

Unknown heights are indicated by the contraction **UNKN**.

Example:

`/SK OVC065-TOPUNKN` ⟶ Base of an overcast layer 6,500 feet MSL, top unknown

If a pilot indicates he/she is in the clouds, **IMC** is entered.

Example:

`/SK OVC065-TOPUNKN /RM IMC` ⟶ Base of an overcast layer 6,500 feet MSL, top unknown, remark, in the clouds

When more than one layer is reported, layers are separated by a solidus (*/*).

3.2.1.7 Flight Visibility and Weather (/WX)
Weather conditions encountered by the pilot are reported as follows:

Flight visibility, when reported, is entered first in the **/WX** field. It is coded as **FV** followed by a two-digit visibility value rounded down, if necessary, to the nearest whole statute mile and appended with **SM** (**FV03SM**). If visibility is reported as unrestricted, **FV99SM** is entered.

Flight weather types are entered using one or more of the standard surface weather reporting symbols contained in Table 3-7.

Table 3-7. PIREP Weather Type and Symbols

Type	METAR Code
Drifting / Blowing Snow	DRSN/BLSN
Drifting Dust	DRDU
Drifting Sand	DRSA
Drizzle/Freezing Drizzle	DZ/FZDZ
Dust / Blowing Dust	DU/BLDU
Duststorm	DS
Fog (visibility less than 5/8SM)	FG
Freezing Fog	FZFG
Freezing Rain	FZRA
Funnel Cloud	FC
Hail (Approximately ¼-inch diameter or more)	GR
Hail Shower	SHGR
Haze	HZ
Ice Crystals	IC
Ice Pellets/Showers	PL/SHPL
Mist (visibility great than or equal to 5/8SM)	BR
Patchy Fog	BCFG
Patchy Fog on part of airport	PRFG
Rain/Showers	RA/SHRA
Sand/Blowing Sand	SA/BLSA
Sandstorms	SS
Shallow Fog	MIFG
Small Hail/Snow Pellet Showers	SHGS
Small Hail/Snow Pellets	GS
Smoke	FU
Snow Grains	SG
Snow / Showers	SN/SHSN
Spray	PY
Squalls	SQ
Thunderstorm	TS
Tornado/Waterspout	+FC
Unknown Precipitation	UP
Volcanic Ash	VA
Well developed Dust/Sand Whirls	PO

Intensity modifiers for precipitation (- for light, no qualifier for moderate, and + for heavy) indicates precipitation type, except ice crystals and hail, including those associated with a thunderstorm and those of a showery nature.

Intensity modifiers for obscurations are ascribed as moderate or heavy (+) for dust and sandstorms only. No intensity modifiers are used for blowing dust, blowing sand, or blowing snow.

Example:

`/WV FV01SM +DS000-TOP083/SKC /RM DURC` ········► Flight visibility 1 statute mile, base heavy duststorm layer at the surface, top 8,300 feet MSL, clear above, remarks, during climb

When more than one form of precipitation is combined in the report, the dominant type is reported first.

Examples:

`/WX FV00SM +TSRAGR` ········► Flight visibility zero statute miles, thunderstorm, heavy rain, hail
`/WX FV02SM BRHZ000-TOP083` ········► Flight visibility 2 statute miles, base of a haze and mist layer at the surface, top 8,300 feet MSL

If a funnel cloud is reported, it is coded as **FC** following **/WX** group and is spelled out as **Funnel Cloud** after **/RM** group. If a tornado or waterspout is reported, it is coded **+FC** following **/WX** group and **TORNADO** or **WATERSPOUT** is spelled out after the **/RM** group.

Examples:

`/WX FC /RM FUNNEL CLOUD` ········► Funnel cloud, remarks, funnel cloud
`/WX +FC /RM TORNADO` ········► Tornado, remark, tornado

When the size of hail is stated, it is coded in 1/4-inch increments in remarks (**/RM**) group.

The proximity qualifier **VC** (vicinity) is only used with **TS**, **FG**, **FC**, **+FC**, **SH**, **PO**, **BLDU**, **BLSA**, and **BLSN**.

Example:

`/WX FV02SM BLDU000-TOP083 VC W` ········► Flight visibility 2 statute miles, base of a blowing dust layer at the surface, top 8,300 feet MSL in the vicinity, west

When more than one type of weather is reported, they are reported in the following order:

- **TORNADO**, **WATERSPOUT**, or **FUNNEL CLOUD**
- Thunderstorm with or without associated precipitation
- Weather phenomena in order of decreasing predominance.

No more than three groups are used in a single PIREP.

Weather layers are entered with the base and/or top of the layer when reported. The same format as in the sky condition (**/SK**) group is used.

Example:

`/WX FU002-TOP030` ········► Base of a smoke layer, 200 feet MSL, top 3,000 feet MSL

3.2.1.8 Air Temperature (/TA)

Outside air temperature (/**TA**) is reported using two digits in degrees Celsius. Negative temperatures is prefixed with an **M**; e.g., /**TA 08** or /**TA M08**.

3.2.1.9 Wind Direction and Speed (/WV)

Wind direction and speed is encoded using three digits to indicate wind direction (magnetic) and two or three digits to indicate reported wind speed. When the reported speed is less than 10 knots, a leading zero is used. The wind group will always have **KT** appended to represent the units in knots.

Examples:

/**WV 02009KT** ┄┄→ Wind 20 degrees (magnetic) at 9 knots
/**WV 28057KT** ┄┄→ Wind 280 degrees (magnetic) at 57 knots
/**WV 350102KT** ┄┄→ Wind 350 degrees (magnetic) at 102 knots

3.2.1.10 Turbulence (/TB)

Turbulence intensity, type, and altitude are reported after wind direction and speed.

Intensity is coded first. Duration is coded next if reported by the pilot (intermittent, occasional, continuous) followed by the intensity using contractions **LGT**, **MOD**, **SEV**, or **EXTRM**. Range or variation of intensity is separated with a hyphen; e.g., MOD-SEV. If turbulence was forecasted, but not encountered, **NEG** is entered.

Type is coded second. **CAT** (Clear Air Turbulence) or **CHOP** is entered if reported by the pilot. High-level turbulence (normally above 15,000 feet AGL) not associated with clouds (including thunderstorms) is reported as CAT.

Altitude is reported (last) only if it differs from value reported in the Altitude/Flight Level (/**FL**) group. When a layer of turbulence is reported, height values are separated with a hyphen. If lower or upper limits are not defined, **BLO** or **ABV** is used.

Examples:

/**TB LGT** ┄┄┄┄→ Light turbulence
/**TB LGT 040** ┄┄┄→ Light turbulence at 4,000 feet MSL
/**TB OCNL MOD-SEV BLO 080** ┄→ Occasional moderate to severe turbulence below 8,000 feet MSL
/**TB MOD-SEV CAT 350** ┄→ Moderate to severe clear air turbulence at 35,000 feet MSL
/**TB NEG 120-180** ┄┄→ Negative turbulence between 12,000 to 18,000 feet MSL
/**TB CONS MOD CHOP 220/NEG 230-280** ┄→ Continuous moderate chop at 22,000 feet MSL, negative turbulence between 23,000 to 28,000 feet MSL
/**TB MOD CAT ABV 290** ┄→ Moderate clear air turbulence above 29,000 feet MSL

Turbulence reports should include location, altitude, or range of altitudes, and aircraft type, and, when reported, whether in clouds or clear air. The pilot determines the degree of turbulence, intensity, and duration (occasional, intermittent, and continuous). The report should be obtained and disseminated, when possible, in conformance with the U.S. Standard Turbulence Criteria Table 3-8.

Table 3-8. PIREP Turbulence Reporting Criteria

Intensity	Aircraft Reaction	Reaction Inside Aircraft	Reporting Term-Definition
Light	Turbulence that momentarily causes slight, erratic changes in altitude and/or attitude (pitch, roll, yaw). Report as **Light Turbulence;**[1] or Turbulence that causes slight, rapid and somewhat rhythmic bumpiness without appreciable changes in altitude or attitude. Report as **Light Chop.**	Occupants may feel a slight strain against belts or shoulder straps. Unsecured objects may be displaced slightly. Food service may be conducted and little or no difficulty is encountered in walking.	Occasional – Less than 1/3 of the time. Intermittent-1/3 to 2/3 Continuous-More than 2/3
Moderate	Turbulence that is similar to Light Turbulence but of greater intensity. Changes in altitude and/or attitude occur but the aircraft remains in positive control at all times. It usually causes variations in indicated airspeed. Report as **Moderate Turbulence;**[1] or Turbulence that is similar to Light Chop but of greater intensity. It causes rapid bumps or jolts without appreciable changes in aircraft or attitude. Report as **Moderate Chop.**[1]	Occupants feel definite strains against seat belts or shoulder straps. Unsecured objects are dislodged. Food service and walking are difficult.	NOTE 1. Pilots should report location(s), time (UTC), intensity, weather in or near clouds, altitude, type of aircraft and, when applicable, duration of turbulence. 2. Duration may be based on time between two locations or over a single location. All locations should be readily identifiable. **EXAMPLES:** Over Omaha. 1232Z, Moderate Turbulence, in cloud, flight Level 310, B737.
Severe	Turbulence that causes large, abrupt changes in altitude and/or attitude. It usually causes large variations in indicated airspeed. Aircraft may be momentarily out of control. Report as **Severe Turbulence.**[1]	Occupants are forced violently against seat belts or shoulder straps. Unsecured objects are tossed about. Food service and walking are impossible.	b. From 50 miles south of Albuquerque to 30 miles north of Phoenix, 1210Z to 1250Z, occasional Moderate Chop, Flight Level 330, DC8.
Extreme	Turbulence in which the aircraft is violently tossed about and is practically impossible to control. It may cause structural damage. Report as **Extreme Turbulence.**[1]		

[1] High level turbulence (normally above 15,000 feet ASL) not associated with clouds, including thunderstorms, should be reported as CAT (clear air turbulence) preceded by the appropriate intensity, or light or moderate chop.

3.2.1.11 Icing (/IC)
Icing intensity, type and altitude is reported after turbulence.

Intensity is coded first using contractions **TRACE**, **LGT** (light), **MOD** (moderate), or **SEV** severe). Reports of a range or variation of intensity is separated with a hyphen. If icing was forecast but not encountered, **NEG** (negative) is coded.

The following table classifies icing intensity according to its operational effects on aircraft.

Table 3-9. Icing Intensities, Contractions, and Airframe Ice Accumulation

Intensity	Contraction	Airframe Ice Accumulation
Trace	TRACE	Ice becomes perceptible. Rate of accumulation slightly greater than rate of sublimation. It is not hazardous even without the use of deicing/anti-icing equipment unless encountered for an extended period of time (over 1 hour).
Light	LGT	The rate of accumulation may create a problem if flight is prolonged in this environment (over 1 hour). Occasional use of deicing/anti-icing equipment removes/prevents accumulation. It does not present a problem if the deicing/anti-icing equipment is used.
Moderate	MOD	The rate of accumulation is such that even short encounters become potentially hazardous and use of deicing/anti-icing equipment or diversion is necessary.
Severe	SEV	The rate of accumulation is such that deicing/anti-icing equipment fails to reduce or control the hazard. Immediate diversion is necessary.

Icing type is reported second. Reportable types are **RIME**, **CLR** (clear), or **MX** (mixed).

The following table classifies icing type according to it description.

Table 3-10. Icing Types, Contractions, and Descriptions

Icing Type	Contraction	Description
Rime	RM	Rough, milky, opaque ice formed by the instantaneous freezing of small super-cooled water droplets.
Clear	CLR	A glossy, clear or translucent ice formed by the relatively slow freezing of large super-cooled water droplets.
Mixed	MX	A combination of both rime and clear.

The reported icing/altitude is coded (last) only if different from the value reported in the altitude/flight level (**/FL**) group. A hyphen is used to separate reported layers of icing. **ABV** (above) or **BLO** (below) is coded when a layer is not defined.

Pilot reports of icing should also include location (**/OV**), type aircraft (**/TP**), and air temperature (**/TA**).

Examples:

`/IC LGT-MOD MX 085` ⋯⋯➤ Light to moderate mixed icing, 8,500 feet MSL
`/IC LGT RIME` ⋯⋯➤ Light rime icing
`/IC MOD RIME BLO 095` ⋯⋯➤ Moderate rime icing below 9,500 feet MSL
`/IC SEV CLR 035-062` ⋯⋯➤ Severe clear icing 3,500 to 6,200 feet MSL

3.2.1.12 Remarks (/RM)

The remarks (**/RM**) group is used to report a phenomenon which is considered important but does not fit in any of the other groups. This includes, but is not limited to, low-level wind shear

(**LLWS**) reports, thunderstorm lines, coverage and movement, size of hail (1/4-inch increments), lightning, clouds observed but not encountered, geographical or local description of where the phenomenon occurred, and contrails. Hazardous weather is reported first. LLWS is described to the extent possible.

3.2.1.12.1 Wind Shear

Ten knots or more fluctuations in wind speed (+/- 10KTS), within 2,000 feet of the surface, require an Urgent (**UUA**) pilot report. When Low Level Wind Shear is entered in a pilot report, **LLWS** is entered as the first remark in the remarks (**/RM**) group.

Example:

`/RM LLWS +/-15 KT SFC-008 DURC RY22 JFK` ┄┄► Remarks, Low Level Wind Shear, air speed fluctuations of plus or minus 15 knots, surface to 800 feet during climb, runway 22, John F. Kennedy International Airport, New York.

3.2.1.12.2 FUNNEL CLOUD, TORNADO, and WATERSPOUT

Funnel cloud, tornado, and waterspout are entered with the direction of movement when reported.

Example:

`/RM TORNADO W MOV E` ┄┄► Remarks, tornado west moving east

3.2.1.12.3 Thunderstorm

Thunderstorm coverage is coded as **ISOL** (isolated), **FEW** (few), **SCT** (scattered), **NMRS** (numerous) followed by description as **LN** (line), **BKN LN** (broken line), **SLD LN** (solid line) when reported. This is followed with **TS**, the location and movement, and the type of lightning when reported.

Example:

`/RM NMRS TS S MOV E GR1/2` ┄┄► Remarks, numerous thunderstorms south moving east, hail 1/2–inch in diameter

3.2.1.12.4 Lightning

Lightning frequency is coded as **OCNL** (occasional) or **FRQ** (frequent), followed by type as **LTGIC** (lightning in cloud), **LTGCC** (lightning cloud to cloud), **LTGCG** (lightning cloud to ground), **LTGCA** (lightning cloud to air), or combinations, when reported.

Example:

`/RM OCNL LTGICCG` ┄┄► Remarks, occasional lighting in cloud, cloud to ground

3.2.1.12.5 Electrical Discharge

For an electrical discharge, **DISCHARGE** is coded followed by the altitude.

Example:

`/RM DISCHARGE 120` ───► Remarks, discharge, 12,000 feet MSL

3.2.1.12.6 Clouds
Remarks are used when clouds can be seen but were not encountered and reported in the sky condition group (**/SK**)

Examples:

`/RM CB E MOV N` ───► Remarks, cumulonimbus east moving north
`/RM OVC BLO` ───► Remarks, overcast below

3.2.1.12.7 Plain Language
If specific phraseology is not adequate, plain language is used to describe the phenomena or local geographic locations. Remarks that do not fit in other groups like **DURC** (during climb), **DURD** (during descent), **RCA** (reach cruising altitude), **TOP**, **TOC** (top of climb), or **CONTRAILS** are included.

Examples:

```
/RM BUMPY VERY ROUGH RIDE
/RM CONTRAILS
/UA/OV BIS270030/TM 1445/FL060/TP CVLT/TB LGT /RM DONNER SUMMIT PASS
```

3.2.1.12.8 Volcanic Eruptions
Volcanic ash alone is an Urgent PIREP. A report of volcanic activity includes as much information as possible including the name of the mountain, ash cloud and movement, height of the top and bottom of the ash, etc., is included. If the report is received from a source other than a pilot, Aircraft **UNKN**, Flight Level **UNKN**, and **/RM UNOFFICIAL** is entered.

Example:

```
/UUA/OV ANC240075/TM 2110/FL370/TP DC10/WX VA/RM VOLCANIC ERUPTION
2008Z MT AUGUSTINE ASH 40S MOV SSE
```

Urgent Pilot Weather Report, 240 degrees at 75 nautical miles from Anchorage International Airport, Alaska, 2110 UTC, flight level 310, a DC10 reported volcanic ash, remarks, volcanic eruption occurred at 2008 UTC Mount Augustine, ash 40 nautical miles south moving south-southeast.

3.2.1.12.9 SKYSPOTTER
The **SKYSPOTTER** program is a result of a recommendation from the Safer Skies FAA/INDUSTRY Joint Safety Analysis and Implementation Teams. The term **SKYSPOTTER** indicates a pilot has received specialized training in observing and reporting in-flight weather phenomenon, pilot weather reports, or PIREPs.

When a PIREP is received from a pilot identifying themselves as a **SKYSPOTTER** aircraft, the additional comment "**/AWC**" is added at the end of the remarks section of the PIREP.

Example:

PIREP TEXT/RM REMARKS/AWC

3.2.2 PIREP Examples

UUA /OV ORD/TM 1235/FLUNKN/TP B727/TB MOD/RM LLWS +/- 20KT BLW 003
DURD RWY27L

Urgent Pilot Weather Report, over Chicago O'Hare Airport, Illinois, 1235 UTC, flight level unknown, from a Boeing 727, moderate turbulence, remarks, Low Level Wind Shear, airspeed fluctuations of plus or minus 20 knots below 300 feet AGL during descent, runway 27 left.

UUA /OV BAM260045/TM 2225/FL180/TP BE20/TB SEV/RM BROKE ALL THE
BOTTLES IN THE BAR

Urgent Pilot Weather Report, 260 degrees at 45 nautical miles from Hazen VOR, Nevada, 2225 UTC, 18,000 feet MSL, Beech Super King Air 200, severe turbulence, remarks, broke all the bottles in the bar.

UA /OV KMRB-KPIT/TM 1600/FL100/TP BE55/SK BKN024-TOP032/BKN-OVC043-
TOPUNKN /TA M12/IC LGT-MOD RIME 055-080

Pilot Weather Report, Martinsburg, West Virginia to Pittsburgh International Airport, Pennsylvania, 1600 UTC, 10,000 feet MSL, Beechcraft Baron, base of a broken layer 2,400 feet MSL, top 3,200 feet MSL, base of a broken to overcast layer 4,300 feet MSL, top unknown, temperature minus 12, light to moderate rime ice between 5,500 to 8,000 feet MSL.

UA /OV IRW090064/TM 1522/FL080/TP C172/SK SCT090-TOPUNKN/WX FV05SM
HZ/TA M04/WV 24040KT/TB LGT/RM IN CLR

Pilot Weather Report, 90 degrees and 64 nautical miles from Will Rogers VORTAC, Oklahoma City, Oklahoma, 1522 UTC, 8,000 feet MSL, Cessna 172, base of a scattered layer 9,000 feet MSL, top unknown, flight visibility 5 statute miles, haze, temperature minus 4, wind 240 degrees at 40 knots, light turbulence, remarks, in clear.

`UA /OV KLIT-KFSM/TM 0310/FL100/TP BE36/SK SCT070-TOP110/TA M03/WV`
`25015KT`

Pilot Weather Report, between Little Rock and Fort Smith, Arkansas, 0310 UTC at 10,000 feet MSL. Beech 36, base of a scattered layer at 7,000 feet MSL, top 11,000 feet MSL, temperature minus 3, wind 250 degrees at 15 knots.

`UA /OV KAEG/TM 1845/FL UNKN/TP UNKN /RM TIJERAS PASS CLSD DUE TO FG`
`AND LOW CLDS UNA VFR RTN KAEG.`

Pilot Weather Report, over Double Eagle II Airport, Albuquerque, New Mexico, 1845 UTC, remarks, Tijeras Pass closed due to fog and low clouds, unable to fly VFR, returned to Double Eagle II Airport.

`UA /OV ENA14520/TM 2200/FL310/TP B737/TB MOD CAT 350-390.`

Pilot Weather Report, 145 degrees at 20 nautical miles from Kenai, Alaska, at 2200 UTC, at flight level 310, Boeing 737, moderate clear air turbulence between 35,000 and 39,000 feet MSL.

3.3 Radar Weather Report (SD/ROB)

A Radar Weather Report (SD/ROB) contains information about precipitation observed by weather radar. This is a textual product derived from the WSR-88D NEXRAD radar without human intervention. *The resolution of this textual product is very coarse, up to 80 minutes old, and should only be used if no other radar information is unavailable.*

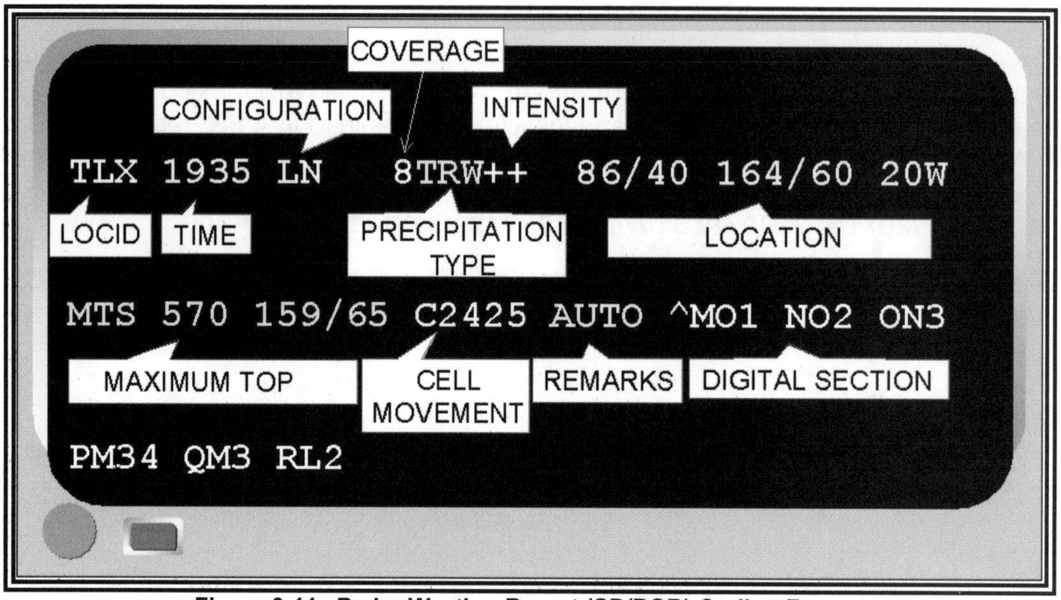

Figure 3-11. Radar Weather Report (SD/ROB) Coding Format

3.3.1 Format
Reports are transmitted hourly from WSR-88D Weather Radar sites (see figure 3-12). The SD/ROB format is presented in Figure 3-8.

3.3.1.1 Location Identifier
The location identifier is reported as the three-letter International Air Transport Association (IATA) code.

Example:

TLX ··········▶ Oklahoma City Twin Lakes, Oklahoma

3.3.1.2 Time
The time of the observation is reported in four-digits Universal Time Coordinated (UTC).

Example:

1935 ···············▶ 1935 UTC

3.3.1.3 Configuration
Three types of configurations can be reported: **CELL**, **LN** (line), and **AREA**. Multiple configurations can be reported within one Weather Radar Report.

A **CELL** is a single, isolated convective echo.

A **LN** (line) is a convective echo that meets the following criteria:

- Contains heavy or greater intensity precipitation
- Is at least 30 miles long
- Length is at least four times greater than width
- Contains at least 25 percent coverage

An **AREA** is a group of echoes of similar type, not classified as a line.

Figure 3-9 illustrates the three configurations that can be reported in a Weather Radar Report.

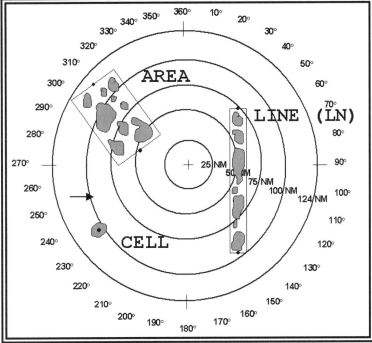

Figure 3-12. Radar Weather Report (SD/ROB) Configurations

3.3.1.4 Coverage
Coverage of precipitation is coded in single digits representing tenths of coverage.

For echo configurations containing multiple precipitation types, coverage is coded for each type. Total coverage is obtained by adding the individual values.

Examples:

2TRW+4R ⋯⋯▸ 2/10 coverage TRW+, 4/10 coverage R, 6/10 total coverage
3R6S– ⋯⋯⋯▸ 3/10 coverage R, 6/10 coverage S-, 9/10 total coverage

3.3.1.5 Precipitation Type
Precipitation type is determined by computer model.

Reportable types are:

- Rain (**R**)
- Rain shower (**RW**)
- Snow (**S**)
- Snow shower (**SW**)
- Thunderstorm (**T**)

Multiple precipitation types can be reported within a configuration.

3.3.1.6 Precipitation Intensity

Four precipitation intensities can be reported as shown in table 3-11.

Table 3-11. SD/ROB Reportable Intensities

Symbol	Intensity	dBZ
-	Light	0-29
(no entry)	Moderate	30-40
+	Heavy	41-45
++	Heavy	46-49
X	Extreme	50-56
XX	Extreme	57 or more

Examples:

7R- → 7/10 coverage of light rain

3R-6S → 3/10 coverage light rain, 6/10 coverage moderate snow, 9/10 total coverage

2TRWX4R- → 2/10 coverage thunderstorms, extreme rain showers, 4/10 coverage light rain, 6/10 total coverage

3.3.1.7 Location

An area is coded with two end points and a width that defines a rectangle. Each end point is defined by an azimuth and range (AZRAN).

A line is also coded with two end points and a width that defines a rectangle. Each end point is defined by an AZRAN.

A cell is coded as a single point with a diameter (**D**). This point is defined by an AZRAN.

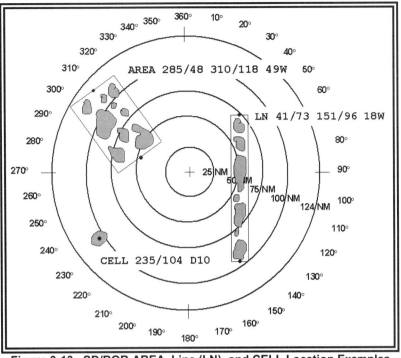

Figure 3-13. SD/ROB AREA, Line (LN), and CELL Location Examples
The "+" denotes the radar location.

3.3.1.8 Maximum Top

Maximum top (**MT** or **MTS**) denotes the altitude and location of the top of the highest precipitation echo.

All radar heights are estimates and assume standard atmosphere conditions and, thus, standard radar wave propagation. **MT** denotes radar data alone was used to determine the maximum top. **MTS** denotes both satellite and radar data were used to estimate the maximum top.

The maximum top is coded as a three-digit number in hundreds of feet MSL. Location is coded as an azimuth and range (AZRAN) relative to the radar site. If echo tops are uniform in altitude, the letter "U" precedes the altitude with no AZRAN provided.

Examples:

MT 150 19/32 ·············▶ Maximum top 15,000 feet MSL at 19 degrees, 32 nautical miles

MT 340 182/98 ·············▶ Maximum top 34,000 feet MSL at 182 degrees, 98 nautical miles

MTS 520 5/121 ·············▶ Maximum top with satellite data 52,000 feet MSL at 5 degrees, 121 nautical miles

3.3.1.9 Cell Movement

Cell movement is the average motion of all the cells within a configuration. It is coded in the following format: the cell movement group is indicated by the letter **C** followed by four digits. The first two digits represent the direction the cell(s) is (are) moving from in tens of degrees referenced to true north. The last two digits represent the speed of the configuration in knots.

Movement of areas and lines is not coded.

Examples:

C0209 ················► Cell movement from 20 degrees at 9 knots
C2043 ················► Cell movement from 200 degrees at 43 knots
C3616 ················► Cell movement from 360 degrees at 16 knots

3.3.1.10 Remarks
Remarks contain information about the radar's status and type of report. Currently, all weather radar reports are automated.

Table 3-12. Weather Radar Report Remarks and Meaning

REMARK	MEANING
PPINE	Equipment normal and operating, but no echoes observed
PPINA	Observation not available
PPIOM	Radar out for maintenance
AUTO	Report derived from an automated weather radar

3.3.1.11 Digital Section
The information contained in the digital section is used primarily to create the Radar Summary Chart. However, with the proper grid overlay chart for the corresponding radar site, the digital section code can also be used to determine precipitation location and intensity. (See Figure 3-11 for an example of a digital code plotted from the Oklahoma City, Oklahoma, Weather Radar Report.)

Each digit represents the maximum precipitation intensity found within a grid box as determined by the weather radar. Light intensity is denoted by a **1**, **2** is for moderate, **3** and **4** is for heavy, **5** and **6** is used for extreme precipitation. These digits were once commonly referred to as VIP levels because precipitation intensity, and therefore the digit, was derived using a video integrator processor (VIP). Whereas the old WSR-57 and WSR-74 weather radar video integrator processors displayed six data levels, the WSR-88D weather radar displays sixteen data levels. The data levels are still converted back to six levels for use in the Radar Weather Report. To avoid confusion, the term VIP should no longer be used to describe precipitation intensity. For example, if a grid box is coded with the number 2, it would be described as "moderate" precipitation," not "VIP 2" or "level 2" precipitation.

A grid box is identified by two letters. The first represents the row in which the box is found and the second letter represents the column. For example **MO1** identifies the box located in row M and column O as containing light precipitation. A code of **MO1234** indicates precipitation in four consecutive boxes in the same row. Working from left to right: box MO = 1, box MP = 2, MQ = 3, and box MR = 4.

A Weather Radar Report contains data about precipitation echoes only. It does not contain information about important non-precipitation echoes such as clouds, fronts, dust, etc., which can be detected by weather radar under certain circumstances.

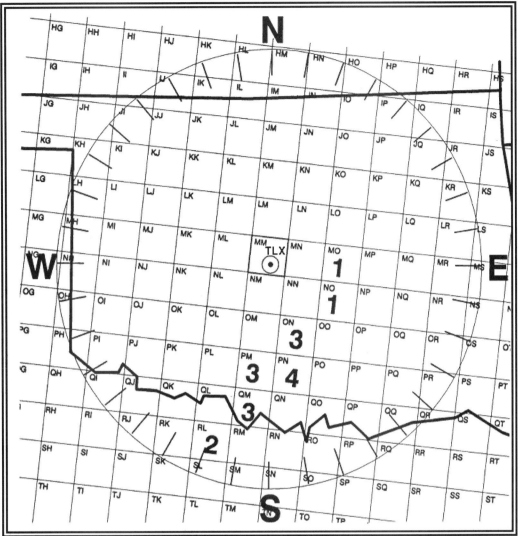

Figure 3-14. SD/ROB Digital Section Information Plotted on a PPI Grid Overlay Example
(See Table 3-11 for Intensity Level Codes 1 through 6.)

3.3.2 Examples

GRB 1135 AREA 4TRW+ 9/101 133/76 54W MT 310 45/47 C2428 AUTO

Green Bay, Wisconsin, automated Radar Weather Report at 1135 UTC. An area of echoes, 4/10 coverage, contained thunderstorms and heavy rain showers. Area is defined by points (referenced from GRB radar site) at 9 degrees, 101 nautical miles and 133 degrees, 76 nautical miles. These points, plotted on a map and connected with a straight line, define the center line of the echo pattern. The width of the area was 54 nautical miles; i.e., 27 nautical miles either side of the center line. Maximum top was 31,000 feet MSL located at 45 degrees and 47 nautical miles from Green Bay. Cell movement was from 240 degrees at 28 knots.

ICT 1935 LN 9TRWX 274/84 216/93 22W MTS 440 260/48 C2131 AUTO

Wichita, Kansas, automated Radar Weather Report at 1935 UTC. A line of echoes, 9/10 coverage, contained thunderstorm with intense rain showers. The center of the line extended from 274 degrees, 84 nautical miles to 216 degrees, 93 nautical miles. The line was 22 nautical miles wide.

To display graphically, plot the center points on a map and connect the points with a straight line; then plot the width. Since the thunderstorm line was 22 nautical miles wide, it extended 11 nautical miles either side of your plotted line.

The maximum top is 44,000 feet MSL at 260 degrees, 48 nautical miles from Wichita. Cell movement was from 210 degrees at 31 knots.

```
GGW 1135 AREA 3S- 95/129 154/81 34W MT 100 130/49 0805 AUTO
```

Glasgow, Montana, automated Radar Weather Report at 1135 UTC. An area, 3/10 coverage, of light snow. The area's centerline extended from points at 95 degrees, 129 nautical miles to 154 degrees, 81 nautical miles from Glasgow. The area was 34 nautical miles wide. The maximum top was 10,000 feet MSL, at 130 degrees, 49 nautical miles from Glasgow. Cell movement was from 80 degrees at 5 knots.

```
JGX 2235 AREA 2TRW++6R- 67/130 308/45 106W MT 380 66/54 C2038 AUTO
```

Atlanta, Georgia, automated Radar Weather Report at 2235 UTC. An area of echoes, total coverage 8/10, with 2/10 of thunderstorms with very heavy rain showers and 6/10 coverage of light rain (This suggests that the thunderstorms were embedded in an area of light rain). The area was 53 nautical miles either side of the line defined by the two points, 67 degrees, 130 nautical miles and 308 degrees, 45 nautical miles from Atlanta. Maximum top was at 38,000 feet and was located on the 66 degree radial of JGX at 54 nautical miles. Cell movement was from 200 degrees at 38 knots.

```
HKM 0235 CELL TRW+ 19/22 D5 MT 270 18/23 C0414 AUTO
```

Kohala, Hawaii, automated Radar Weather Report at 0235 UTC. A cell, containing thunderstorms with very heavy rain showers, 5 miles in diameter, was located 19 degrees, 22 nautical miles from Kohala. Maximum top was 27,000 feet located at 18 degrees, 23 nautical miles from Kohala. Movement was from 40 degrees at 14 knots.

```
TLX 0435 PPINE AUTO
```

Oklahoma City, Oklahoma, automated Radar Weather Report at 0435 UTC, detected no echoes.

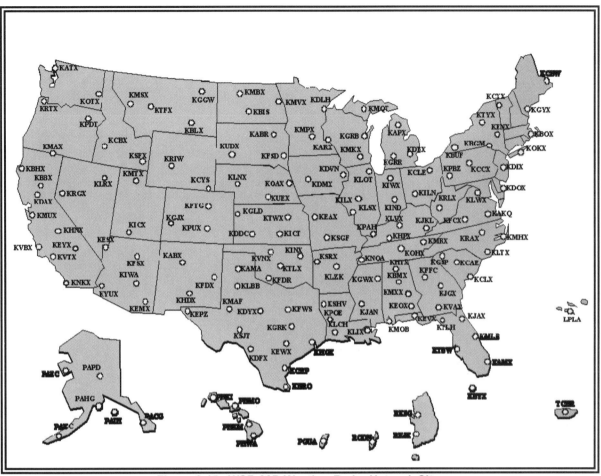

Figure 3-15. WSR-88D Weather Radar Network Sites

4 RADAR AND SATELLITE IMAGERY

4.1 Radar

4.1.1 Description
Radar images are graphical displays of precipitation and non-precipitation targets detected by weather radar. WSR-88D Doppler radar displays these targets on a variety of products which can be found on the internet on the National Weather Service (NWS) Doppler Radar Images web site at: http://radar.weather.gov/ridge/

4.1.2 Modes of Operation
The WSR-88D Doppler radar has **two** operational modes, **Clear Air** and **Precipitation**.

4.1.2.1 Clear Air Mode
In Clear Air Mode, the radar is in its most sensitive operation. This mode has the slowest antenna rotation rate which permits the radar to sample the atmosphere longer. This slower sampling increases the radar's sensitivity and ability to detect smaller objects in the atmosphere. The term "clear air" does not imply "no-precipitation" mode. Even in Clear Air Mode, the WSR-88D can detect light, stratiform precipitation (e.g., snow) due to the increased sensitivity.

Many of the radar returns in Clear Air Mode are airborne dust and particulate matter. The WSR-88D images are updated every 10 minutes when operating in this mode.

4.1.2.2 Precipitation Mode
Precipitation targets typically provide stronger return signals to the radar than non-precipitation targets. Therefore, the WSR-88D is operated in Precipitation Mode when precipitation is present although some non-precipitation echoes can still be detected in this operating mode.

The faster rotation of the WSR-88D in Precipitation Mode allows images to update at a faster rate approximately every 4 to 6 minutes.

4.1.3 Echo Intensities

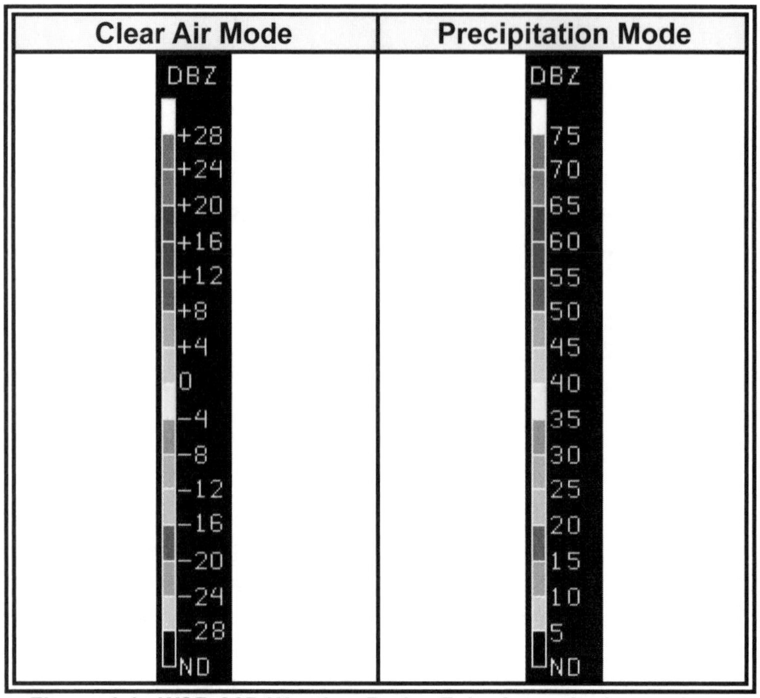

Figure 4-1. WSR-88D Weather Radar Echo Intensity Legend

The colors on radar images represent different echo reflectivities (intensities) measured in dBZ (decibels of Z). The dBZ values increase based on the strength of the return signal from targets in the atmosphere. Each reflectivity image includes a color scale that represents a correlation between reflectivity value and color on the radar image. Figure 4-1 depicts these correlations for both Clear Air and Precipitation Mode. For Clear Air Mode the scale ranges from -28 to +28 dBZ, for Precipitation Mode the scale ranges from 5 to 75 dBZ. *The color on each scale remains the same in both operational modes, only the dBZ values change.* The scales also include **ND** correlated to black which indicates no data was measured.

Reflectivity is correlated to intensity of precipitation. For example, in Precipitation Mode, when the dBZ value reaches 15, light precipitation is present. The higher the indicated reflectivity value, the higher the rainfall rate. The interpretation of reflectivity values is the same for both Clear Air and Precipitation Modes.

Reflectivity is also correlated with intensity terminology (phraseology) for air traffic control purposes. Table 4-1 defines this correlation.

Table 4-1. WSR-88D Weather Radar Precipitation Intensity Terminology

Reflectivity (dBZ) Ranges	Weather Radar Echo Intensity Terminology
<30 dBZ	Light
30-40 dBZ	Moderate
>40-50 dBZ	Heavy
50+ dBZ	Extreme

Values below 15 dBZ are typically associated with clouds. However, they may also be caused by atmospheric particulate matter such as dust, insects, pollen, or other phenomena. The scale **cannot** be used to determine the intensity of snowfall. However, snowfall rates generally increase with increasing reflectivity.

4.1.4 Products

The NWS produces numerous radar products of interest to the aviation community. The next section will discuss Base Reflectivity and Composite Reflectivity both available through National Weather Service (NWS) Doppler Radar Images web site at: http://radar.weather.gov/ridge/

4.1.4.1 Base Reflectivity

Base Reflectivity is a display of both the location and intensity of reflectivity data. Base Reflectivity images encompass several different elevation angles (tilts) of the antenna. The Base Reflectivity image currently available on the ADDS website begins at the lowest tilt angle (0.5°), more specifically 0.5° above the horizon.

Both a short range (Figure 4-2) and long range (Figure 4-3) image is available from the 0.5° Base Reflectivity product. The maximum range of the short range Base Reflectivity product is 124 NM from the radar location. This view will not display echoes farther than 124 NM from the radar site, although precipitation may be occurring at these greater distances. Other options to view precipitation beyond 124 NM from the radar site include selecting the long-range view which increases coverage out to 248 NM, selecting adjacent radars, or viewing a radar mosaic.

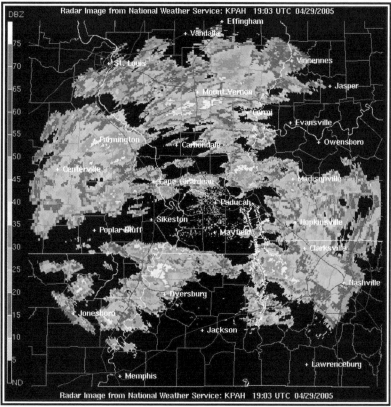

Figure 4-2. WSR-88D Weather Radar Short Range (124 NM) Base Reflectivity Product Example

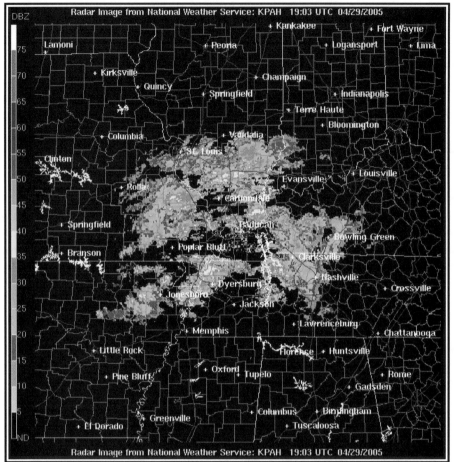

Figure 4-3. WSR-88D Weather Radar Long Range (248 NM) Base Reflectivity Product Example

4.1.4.1.1 Base Reflectivity Use

The Base Reflectivity product can be used to determine the location of precipitation and non-precipitation echoes, the intensity of liquid precipitation, and the general movement of precipitation when animating the image.

If the echo is precipitation, the product can be used to determine if it is convective or stratiform in nature. Stratiform precipitation (Figure 4-4) has the following characteristics:

- Widespread in areal coverage,
- Weak reflectivity gradients,
- Precipitation intensities are generally light or moderate (39 dBZ or less),
    - Occasionally, precipitation intensities can be stronger
- Echo patterns change slowly when animating the image.

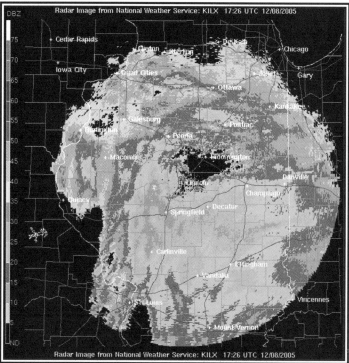

Figure 4-4. WSR-88D Weather Radar Stratiform Precipitation on the 0.5°Base Reflectivity Product Example

Hazards associated with stratiform precipitation include possible widespread icing above the freezing level, low ceilings and reduced visibilities.

Convective precipitation (Figure 4-5) can be described using the following characteristics:

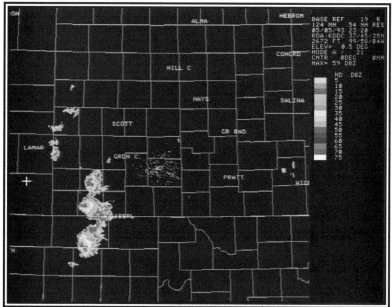

Figure 4-5. WSR-88D Weather Radar Convective Precipitation on the 0.5°Base Reflectivity Product Example

- Echoes tend to form as lines or cells,
- Reflectivity gradients are strong,

- Precipitation intensities generally vary from moderate to extreme,
  - Occasionally precipitation intensities can be light
- Echo patterns change rapidly when animating the image

Numerous hazards are associated with convective precipitation. They include: turbulence, low-level wind shear, strong and gusty surface winds, icing above the freezing level, hail, lightning, tornadoes and localized IFR conditions with heavy precipitation.

4.1.4.1.2 Strengths of Base Reflectivity
The strengths of the Base Reflectivity product include:

- The location of precipitation and non-precipitation echoes is depicted, and
- The intensity and movement of precipitation is relatively easy and straight forward to determine.

4.1.4.1.3 Limitations of Base Reflectivity
Limitations associated with the Base Reflectivity product include:

- The radar beam may overshoot targets, and
- The image may be contaminated by:
  - Beam blockage
  - Ground clutter
  - Anomalous Propagation (AP) and
  - Ghosts.

4.1.4.1.3.1 Radar Beam Overshooting
Radar beam overshooting may occur because the radar beam (typically the 0.5 degree slice) can be higher than the top of precipitation. This will most likely occur with stratiform precipitation and low-topped convection. For example, at a distance of 124 NM from the radar, the 0.5° Base Reflectivity radar beam is at an altitude of approximately 18,000 feet; at 248 NM the beam height is approximately 54,000 feet. Any precipitation with tops below these altitudes and distances will **not** be displayed on the image. Therefore, it is quite possible that precipitation may be occurring where none appears on the radar image.

4.1.4.1.3.2 Beam Blockage
Beam blockage (Figure 4-6) occurs when the radar beam is blocked by terrain and is particularly predominant in mountainous terrain. This impacts both the Composite Reflectivity and Base Reflectivity images.

Beam blockage is most easily seen on the 0.5° Base Reflectivity images where it appears as a pie-shaped area (or areas) perpetually void of echoes. When animating the imagery, the beam blockage area will remain clear of echoes even as precipitation and other targets pass through.

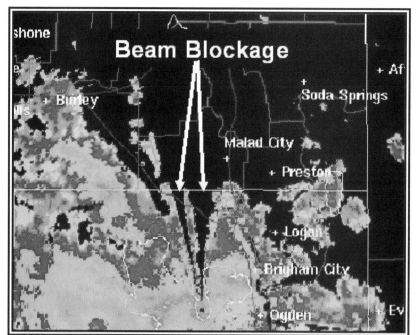

Figure 4-6. WSR-88D Weather Radar Beam Blockage on Base Reflectivity Product Example

4.1.4.1.3.3 Ground Clutter

Ground clutter (Figure 4-8) is radar echoes returns from trees, buildings, or other objects on the ground. It appears as a roughly circular region of high reflectivities at ranges close to the radar. Ground clutter appears stationary when animating images and can mask precipitation located near the radar. Most ground clutter is automatically removed from WSR-88D imagery, so typically it is does not interfere with image interpretation.

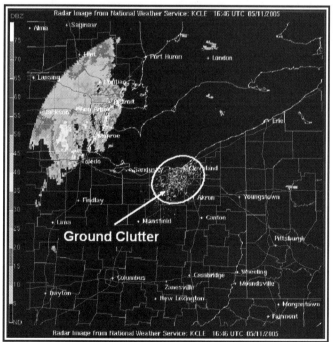

Figure 4-8. WSR-88D Weather Radar Ground Clutter Example

4.1.4.1.3.4 Ghost

A Ghost (Figure 4-9) is a diffused echo in apparently clear air caused by a "cloud" of point targets such as insects or by refraction returns of the radar beam in truly clear air.

The latter case commonly develops at sunset due to superrefraction during the warm season. The ghost develops as an area of low reflectivity echoes (typically less than 15dBZ) near the radar site and quickly expands. When animating the imagery, the ghost echo shows little movement.

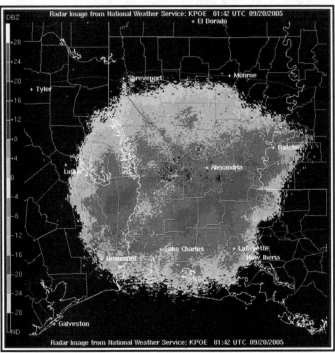

Figure 4-9. WSR-88D Weather Radar Ghost Example

4.1.4.1.3.5 Angels

Angels are echoes caused by a physical phenomenon not discernible by the eye at the radar site. They are usually causes by bats, birds or insects. Angels typically appear as a donut-shaped echo with low reflectivity values (Figure 4-10). When animated, the echo expands and becomes more diffuse with time.

Angels typically only appear only when the radar is in Clear Air Mode because of their weak reflectivity. Echoes caused by birds are typically detected in the morning when they take flight for the day. Echoes caused by bats are typically detected in the evening, when they are departing from caves.

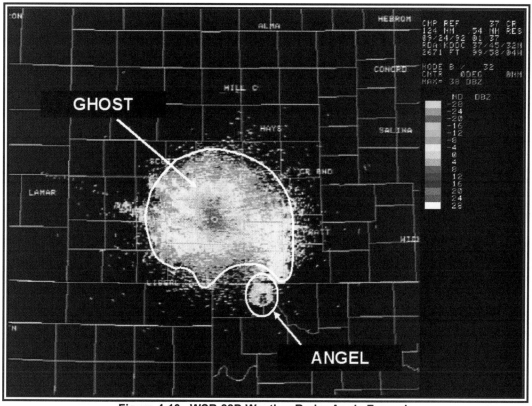

Figure 4-10. WSR-88D Weather Radar Angle Example
This angel was caused by bats departing Selman Bat Cave at Alabaster Caverns State Park, Oklahoma around sunset

4.1.4.1.3.6 Anomalous Propagation (AP)

Anomalous propagation (AP) (Figure 4-11) is an en extended pattern of ground echoes caused by superrefraction of the radar beam. Superrefraction causes the radar beam to bend downward and strike the ground. It differs from ground clutter because it can occur anywhere within the radar's range, not just at ranges close to the radar.

AP typically appears as speckled or blotchy, high reflectivity echoes. When animating images, AP tends to "bloom up" and dissipate and has no continuity of motion. AP can sometimes be misinterpreted as thunderstorms; differentiating between to two is determined by animating images. Thunderstorms move with a smooth, continuous motion while AP appears to "bloom up" and dissipate randomly.

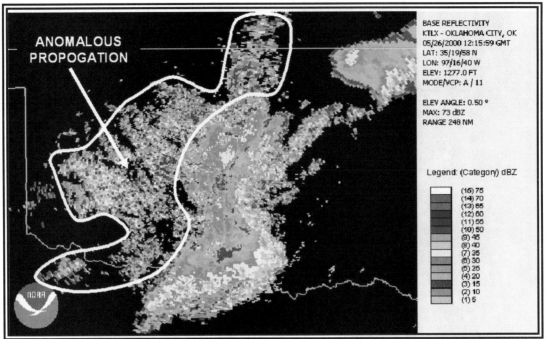

Figure 4-11. WSR-88D Weather Radar Anomalous Propagation (AP) Example

4.1.4.2 Composite Reflectivity

Composite reflectivity is the maximum echo intensity (reflectivity) detected within a column of the atmosphere above a location. The radar scans through all of the elevation slices to determine the highest dBZ value in the vertical column (Figure 4-12) then displays that value on the product. When compared with Base Reflectivity, the Composite Reflectivity can reveal important storm structure features and intensity trends of storms.

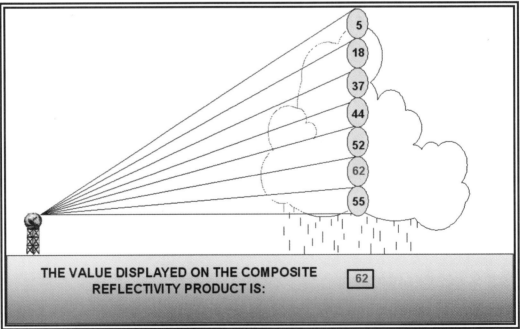

Figure 4-12. Creation of a Composite Reflectivity product

The maximum range of the long range Composite Reflectivity product (Figure 4-13) is 248 NM from the radar. The "blocky" appearance of this product is due to its lower spatial resolution as it has one-fourth the resolution of the Base Reflectivity product.

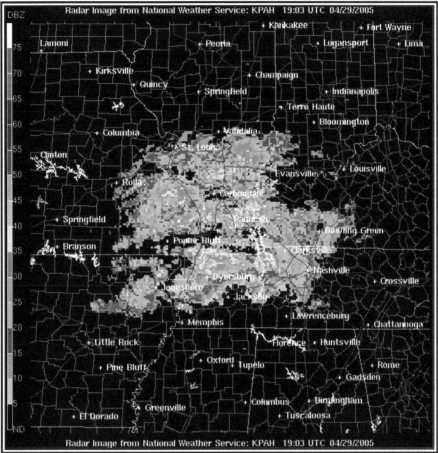

Figure 4-13. WSR-88D Weather Radar Long Range (248 NM) Composite Reflectivity Product Example

For a higher resolution (1.1 x 1.1 NM grid) Composite Reflectivity image, users must select the short range view (Figure 4-14). The image is less "blocky" as compared to the long range image. However, the maximum range is reduced to 124 NM from the radar location.

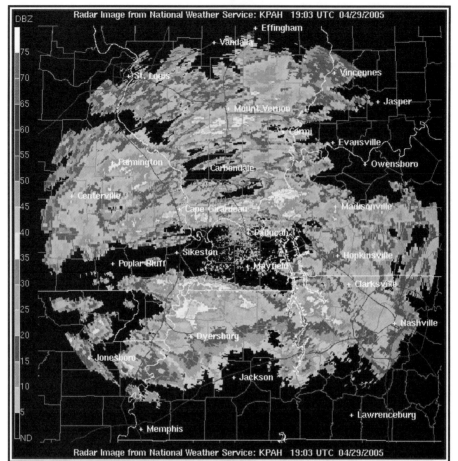

Figure 4-14. WSR-88D Weather Radar Short Range (124 NM) Composite Reflectivity Product Example

4.1.4.2.1 Composite Reflectivity Use

The primary use of the Composite Reflectivity product, which offers the highest reflectivity value in a vertical column, is to determine the vertical structure of the precipitation. The image must be compared with the Base Reflectivity image to determine the vertical structure of the precipitation. Figure 4-15 includes the 0.5° Base Reflectivity and Composite Reflectivity images for the same location and period of time.

In Figure 4-15, within location A, the intensity of the echoes is higher on the Composite Reflectivity image. Also, within area B, many more echoes present on the Composite Reflectivity. Since the Composite Reflectivity product displays the highest reflectivity of **all** elevation scans, it is detecting these higher reflectivities at some higher altitude/elevation than the Base Reflectivity product, which is sampling closer to the ground. This often occurs when precipitation and especially thunderstorms are developing.

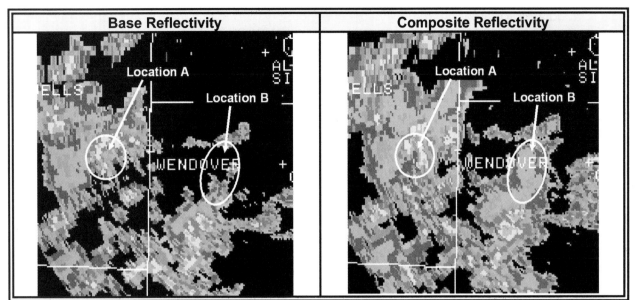

Figure 4-15. WSR-88D Weather Radar 0.5° Base Reflectivity Versus Composite Reflectivity Comparison

4.1.4.2.2 Strengths of Composite Reflectivity

The primary strength of the Composite Reflectivity product is its three-dimensional view of reflectivity. The method used to determine this three-dimensional view is described in section 4.1.4.2.1.

4.1.4.2.3 Limitations of Composite Reflectivity

Limitations associated with the Composite Reflectivity product include:

- The radar beam may overshoot targets, and
- The image may be contaminated by:
  - Beam blockage
  - Ground clutter
  - Anomalous Propagation (AP) and
  - Ghosts.

4.1.4.2.3.1 Radar Beam Overshooting

Radar beam overshooting may occur because the lowest base reflectivity tilt (0.5) can be higher than the top of precipitation. This will most likely occur with stratiform precipitation and low-topped convection. For example, at a distance of 124 NM from the radar, the radar beam is at an altitude of approximately 18,000 feet above the radar; at 248 NM the beam height is approximately 54,000 feet. Any precipitation with tops below these altitudes and distances will **not** be displayed on the image. Therefore, it is quite possible that precipitation may be occurring where none appears on the radar image.

4.1.4.2.3.2 Beam Blockage

Beam blockage (Figure 4-6) occurs when the radar beam is blocked by terrain and is particularly predominant in mountainous terrain. This impacts both the Composite Reflectivity and Base Reflectivity images.

Beam blockage is most easily seen on the 0.5° Base Reflectivity images where it appears as a pie-shaped area (or areas) perpetually void of echoes. When animating the imagery, the beam blockage area will remain clear of echoes even as precipitation and other targets pass through.

4.1.4.2.3.3 Ground Clutter

Ground clutter (Figure 4-8) is radar echoes returns from trees, buildings, or other objects on the ground. It appears as a roughly circular region of high reflectivities at ranges close to the radar. Ground clutter appears stationary when animating images and can mask precipitation located near the radar. Most ground clutter is automatically removed from WSR-88D imagery, so typically it is does not interfere with image interpretation.

4.1.4.2.3.4 Ghost

A Ghost (Figure 4-9) is a diffused echo in apparently clear air that is caused by a "cloud" of point targets such as insects or by refraction returns of the radar beam in truly clear air.

The latter case commonly develops at sunset due to superrefraction during the warm season. The ghost develops as an area of low reflectivity echoes (typically less than 15 dBZ) near the radar site and quickly expands. When animating the imagery, the ghost echo shows little movement.

4.1.4.2.3.5 Angels

Angels are echoes caused by a physical phenomenon not discernible by the eye at the radar site. They are usually causes by bats, birds or insects. Angels typically appear as a donut-shaped echo with low reflectivity values (Figure 4-10). When animating, the echo expends and becomes more diffuse with time.

Angels typically only appear only when the radar is in clear air mode because of their weak reflectivity. Echoes caused by birds are typically detected in the morning when they take flight for the day. Echoes caused by bats are typically detected in the evening when they take flight from caves.

4.1.4.2.3.6 Anomalous Propagation (AP)

Anomalous propagation (AP) (Figure 4-11) is an en extended pattern of ground echoes caused by superrefraction of the radar beam. Superrefraction causes the radar beam to bend downward and strike the ground. It differs from ground clutter because it can occur anywhere within the radar's range, not just at ranges close to the radar.

AP typically appears as speckled or blotchy, high reflectivity echoes. When animating images, AP tends to "bloom up" and dissipate and has no continuity of motion. AP can sometimes be misinterpreted as thunderstorms; differentiating between to two is determined by animating images. Thunderstorms move with a smooth, continuous motion while AP appears to "bloom up" and dissipate randomly.

4.1.5 Radar Mosaics

A radar mosaic consists of multiple single site radar images combined to produce a radar image on a regional or national scale. Regional and national mosaics can be found at the National Weather Service (NWS) Doppler Radar Images web site: http://radar.weather.gov/ridge/

The mosaics are located toward the bottom of the page.

4.1.5.1   0.5° Mosaics - Contiguous U.S. and Hawaii

The NWS produces a set of regional and national mosaics (Table 4-2) in the contiguous U.S. using the 124 NM 0.5° Base Reflectivity product (Figure 4-16).

Table 4-2. NWS Radar Mosaic Products

Pacific Northwest	Pacific Southwest
Upper Mississippi Valley	Southern Mississippi Valley
Northeast	Southeast
Southern Rockies	Northern Rockies
Southern Plains	Great Lakes
Low Resolution National	High Resolution National

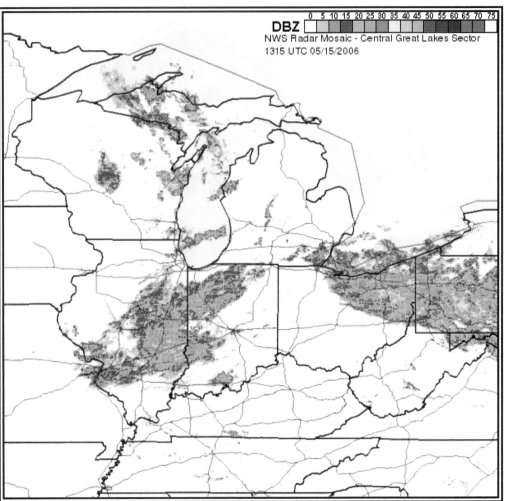

Figure 4-16. Great Lakes Regional Radar Mosaic Example

The most recent image from single site radars is used to create the product. Single site data older than 15 minutes from the current time of the product are excluded from the image. Therefore, data on the mosaics will be no greater than 15 minutes old. Where radar coverage overlaps, the highest dBZ value will be plotted on the image.

4.1.5.2 0.5º Mosaics - Alaska

The Alaskan mosaic (Figure 4-17) differs from the contiguous product in only one way: it is created using the 248 NM 0.5º Base Reflectivity single site product.

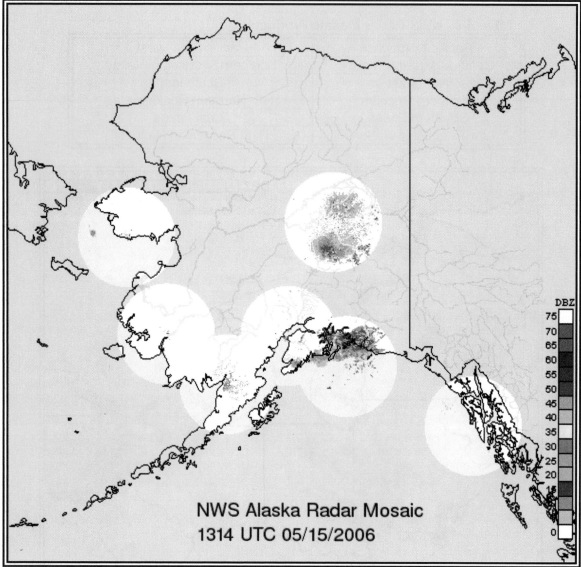

Figure 4-17. Alaskan Regional Radar Mosaic Example

The long range Base Reflectivity product is used because the radar sites are located at greater distances from each other. Even with the use of the long range product, many areas of Alaskan do not have radar coverage. These areas are shaded gray on Figure 4-17.

4.2 Satellite

4.2.1 Description

Satellite is perhaps the single most important source of weather data world wide, particularly over data sparse regions such as countries without organized weather data collection and the oceans.

GOES satellite imagery can be found on the NWS Aviation Digital Data Service (ADDS) website at: http://adds.aviationweather.noaa.gov/satellite/. Additional satellite imagery for Alaska can be found on the Alaska Aviation Weather Unit (AAWU) website at: http://aawu.arh.noaa.gov/sat.php

4.2.2 Imagery Types

Three types of satellite imagery are commonly used: Visible, infrared (IR), and water vapor. Visible imagery is available only during daylight hours. IR and water vapor imagery are available day or night.

4.2.2.1 Visible Imagery

Visible imagery (Figures 4-16 and 4-17) displays reflected sunlight from the Earth's surface, clouds, and particulate matter in the atmosphere. These images are simply black and white pictures of the Earth from space. During daylight hours, visible imagery is the most widely used image type because it has the highest resolution and approximates what is seen with the human eye.

Gray shades displayed on visible imagery can be correlated with particular features. Assuming a high sun angle, thick clouds and snow will appear white, thin clouds will appear translucent gray, land appears gray, and deep bodies of water such as lakes and oceans will appear black.

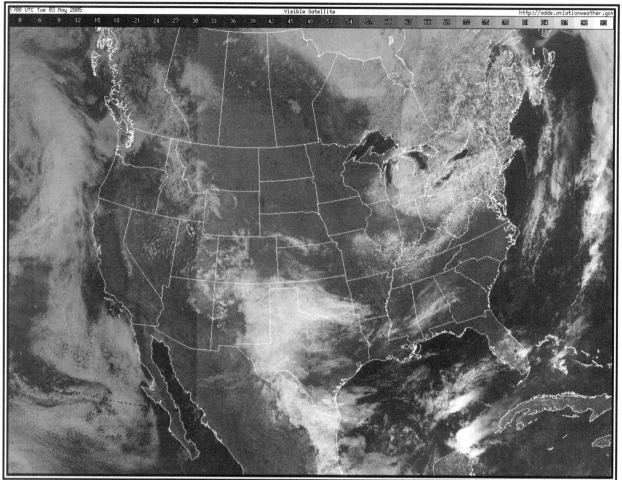

Figure 4-18. Visible Satellite Image – U.S. Example

Figure 4-19. Visible Satellite Image – Regional-Scale Example

4.2.2.1.1 Visible Image Data Legend

The data legend (Figure 4-18) on a visible image displays albedo, or reflectance, expressed as a percentage. For example, an albedo of 72 means 72 percent of the sunlight which struck a feature was reflected back to space.

Figure 4-20. Visible Satellite Data Legend.
The gray shades (values) represent albedo or reflectance expressed as a percentage.

4.2.2.2 Infrared (IR) Imagery

Infrared (IR) images (Figure 4-19 through 4-22) display temperatures of the Earth's surface, clouds, and particulate matter. Generally speaking, the warmer an object, the more infrared energy it emits. The satellite sensor measures this energy and calibrates it to temperature using a very simple physical relationship.

Clouds that are very high in the atmosphere are generally quite cold (perhaps -50°C) whereas clouds very near the earth's surface are generally quite warm (perhaps +5°C). Likewise, land may be even warmer than the lower clouds (perhaps +20°C). Those colder clouds emit much less infrared energy than the warmer clouds and the land emits more than those warm clouds.

The data measured by satellite is calibrated and colorized according to the temperature. If the temperature of the atmosphere decreases with height (which is typical), cloud-top temperature can be used to roughly determine which clouds are high-level and which are low-level.

When clouds are present, the temperature displayed on the infrared images is that of the tops of clouds. When clouds are not present, the temperature is that of the ground or the ocean.

Figure 4-21. Infrared (Color) Satellite Image – U.S. Example
The scale is in degrees Celsius. Blue/purple colors indicate colder temperatures, while orange/red colors indicate warmer temperatures.

Figure 4-22. Infrared (Color) Satellite Image – Regional-Scale Example
The scale is in degrees Celsius. Blue/purple colors indicate colder temperatures, while orange/red colors indicate warmer temperatures.

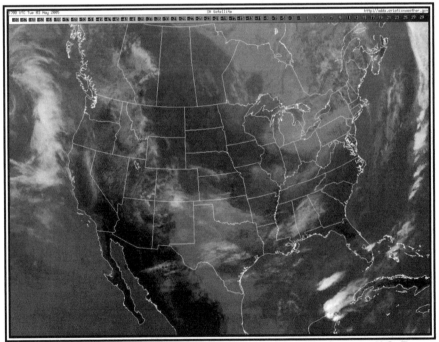

Figure 4-23. Unenhanced Infrared (black and white) Satellite Image – U.S. Example
The scale is in degrees Celsius. Lighter gray shades indicate colder temperatures, while darker gray shades indicate warmer temperatures.

Figure 4-24. Unenhanced Infrared (black and white) Satellite Image – Regional- Scale Example
The scale is in degrees Celsius. Lighter gray shades indicate colder temperatures, while darker gray shades indicate warmer temperatures.

4.2.2.2.1 Infrared Image Data Legends

The data legend (Figure 4-23 and Figure 4-24) on an infrared image is calibrated to temperature expressed in degrees Celsius. The legend may vary based on the satellite image provider.

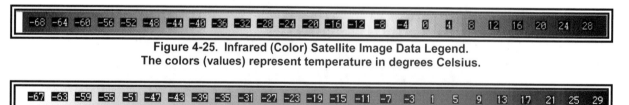

Figure 4-25. Infrared (Color) Satellite Image Data Legend.
The colors (values) represent temperature in degrees Celsius.

Figure 4-26. Unenhanced Infrared (black and while) Satellite Image Data Legend.
The gray shades (values) represent temperature in degrees Celsius.

4.2.3.3 Water Vapor Imagery

The water vapor imagery (Figure 4-25 and Figure 4-26) displays the quantity of water vapor generally located in the middle and upper troposphere within the layer between 700 millibars (FL100) to 200 millibars (FL390). The actual numbers displayed on the water vapor images correspond to temperature in degrees Celsius. No direct relationship exists between these values and the temperatures of clouds, unlike IR imagery. Water Vapor imagery does not really "see" clouds but "sees" high-level water vapor instead.

The most useful information to be gained from the water vapor images is the locations and movements of weather systems, jet streams, and thunderstorms. Another useful tidbit is aided by the color scale used on the images. In general, regions displayed in shades of red are VERY dry in the upper atmosphere and MAY correlate to crisp blue skies from a ground perspective. On the contrary, regions displayed in shades of blue or green are indicative of lots of high-level moisture and may also indicate cloudiness. This cloudiness could simply be high-level cirrus types or thunderstorms. That determination cannot be gained from this image by itself but could easily be determined when used in conjunction with corresponding visible and infrared satellite images.

Figure 4-27. Water Vapor Satellite Image – U.S. Example
The scale is in degrees Celsius. Blue/green colors indicate moisture and/or clouds in the mid/upper troposphere, while dark gray/orange/red colors indicate dry air in the mid/upper troposphere.

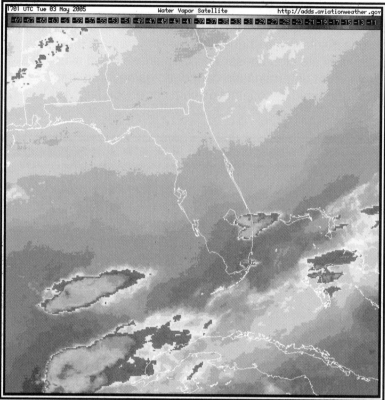

Figure 4-28. Water Vapor Satellite Image – Regional-Scale Example
The scale is in degrees Celsius. Blue/green colors indicate moisture and/or clouds in the mid/upper troposphere, while dark gray/orange/red colors indicate dry air in the mid/upper troposphere.

4.2.3.3.1   Water Vapor Image Data Legend

The data legend (Figure 4-27) on a water vapor images is calibrated to temperature expressed in degrees Celsius. The actual data values on the water vapor images are not particularly useful. Interpretation of the patterns and how they change over time is more important. The legend may vary depending on the satellite image provider.

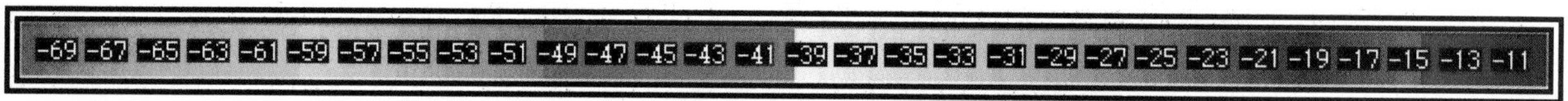

Figure 4-29. Water Vapor Satellite Image Data Legend.
The colors (values) represent temperature in degrees Celsius.

5 GRAPHICAL OBSERVATIONS AND DERIVED PRODUCTS

5.1 Surface Analysis Charts

Surface Analysis charts are analyzed charts of surface weather observations. The chart depicts the distribution of several items including sea level pressure, the positions of highs, lows, ridges, and troughs, the location and character of fronts, and the various boundaries such as drylines, outflow boundaries, sea-breeze fronts, and convergence lines. Other symbols are often added depending upon the intended use of the chart. Pressure is referred to in mean sea level (MSL) on the surface analysis chart while all other elements are presented as they occur at the surface point of observation. A chart in this general form is commonly referred to as the weather map.

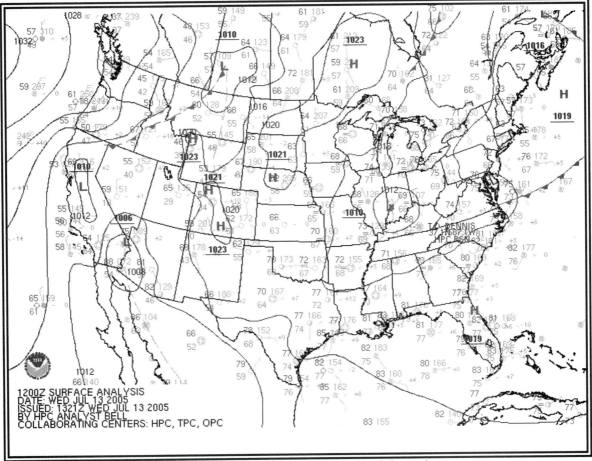

Figure 5-1. HPC Surface Analysis Chart Example

5.1.1 Issuance
Five National Weather Service (NWS) offices issue surface analysis charts:

- The Hydrometeorological Prediction Center (HPC) in Camp Springs, Maryland is responsible for land areas of North America. The charts are available at
  - http://www.hpc.ncep.noaa.gov/html/sfc2.shtml
  - http://www.hpc.ncep.noaa.gov/html/avnsfc.shtml

- The Ocean Prediction Center (OPC) in Camp Spring, Maryland is responsible for the Atlantic and Pacific Oceans north of 30°N latitude. The charts are available at:
  - http://www.opc.ncep.noaa.gov/

- The Tropical Prediction Center (TPC) in Miami, Florida is responsible for the tropical regions of the western hemisphere south of 30°N latitude and east of 160°E longitude. The surface analysis charts are available at:
  - http://www.nhc.noaa.gov/marine_forecasts.shtml

- The Alaskan Aviation Weather Unit (AAWU) in Anchorage, Alaska is responsible for the state of Alaska. The surface analysis chart is available at:
  - http://aawu.arh.noaa.gov/surface.php

- The Weather Forecast Office in Honolulu, Hawaii (WFO HNL) is responsible for the tropical Pacific Ocean, south of 30°N latitude and west of 160°E longitude. The charts are available at:
  - http://www.prh.noaa.gov/hnl/pages/analyses.php

Each office produces multiple versions of the surface analysis chart with varying formats.

5.1.2 HPC Surface Analysis Charts

- The Hydrometeorological Prediction Center (HPC) in Camp Springs, Maryland is responsible for land areas of North America. The charts are available at
  - http://www.hpc.ncep.noaa.gov/html/sfc2.shtml
  - http://www.hpc.ncep.noaa.gov/html/avnsfc.shtml

5.1.2.1 Issuance

The Hydrometeorological Prediction Center (HPC) issues Surface Analysis Charts for North America eight times daily (Table 5-1).

Table 5-1. HPC Surface Analysis Charts Issuance Schedule

Valid Time (UTC)	00	03	06	09	12	15	18	21

5.1.2.2 Analysis Symbols

Figure 5-2 shows analysis symbols used on HPC surface analysis charts:

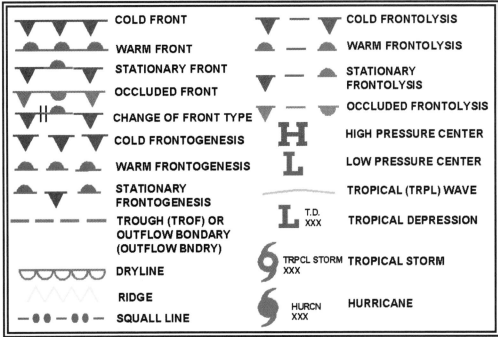

Figure 5-2. HPC Surface Analysis Chart Symbols

5.1.2.3 Station Plot Models

Land, ship, buoy, and C-MAN stations are plotted on the chart to aid in analyzing and interpreting the surface weather features. These plotted observations are referred to as station models. Some stations may not be plotted due to space limitations. However, all reporting stations are used in the analysis.

Figure 5-3 and 5-4 contain the most commonly used station plot models used in surface analysis charts:

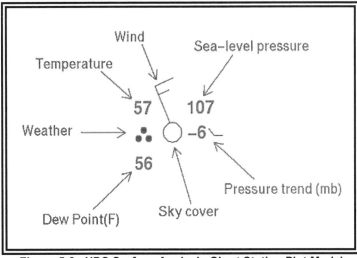

Figure 5-3. HPC Surface Analysis Chart Station Plot Model

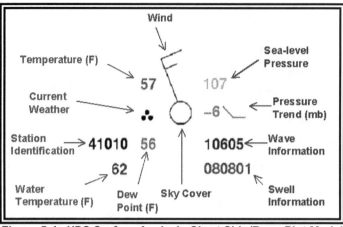

Figure 5-4. HPC Surface Analysis Chart Ship/Buoy Plot Model

HPC also produces surface analysis charts specifically for the aviation community. Figure 5-5 contains the station plot model for these charts:

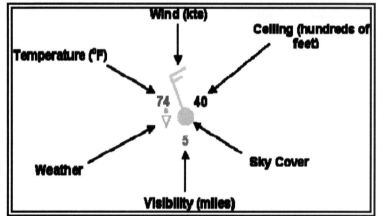

Figure 5-5. HPC Surface Analysis Chart for Aviation Interests Station Plot Model

5.1.2.3.1 Station Identifier
The format of the station identifier depends on the observing platform.

- Ship -- Typically 4 or 5 characters. If 5 characters, then the fifth will usually be a digit.

- Buoy -- Whether drifting or stationary, a buoy will have a 5-digit identifier. The first digit will always be a **4**.

- C-MAN -- Stands for Coastal-Marine Automated Network, and are usually close to coastal areas. Their identifier will appear like a 5-character ship identifier, however the 4th character will identify off which state the platform is located.

- Land -- Land stations will always be 3 characters, making them easily distinguishable from ship, buoy, and C-MAN observations.

5.1.2.3.2 Temperature
The air temperature is plotted in whole degrees Fahrenheit.

5.1.2.3.3 Dew Point
The dew point temperature is plotted in whole degrees Fahrenheit.

5.1.2.3.4 Weather
A weather symbol is plotted if, at the time of observation, precipitation is either occurring or a condition exists causing reduced visibility.

Figure 5-6 contains a list of the most common weather symbols:

Figure 5-6. HPC Surface Analysis Chart Common Weather Symbols

A complete list of weather symbols can be found at in Appendix I.

5.1.2.3.5 Wind
Wind is plotted in increments of 5 knots (kts). The wind direction is in "true" degrees and is depicted by a stem (line) pointed in the direction from which the wind is blowing. Wind speed is determined by adding the values of the flags (50kts), barbs (10kts), and half-barbs (5kts) found on the stem.

If the wind is calm at the time of observation, only a single circle over the station is depicted.

Figure 5-7 are some sample wind symbols:

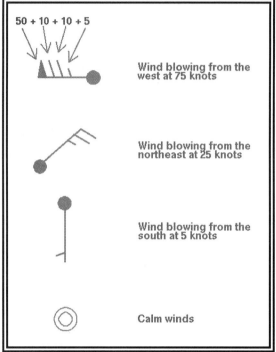

Figure 5-7. HPC Surface Analysis Chart Wind Plotting Model

5.1.2.3.6 Ceiling
Ceiling is plotted in hundreds of feet above ground level.

5.1.2.3.7 Visibility
Surface visibility is plotted in whole statute miles.

5.1.2.3.8 Pressure
Sea-level pressure is plotted in tenths of millibars (mb), with the first two digits (generally 10 or 9) omitted. For reference, 1013 mb is equivalent to 29.92 inches of mercury. Below are some sample conversions between plotted and complete sea-level pressure values:

410: 1041.0 mb
103: 1010.3 mb
987: 998.7 mb
872: 987.2 mb

5.1.2.3.9 Pressure Trend
The pressure trend has two components, a number and a symbol, to indicate how the sea level pressure has changed during the past three hours. The number provides the 3-hour change in tenths of millibars while the symbol provides a graphic illustration of how this change occurred.

Figure 5-8 contains the meanings of the pressure trend symbols:

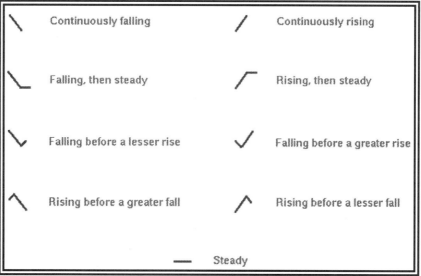

Figure 5-8. HPC Surface Analysis Chart Pressure Trends

5.1.2.3.10 Sky Cover

The approximate amount of sky cover can be determined by the circle at the center of the station plot. The amount the circle is filled reflects the amount of sky covered by clouds. Figure 5-9 contains the common cloud cover depictions:

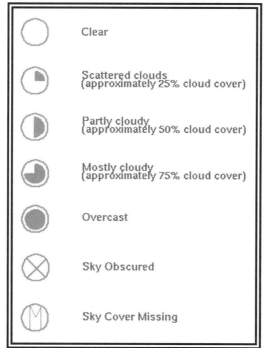

Figure 5-9. HPC Surface Analysis Chart Sky Cover Symbols

5.1.2.3.11 Water Temperature

Water temperature is plotted in whole degrees Fahrenheit.

5.1.2.3.12 Swell Information
Swell direction, period, and height are represented in the surface observations by a 6-digit code. The first two digits represent the swell direction, the middle digits describe the swell period (in seconds), and the last two digits are the swell's height (in half meters).

090703

09 - The swell direction is from 90 degrees (i.e. it is coming from due east).
07 - The period of the swell is 7 seconds.
03 - The height of the swell is 3 half meters.

271006

27 - The swell direction is from 270 degrees (due west).
10 - The period is 10 seconds.
06 - The height of the swell is 6 half meters.

5.1.2.3.13 Wave Information
Period and height of waves are represented by a 5-digit code. The first digit is always **1**. The second and third digits describe the wave period (in seconds), and the final two digits give the wave height (in half meters).

10603

1 - A group identifier. The first digit will always be **1**.
06 - The wave period is 6 seconds.
03 - The wave height is 3 half meters.

10515

1 - The group identifier again.
05 - The wave period is 5 seconds.
15 - Wave height is 15 half meters.

In some charts by the OPC, only the wave height (in feet) is plotted.

5.1.2.4 Analyses
Isobars, pressure systems, and fronts are the most common analyses depicted on surface analysis charts.

5.1.2.4.1 Isobars
An isobar is a line of equal or constant pressure commonly used in the analysis of pressure patterns.

On a Surface Analysis Chart, isobars are solid lines usually spaced at intervals of 4 millibars (mb). Each isobar is labeled. For example, **1032** signifies 1,032.0 mb and **992** signify 992.0 mb.

METAR/SPECI (Section 2.1) reports pressure in hectopascals. However, one millibar is equivalent to one hectopascal, so no conversion is required.

5.1.2.4.2 Pressure Systems

On a Surface Analysis Chart, a High (**H**) is a maximum of atmospheric pressure, while a Low (**L**) is a minimum of atmospheric pressure. Central pressure is the atmospheric pressure located at the center of a High or Low. In general, the central pressure is the highest pressure in the center of a High and the lowest pressure at the center of a Low. The central pressure is denoted near each pressure center.

A trough or an elongated area of low atmospheric pressure is denoted by dashed lines and identified with the word **TROF**. A ridge or an elongated area of high atmospheric pressure is denoted by saw-toothed symbols. Ridges are rarely denoted on charts produced by the HPC.

Tropical storms, hurricanes, and typhoons (See Figure 5-2) are low-pressure systems with their names and central pressures denoted.

5.1.2.4.3 Fronts

The analysis shows positions and types of fronts by the symbols in Figure 5-2. The symbols on the front indicate the type of front and point in the direction toward which the front is moving. Two short lines across a front indicate a change in front type.

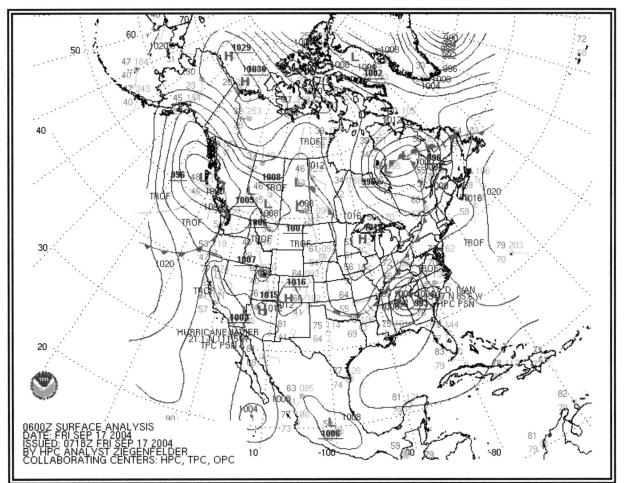

Figure 5-10. HPC Surface Analysis Chart – North America Example

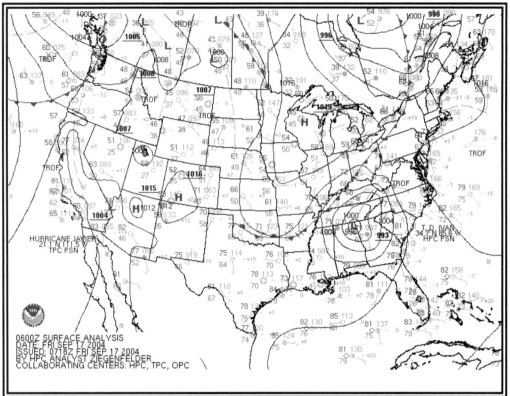

Figure 5-11. HPC Surface Analysis Chart - Contiguous U.S. Example

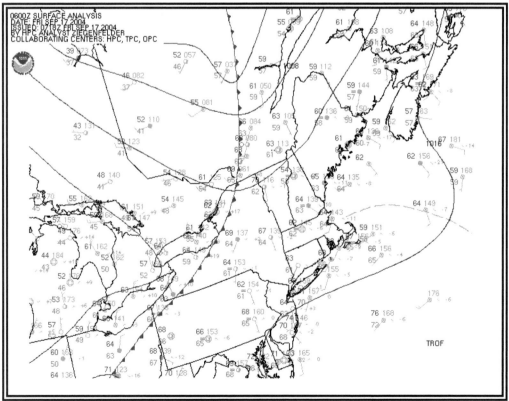

Figure 5-12. HPC Surface Analysis Chart - Northeast U.S. Example

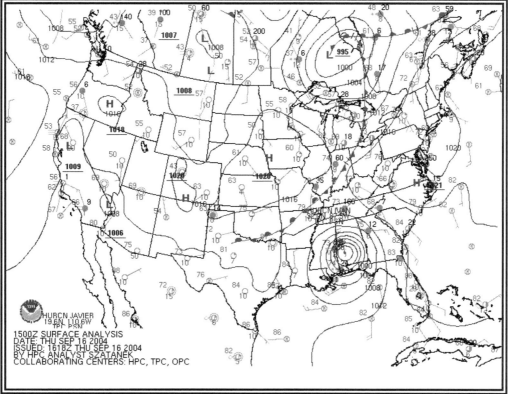

Figure 5-13. HPC Surface Analysis Chart for Aviation Interests – Contiguous U.S. Example

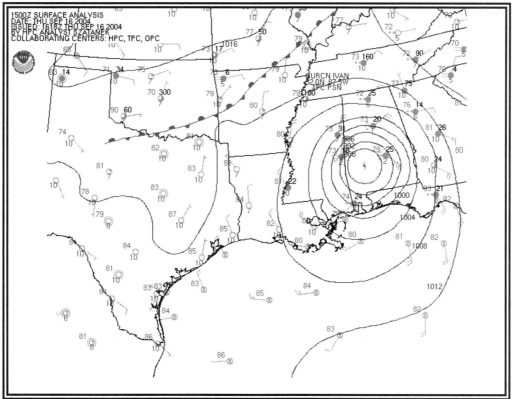

Figure 5-14. HPC Surface Analysis Chart for Aviation Interests – South central U.S. Example

5.1.3 OPC and WFO Honolulu Surface Analysis Charts

The Ocean Prediction Center (OPC) in Camp Spring, Maryland is responsible for the Atlantic and Pacific Oceans north of 30°N latitude. The charts are found at:

- http://www.opc.ncep.noaa.gov/

The Weather Forecast Office in Honolulu, Hawaii (WFO HNL) is responsible for the tropical Pacific Ocean, south of 30°N latitude and west of 160°E longitude. The charts are found at:

- http://www.prh.noaa.gov/hnl/pages/analyses.php

5.1.3.1 Issuance

The Ocean Prediction Center (OPC) produces surface analysis charts for the Atlantic and Pacific Oceans north of 30°N latitude four times daily (Table 5-2). The Weather Forecast Office in Honolulu, Hawaii (WFO HNL) issues surface analysis charts for the tropical Pacific Ocean, south of 30°N latitude and west of 160°E longitude four times daily. Surface analysis charts for the North Pacific are jointly issued by both offices.

Table 5-2. OPC and WFO Honolulu Surface Analysis Charts Issuance Schedule

UTC	00	06	12	18

5.1.3.2 Analysis Symbols

Figure 5-15 shows analysis symbols used on OPC and WFO HNL surface analysis charts.

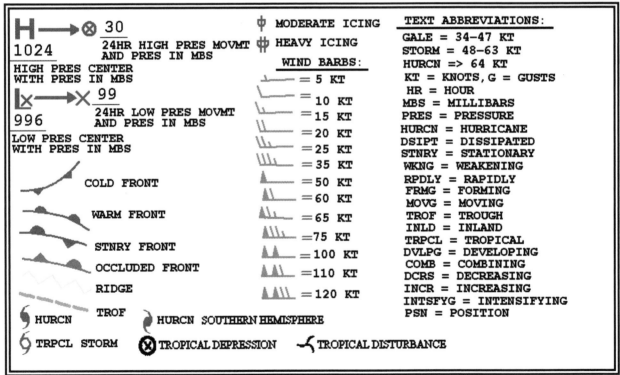

Figure 5-15. OPC and WFO HNL Surface Analysis Chart Symbols

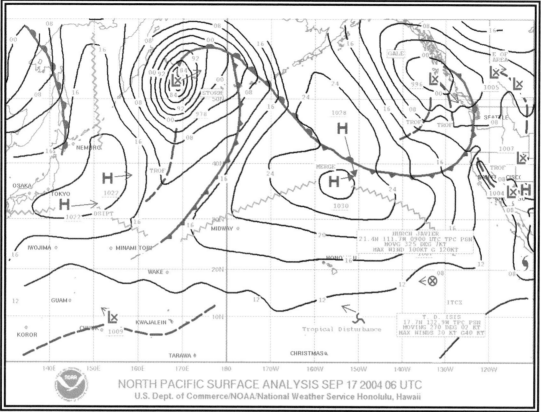

Figure 5-16. WFO HNL Surface Analysis Chart – North Pacific Example

5.1.3.3 Wave Information

Period and height of waves are represented by a 5-digit code. The first digit is always **1**. The second and third digits describe the wave period (in seconds), and the final two digits give the wave height (in half meters). Below are two examples:

10603

1 - A group identifier. The first digit will always be **1**.
06 - The wave period is 6 seconds.
03 - The wave height is 3 half meters.

10515

1 - The group identifier again.
05 - The wave period is 5 seconds.
15 - Wave height is 15 half meters.

In certain charts by the OPC, only the wave height (in feet) is plotted.

5.1.3.4 Analyses

Isobars, pressure systems, and fronts are the most common analyses depicted on the surface analysis charts.

5.1.3.4.1 Isobars

An isobar is a line of equal or constant pressure commonly used to analyze pressure patterns.

On a Surface Analysis Chart, isobars are solid lines usually spaced at intervals of 4 millibars (mb). Each isobar is labeled. For example, **1032** signifies 1,032.0 mb and **992** signify 992.0 mb.

METAR/SPECI (Section 2.1) reports pressure in hectopascals. However, one millibar is equivalent to one hectopascal, so no conversion is required.

5.1.3.4.2 Pressure Systems
On a Surface Analysis Chart, a High (**H**) is a maximum of atmospheric pressure, while a Low (**L**) is a minimum of atmospheric pressure. Central pressure is the atmospheric pressure at the center of a High or Low -- the highest pressure in a High and the lowest pressure in a Low. The central pressure is denoted near each pressure center. Tropical storms, hurricanes, and typhoons (See Figure 5-15) are low-pressure systems with their names and central pressures denoted.

A trough or an elongated area of low atmospheric pressure is denoted by dashed lines and identified with the word **TROF**. A ridge or an elongated area of high atmospheric pressure is denoted with saw-toothed symbols. Ridges are rarely denoted on charts produced by the HPC.

5.1.3.4.3 Fronts
The analysis shows positions and types of fronts by the symbols in Figure 5-15. The symbols on the front indicate the type of front and point in the direction toward which the front is moving. Two short lines across a front indicate a change in front type.

5.1.4 TPC Surface Analysis Charts
The Tropical Prediction Center (TPC) in Miami, Florida is responsible for the tropical regions of the western hemisphere south of 30°N latitude and east of 160°E longitude. The surface analysis chart is located at:
- http://www.nhc.noaa.gov/marine_forecasts.shtml

5.1.4.1 Issuance
The Tropical Prediction Center (TPC) issues Surface Analysis Charts for tropical regions of the western hemisphere south of 30°N latitude and east of 160°E longitude four times a day (Table 5-3).

Table 5-3. TPC Surface Analysis Charts Issuance Schedule

Valid Time (UTC)	00	06	12	18

5.1.4.2 Analysis Symbols
Figure 5-17 shows analysis symbols used on TPC surface analysis charts.

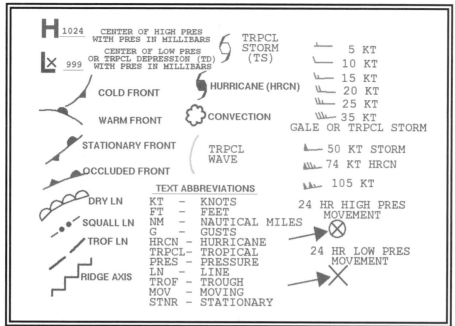

Figure 5-17. TPC Surface Analysis Chart Symbols

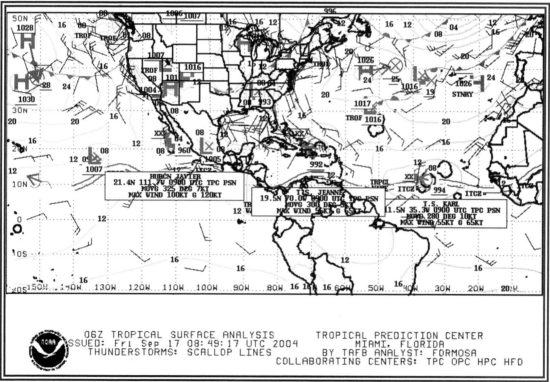

Figure 5-18. TPC Tropical Surface Analysis Chart Example

5.1.4.2.1 Wind

Wind is plotted in increments of 5 knots (kts). The wind direction is in "true" degrees and is depicted by a stem (line) pointed in the direction from which the wind is blowing. The wind speed is determined by adding up the value of the flags (50 kts), lines (10 kts), and half-lines (5 kts), each of which has the following individual values:

A single circle over the station with no wind symbol indicates a calm wind.

Figure 5-19 shows some sample wind symbols:

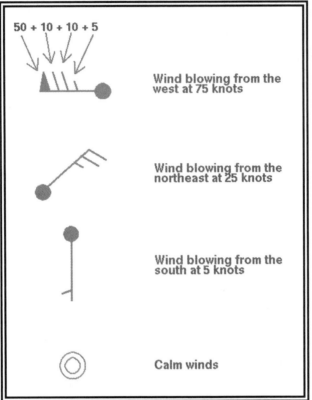

Figure 5-19: TPC Surface Analysis Chart Wind Plotting Model

5.1.4.3 Analyses
Isobars, pressure systems, and fronts are the most common analyses depicted on the surface analysis charts.

5.1.4.3.1 Isobars
An isobar is a line of equal or constant pressure commonly used to analyze pressure patterns.

On a Surface Analysis Chart, isobars are solid lines usually spaced at intervals of 4 millibars (mb). Each isobar is labeled. For example, **1032** signifies 1,032.0 mb and **992** signify 992.0 mb.

METAR/SPECI (Section 2.1) reports pressure in hectopascals. However, one millibar is equivalent to one hectopascal, so no conversion is required.

5.1.4.3.2 Pressure Systems
On a Surface Analysis Chart, a High (**H**) is a maximum of atmospheric pressure, while a Low (**L**) is a minimum of atmospheric pressure. Central pressure is the atmospheric pressure at the center of a High or Low -- the highest pressure in a High and the lowest pressure in a Low. The central pressure is denoted near each pressure center. Tropical storms, hurricanes, and

typhoons (See Figure 5-17) are low-pressure systems with their names and central pressures denoted.

A trough or an elongated area of low atmospheric pressure is denoted by dashed lines and identified with the word **TROF**. A ridge or an elongated area of high atmospheric pressure is denoted by saw-toothed symbols. Ridges are rarely denoted on charts produced by the TPC.

5.1.4.3.3 Fronts
The analysis shows positions and types of fronts by the symbols in Figure 5-17. The symbols on the front indicate the type of front and point in the direction toward which the front is moving. Two short lines across a front indicate a change in front type.

5.1.5 Unified Surface Analysis Chart
The Unified Surface Analysis Chart is a composite of all the surface analysis charts produced by HPC, OPC, TPC and WFO Honolulu. It contains an analysis of isobars, pressure systems and fronts.

5.1.5.1 Issuance
The chart is issued four times daily by the OPC (see Table 5-4).

Table 5-4. Unified Surface Analysis Chart Issuance Schedule

Valid Time (UTC)	00	06	12	18

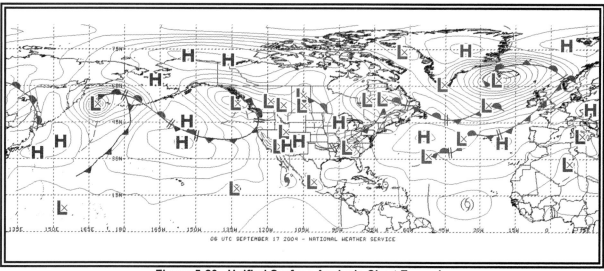

Figure 5-20. Unified Surface Analysis Chart Example

5.1.5.2 Analysis Symbols
Figure 5-21 shows analysis symbols used on the Unified Surface Analysis charts.

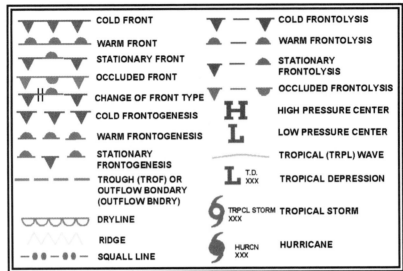

Figure 5-21. Unified Surface Analysis Chart Symbols

5.1.5.3 Analyses
Isobars, pressure systems, and fronts are the most common analyses depicted on the surface analysis charts.

5.1.5.3.1 Isobars
An isobar is a line of equal or constant pressure commonly used to analyze pressure patterns.

On a Surface Analysis Chart, isobars are solid lines usually spaced at intervals of 4 millibars (mb). Each isobar is labeled. For example, **1032** signifies 1032.0 mb and **992** signify 992.0 mb.

METAR/SPECI (Section 2.1) reports pressure in hectopascals. However, one millibar is equivalent to one hectopascal, so no conversion is required.

5.1.5.3.2 Pressure Systems
On a Surface Analysis Chart, a High (**H**) is a maximum of atmospheric pressure while a Low (**L**) is a minimum of atmospheric pressure. Central pressure is the atmospheric pressure at the center of a High or Low -- the highest pressure in a High and the lowest pressure in a Low. The central pressure is denoted near each pressure center. Tropical storms, hurricanes, and typhoons (See Figure 5-21) are low-pressure systems with their names and central pressures denoted.

On a Surface Analysis Chart, a trough is an elongated area of relatively low atmospheric pressure, while a ridge is an elongated area of relatively high atmospheric pressure. Troughs are denoted by dashed lines and identified with the word **TROF**. Ridges are denoted by saw-toothed symbols. Ridges are rarely denoted on charts produced by the HPC.

5.1.5.3.3 Fronts
The analysis shows positions and types of fronts by the symbols in Figure 5-21. The symbols on the front indicate the type of front and point in the direction toward which the front is moving. Two short lines across a front indicate a change in front type.

AC 00-45F   Section 5: Graphical Observations and Derived Products

5.1.6 AAWU Surface Charts

The Alaskan Aviation Weather Unit (AAWU) in Anchorage, Alaska is responsible for the state of Alaska. The surface analysis chart is located at:

- http://aawu.arh.noaa.gov/surface.php

5.1.6.1 Issuance

The AAWU issues Surface Analysis Charts 4 times daily for the state of Alaska. The valid times are shown in Table 5-5.

Table 5-5. AAWU Surface Analysis Issuance Schedule

Valid Time (UTC)	00	06	12	18

5.1.6.2 Analysis Symbols

The symbols (Figure 5-22) used on the Alaskan Surface Analysis Chart are similar to those found on the HPC Surface Analysis chart. However, since the Alaskan Surface Analysis chart is in black and white all of the symbols are black and white as well.

A Key to Symbols used on the AAWU Graphic Products

- Cold Front
- Warm Front
- Occluded Front
- Stationary Front
- Trough
- Ridge

H 1032 – High Pressure Center Pressure in millibars

L 988 – Low Pressure Center Pressure in millibars

– Occasional or greater Precipitation

= – Fog	⌐υ – Freezing Rain	– Mixed Rain/Snow
∞ – Haze	ϡυ – Freezing Drizzle	– Rain Showers
– Smoke	Ш – Light Icing	– Snow Showers
⊹ – Drifting Snow	Ш – Moderate Icing	– Rain/Snow Showers
– Sandstorm	Ш – Severe Icing	Ҟ – Thunderstorm
, – Drizzle	✳ – Snow	∧ – Light Turbulence
• – Rain	↔ – Ice Crystals	∧ – Moderate Turbulence
	△ – Ice Pellets	⋀ – Severe Turbulence

Figure 5-22. AAWU Surface Analysis Chart Symbols

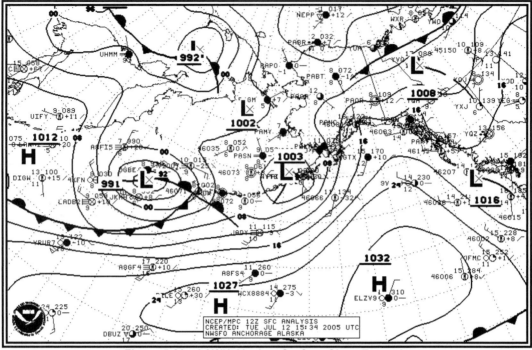

Figure 5-23. AAWU Alaskan Surface Chart Example

5.1.6.3 Station Plot Models

Land, ship, buoy, and C-MAN stations are plotted on the chart to aid in analyzing and interpreting the surface weather features. These plotted observations are referred to as station models. Some stations may not be plotted due to space limitations. However, all reporting stations are used in the analysis.

Figures 5-24 and 5-25 show the most commonly used station plot models used in surface analysis charts.

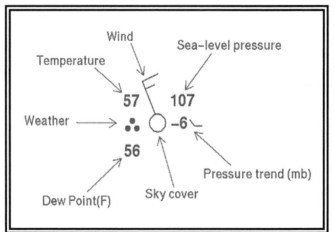

Figure 5-24. AAWU Surface Analysis Chart Station Plot Model

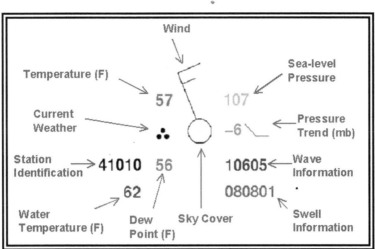

Figure 5-25. AAWU Surface Analysis Chart Ship/buoy Plot Model

5.1.6.3.1 Station Identifier

The format of the station identifier depends on the observing platform.

- Ship -- Typically 4 or 5 characters. If 5 characters, then the fifth will usually be a digit.

- Buoy -- Whether drifting or stationary, a buoy will have a 5-digit identifier. The first digit will always be a **4**.

- C-MAN -- Stands for Coastal-Marine Automated Network, and are usually close to coastal areas. Their identifier will appear like a 5-character ship identifier, however the 4th character will identify off which state the platform is located.

- Land -- Land stations will always be 3 characters, making them easily distinguishable from ship, buoy, and C-MAN observations.

5.1.6.3.2 Temperature

Air temperature is plotted in whole degrees Celsius on large-scale charts. Hourly surface charts may have temperatures using whole degrees Fahrenheit.

5.1.6.3.3 Dew Point

Dew point temperature is plotted in whole degrees Celsius on large-scale charts. Hourly surface charts may have dew point temperatures using whole degrees Fahrenheit.

5.1.6.3.4 Wind

Wind is plotted in increments of 5 knots (kts). The wind direction is in "true" degrees and is depicted by a stem (line) pointed in the direction from which the wind is blowing. Wind speed is determined by adding the values of the flags (50 kts), barbs (10kts), and half barbs (5 kts) found on the stem.

A single circle over the station with no wind symbol indicates a calm wind.

Figure 5-26 shows some sample wind symbols.

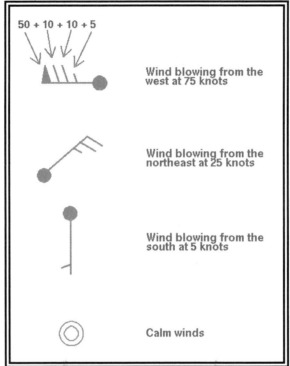

Figure 5-26: AAWU Surface Analysis Chart Wind Plotting Model

5.1.6.3.5 Ceiling
Ceiling is plotted in hundreds of feet above ground level.

5.1.6.3.6 Visibility
Surface visibility is plotted in whole statute miles.

5.1.6.3.7 Pressure
Sea-level pressure is plotted in tenths of millibars (mb), with the first two digits omitted (generally a 10 or 9). For reference, 1013 mb is equivalent to 29.92 inches of mercury. Below are some sample conversions between plotted and complete sea-level pressure values:

410: 1041.0 mb
103: 1010.3 mb
987: 998.7 mb
872: 987.2 mb

5.1.6.3.8 Pressure Trend
The pressure trend has two components, a number and symbol, to indicate how the sea-level pressure has changed during the past three hours. The number provides the 3-hour change in tenths of millibars, while the symbol provides a graphic illustration of how this change occurred.

Figure 5-27 shows the meanings of the pressure trend symbols.

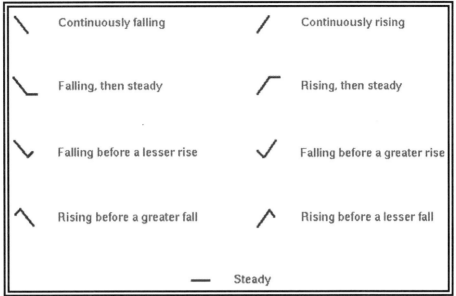

Figure 5-27. AAWU Surface Analysis Chart Pressure Trends

5.1.6.3.9 Sky Cover
The approximate amount of sky cover can be determined by the circle at the center of the station plot. The amount the circle is filled reflects the amount of sky covered by clouds. Figure 5-28 shows the common cloud cover depictions:

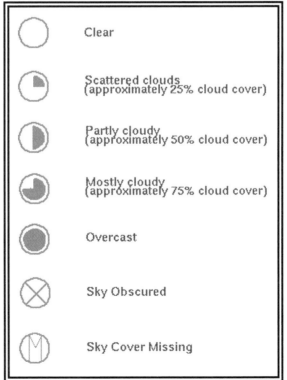

Figure 5-28. AAWU Surface Analysis Chart Sky Cover Symbols

5.1.6.3.10 Water Temperature
Water temperature is plotted in whole degrees Fahrenheit.

5.1.6.3.11 Swell Information

Swell direction, period, and height are represented in the surface observations by a 6-digit code. The first two digits represent the swell direction, the middle digits describe the swell period (in seconds), and the last two digits are the swell's height (in half meters).

090703

09 - The swell direction is from 90 degrees (i.e. it is coming from due east).

07 - The period of the swell is 7 seconds.

03 - The height of the swell is 3 half meters.

271006

27 - The swell direction is from 270 degrees (due west).

10 - The period is 10 seconds.

06 - The height of the swell is 6 half meters.

5.1.6.3.12 Wave Information

The period and height of waves are represented by a 5-digit code. The first digit is always **1**. The second and third digits describe the wave period (in seconds), and the final two digits give the wave height (in half meters). Below are two examples:

10603

1   - A group identifier. The first digit will always be **1**.

06 - The wave period is 6 seconds.

03 - The wave height is 3 half meters.

10515

1   - The group identifier again.

05 - The wave period is 5 seconds.

15 - Wave height is 15 half meters.

5.1.6.4 Analyses

Isobars, pressure systems, and fronts are the most common analyses depicted on surface analysis charts.

5.1.6.4.1 Isobars

An isobar is a line of equal or constant pressure commonly used to analyze pressure patterns.

On a Surface Analysis Chart, isobars are solid lines usually spaced at intervals of 4 millibars (mb). Each isobar is labeled. For example, **1032** signifies 1032.0 mb and **992** signify 992.0 mb.

METAR/SPECI (Section 2.1) reports pressure in hectopascals. However, one millibar is equivalent to one hectopascal, so no conversion is required.

5.1.6.4.2 Pressure Systems

On a Surface Analysis Chart, a High (**H**) is a maximum of atmosphere pressure, while a Low (**L**) is a minimum of atmospheric pressure. Central pressure is the atmospheric pressure at the center of a High or Low – the highest pressure in a High and the lowest pressure in a Low. The central pressure is denoted near each pressure center.

A trough or an elongated area of relatively low atmospheric pressure is denoted by dashed lines and identified with the word **TROF**. A ridge or an elongated area of relatively high atmospheric pressure is denoted by saw-toothed symbols. Ridges are rarely denoted on charts produced by the AAWU.

5.1.6.4.3 Fronts
The analysis shows positions and types of fronts by the symbols in Figure 5-22. The symbols on the front indicate the type of front and point in the direction toward which the front is moving. Two short lines across a front indicate a change in front type.

5.2 Constant Pressure Charts

Constant pressure charts are maps of selected conditions along specified constant pressure surfaces (pressure altitudes) and depict observed weather.

Constant pressure charts help to determine the three-dimensional aspect of depicted pressure systems. Each chart provides a plan-projection view at a specified pressure altitude.

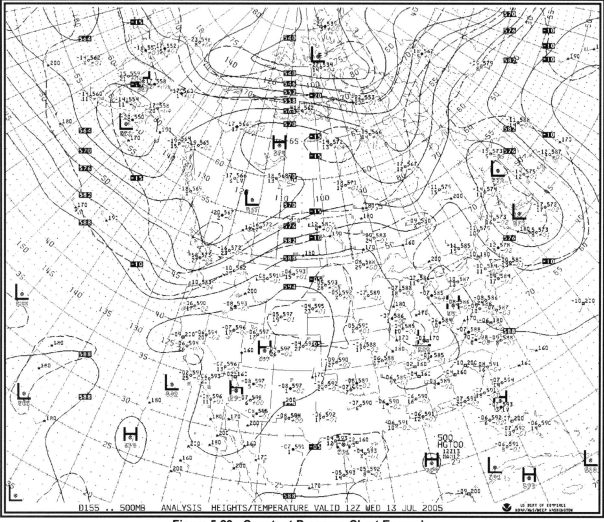

Figure 5-29. Constant Pressure Chart Example

5.2.1 Issuance
Constant pressure charts are issued twice per day from observed data obtained at 00Z and 12Z. Charts are available at the NWS Fax Chart web site at: http://weather.noaa.gov/fax/barotrop.shtml.

5.2.2 Observational Data
Observational data is plotted according to priority with some data deleted to prevent overlap. The retention priority is:

- Radiosonde observations (see Figure 5-30)
- Weather reconnaissance aircraft observations
- Aircraft observations on-time and on-level
- Aircraft observations off-time or off-level
- Satellite wind estimates

Many other data sources are used in the analysis but are not plotted. These include:

- Ships
- Buoys
- Tide gauges
- Wind profilers
- WSR-88D weather radar VAD wind profiles
- Satellite sounder

Figure 5-30. U.S. Radiosonde Network

Table 5-6 Features of Constant Pressure Charts

Constant Pressure Chart	Pressure Altitude (MSL)		Isotachs	Contour Interval (meters)	Height Plot Decoding		Examples of Height Plotting	
					Prefix	Suffix	Plotted	Height (meters, MSL)
850MB	5,000 ft	1,500 m	No	30 m	1	-	530	1,530 m
700MB	10,000 ft	3,000 m	No	30 m	2 or 3*	-	180	3,180 m
500MB	18,000 ft	5,500 m	No	60 m	-	0	582	5,820 m
300MB	30,000 ft	9,000 m	Yes	120 m	-	0	948	9,480 m
200MB	39,000 ft	12,000 m	Yes	120 m	1	0	164	11,640 m

Note: Pressure altitudes are rounded to the nearest 1,000 for feet and 500 for meters.
* Prefix a "2" or "3," whichever brings the height closer to 3,000 meters.

Table 5-7. Constant Pressure Chart Plotting Models

Radiosonde	Reconnaissance Aircraft (RECCO)	Aircraft Report (AIREP)	Satellite Wind Estimate
TT hhh DD h_ch_c	TT hhh DD R	TT $P_aP_aP_a$	$P_aP_aP_a$

SYMBOL	MEANING
	Wind plotted in standard notation. The stem points in the direction from which the wind is blowing plotted to 36 compass points, relative to true north. Wind speed is denoted by a combination of flags (50 knots), barbs (10 knots), and half barbs (5 knots).
TT	Temperature rounded to the nearest whole degree Celsius, with minus sign prefixed if negative.
hhh	Height of the constant pressure surface in meters MSL. See the Table 5-8 to see how each chart abbreviates height.
$P_aP_aP_a$	Pressure altitude in hundreds of feet MSL.
DD	Temperature-dew point spread (depression of the dew point temperature) rounded to the nearest whole degree Celsius. When DD is less than or equal to 5 degrees Celsius, the station circle is darkened so a region of high moisture content will stand out. If DD is greater than 5 degrees Celsius, the station circle is not shaded. If the DD is too large to measure, an X is plotted. For RECCO reports, DD will be missing when the temperature is colder than -41°C.
R	Reconnaissance aircraft type.

Table 5-8. Radiosonde Plotting Model Examples

TT ⟋hhh ● DD $h_c h_c$	$20 \,\bullet\, 504$ $5 \rvert\! \bar{} \, -01$	$08 \,\bullet\, 156$ $6 \rvert\! \bar{} \, -02$	$\llcorner -11 \,\bullet\, 582$ $\mathsf{X} \,\bullet\, +01$	$-36 \,\bullet\, 956$ $\underset{\underline{\quad}}{\mathtt{21}} \,\bullet\, -05$	$-56 \,\bullet\, 214$ $\underset{\underline{\quad}}{\mathtt{7}} \,\bullet\, -02$
Plotting Model	**850MB**	**700MB**	**500MB**	**300MB**	**200MB**
⟋ **Wind***	190°/20 kt	190°/25 kt	270°/15 kt	240°/70 kt	250°/115 kt
TT **Temperature**	20°C	8°C	-11°C	-36°C	-56°C
hhh **Height**	1,504 m	3,156 m	5,820 m	9,560 m	12,140 m
DD **Temperature -Dew Point Spread**	5°C	6°C	Too dry to measure	21°C	7°C
$h_c h_c$ **Height Change**	-10 m	-20 m	+10 m	-50 m	-20 m

* Wind direction assumes that north is at the top of the page. Latitude and longitude lines, as well as other geographical references, must be used to determine actual compass direction.

5.2.3 Analyses

All constant pressure charts contain analyses of height and temperature. Selected charts have an analysis of wind speed as well.

5.2.3.1 Height

Heights are analyzed with contours. Contours are lines of constant height in MSL and are used to map height variations of constant pressure surfaces. They identify and characterize pressure systems on constant pressure charts.

Contours are drawn as solid lines labeled with 3-digit numbers in decameters. Intervals at which the contours are drawn at: 30 meters for the 850 mb and 700 mb charts, 60 meters for the 500-mb chart, and 120 meters for the 300-mb and 200-mb charts. The location of a High or Low is marked with a ⊗ symbol together with a larger **H** or **L**, and the central value in decameters printed under the center location.

Contour gradient is the amount of height change over a specified horizontal distance. Gradients identify slopes of constant pressure surfaces that fluctuate in altitude. Strong gradients are denoted by closely-spaced contours which identify steep slopes. Weak gradients are denoted by widely-spaced contours which identify shallow slopes.

Wind speeds are directly proportional to contour gradients. Faster wind speeds are associated with strong contour gradients and slower wind speeds are associated with weak contour gradients. In mountainous areas, winds are often variable on constant pressure charts with altitudes near terrain elevation due to friction.

5.2.3.2 Temperature

Temperature is analyzed with isotherms which are lines of constant temperature. They are drawn as long dashed lines at intervals of 5° Celsius. They are given a two-digit label in whole degrees Celsius and are preceded with a **+** (positive) or **–** (negative) sign. The zero degree isotherm denotes the freezing level.

Temperature gradient is the amount of temperature change over a specified distance. Isotherm gradients identify the magnitude of temperature variations. Strong gradients are denoted by closely spaced isotherms and identify large temperature variations. Weak gradients are denoted by loosely spaced isotherms and identify small temperature variations.

5.2.3.3 Wind Speed
Wind speed is analyzed with isotachs which are lines of constant wind speed. They are drawn on the 300-mb and 200-mb charts with short-dashed lines at 20-knot intervals beginning with10 knots. They are labeled with a two- or three-digit number followed by a **K** for knots. Regions of high wind speeds are highlighted by alternate bands of shading and no-shading at 40-knot intervals beginning at 70 knots. A jet stream axis is the axis of maximum wind speed in a jet stream. Jet axes are not explicitly indicated, but their positions can be inferred from the isotach pattern and plotted winds.

5.2.3.4 Use
Constant pressure charts are used to provide an overview of selected observed weather conditions at specified pressure altitudes.

Pressure patterns cause and characterize much of the weather. Typically, lows and troughs are associated with bad weather, clouds and precipitation, while highs and ridges are associated with good weather.

Table 5-9. Reconnaissance Aircraft (RECCO) Plotting Model Examples

Plotting Model (diagram)	19₀366 1 AA329A	09₀146 1 AA921A	-05₀580 2 AA923A	-28₀966 3 AA924A	-53₀242 AA916A
Plotting Model	**850MB**	**700MB**	**500MB**	**300MB**	**200MB**
Wind*	150°/90 kt	130°/35 kt	180°/60 kt	240°/30 kt	110°/30 kt
TT Temperature	19°C	9°C	-5°C	-28°C	-53°C
hhh Height	1,366 m	3,146 m	5,800 m	9,660 m	12,420 m
DD Temperature -Dew Point Spread	1°C	1°C	2°C	3°C	Missing
R RECCO Type	AA329A	AA921A	AA923A	AA924A	AA916A
* Wind direction assumes that north is at the top of the page. Latitude and longitude lines, as well as other geographical references, must be used to determine actual compass direction.					

Table 5-10. Aircraft Report (AIREP) Plotting Model Examples

TT ▢ PₐPₐPₐ ▽	-05 060	12 100 LV	-10 180	-38 330	-54 360
Plotting Model	**850MB**	**700MB**	**500MB**	**300MB**	**200MB**
Wind*	20°/10 kt	Light and Variable	300°/30 kt	190°/5 kt	290°/50 kt
TT — Temperature	-5°C	12°C	-10°C	-38°C	-54°C
PₐPₐPₐ — Pressure Altitude (MSL)	6,000 ft	10,000 ft	18,000 ft	33,000 ft	36,000 ft

* Wind direction assumes that north is at the top of the page. Latitude and longitude lines, as well as other geographical references, must be used to determine actual compass direction.

Table 5-11. Satellite Wind Estimate Plotting Model Examples

★ PₐPₐPₐ	070	110	170	330	360
Plotting Model	**850MB**	**700MB**	**500MB**	**300MB**	**200MB**
Wind*	290°/30 kt	360°/20 kt	240°/10 kt	140°/165 kt	310°/60 kt
PₐPₐPₐ — Pressure Altitude (MSL)	7,000 ft	11,000 ft	17,000 ft	33,000 ft	36,000 ft

* Wind direction assumes that north is at the top of the page. Latitude and longitude lines, as well as other geographical references, must be used to determine actual compass direction.

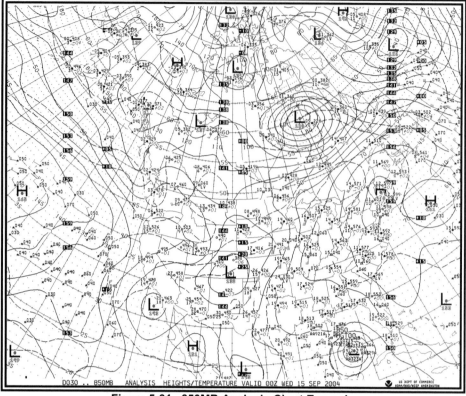

Figure 5-31. 850MB Analysis Chart Example

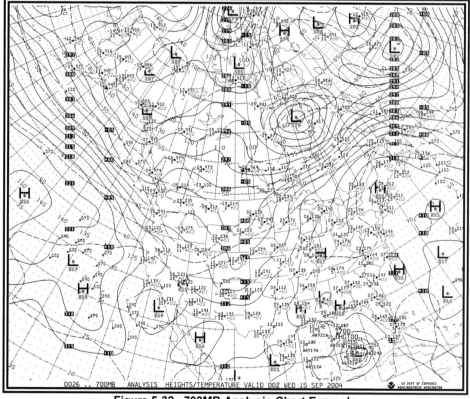

Figure 5-32. 700MB Analysis Chart Example

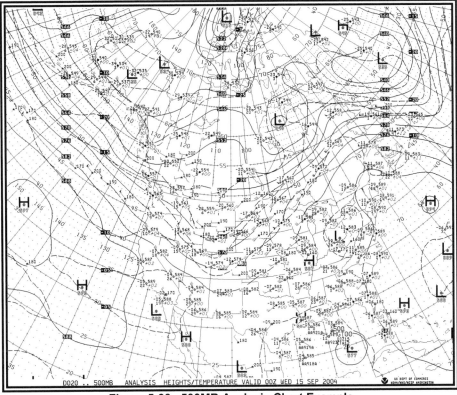

Figure 5-33. 500MB Analysis Chart Example

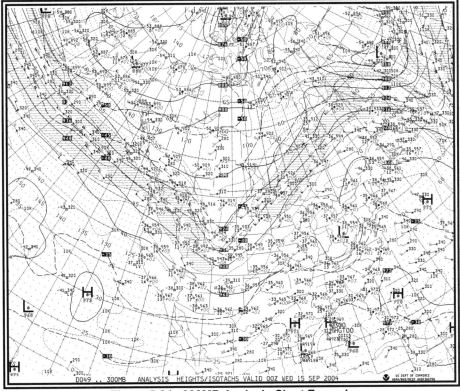

Figure 5-34. 300MB Analysis Chart Example

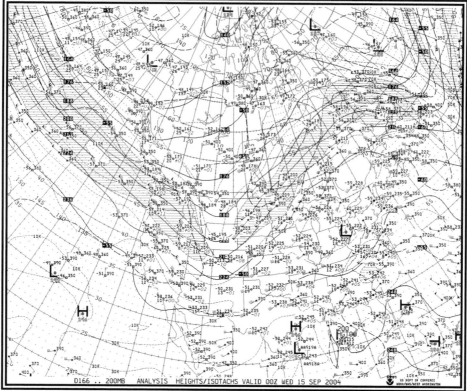

Figure 5-35. 200MB Analysis Chart Example

5.3 Freezing-level Graphics

The freezing level is the lowest altitude in the atmosphere over a given location at which the air temperature reaches $0°C$. This altitude is also known as the height of the $0°C$ constant-temperature surface. A freezing level chart shows the height of the $0°C$ constant-temperature surface.

The concept of freezing level becomes slightly more complicated when more than one altitude is determined to be at a temperature of $0°C$. These "multiple freezing layers" occur when a temperature inversion at altitudes above the defined freezing level are present. For example, if the first freezing level is at 3000 ft MSL and the second is at 7000 ft MSL, a temperature inversion is between these two altitudes. This would indicate temperatures rising above freezing above 3000 ft MSL and then back below freezing at 7000 ft MSL.

The Aviation Weather Center (AWC) provides freezing level graphics available on the Aviation Digital Data Service (ADDS) web site at:
http://adds.aviationweather.noaa.gov/icing/frzg_nav.php

The ADDS Freezing Level graphics provide an initial analysis and forecasts at 3-, 6-, 9-, and 12-hours into the future. The forecasts are based on output from the National Weather Service's (NWS) Rapid Update Cycle (RUC) numerical forecast model.

5.3.1 Issuance
The initial analysis and 3-hour forecast graphics are updated hourly. The 6-, 9-, and 12-hour forecast graphics are updated every three hours.

5.3.2 Observational Data
The RUC forecast model incorporates all of the latest weather observations in order to produce the best available analysis and forecast. These observations include:

- Commercial aircraft

- Profiler related:

    o Wind profilers (404 and boundary layer 915 MHz)

    o VAD (Velocity Azimuth Display) winds from WSR-88D radars

    o RASS (Radio Acoustic Sounding System)

- Rawinsondes and special dropwinsondes

- Surface:

    o GPS total precipitable water estimates

    o GOES cloud-top data (pressure and temperature)

    o GOES total precipitable water estimates

  ○ SSM/I total precipitable water estimates

  ○ GOES high-density visible and infrared (IR) cloud drift winds

- Experimental:

  ○ Radar reflectivity (3-d)

  ○ Lightning

  ○ Regional aircraft data with moisture (TAMDAR)

5.3.3 Format

The colors represent the height in hundreds of feet above mean sea level (MSL) of the lowest freezing level.

- Regions with white indicate the surface and the entire depth of the atmosphere are below freezing.
- Hatched regions represent areas where the surface temperature is below freezing with multiple freezing levels aloft.

- Areas where the surface temperature is above freezing with multiple freezing levels aloft are in regions where adjacent pixels change by more than one color when compared against the color scale (e.g., orange to dark blue).

The following cases illustrate the interpretation of the graphic.

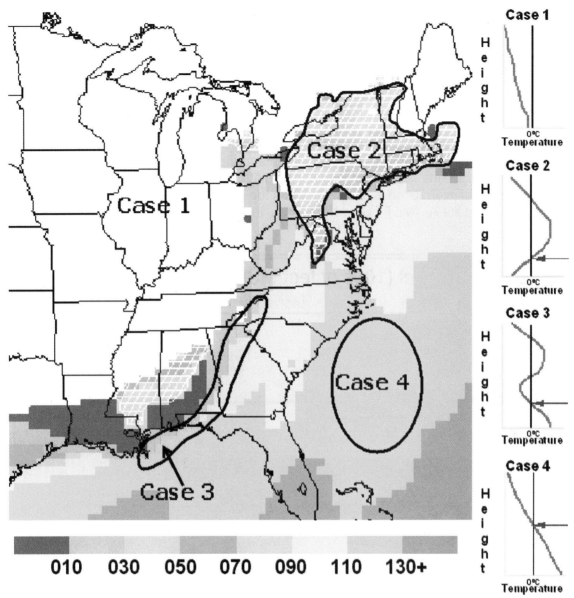

Figure 5-36. RUC 00-hour Freezing Level Graphic Example

Case 1 (Figure 5-36, Case 1) represents the condition where temperature is below freezing at the surface and all levels above the surface (represented in the graphic above by white-colored pixels).

Case 2 (Figure 5-36, Case 2) represents the condition where the temperature goes above and below freezing two or more times vertically through the atmosphere while the surface temperature is less than 0°C. These regions are hatched with white. The underlying color represents the lowest height where the temperature crosses the 0°C line as shown by the blue arrow on the vertical temperature graphic.

Case 3 (Figure 5-36, Case 3) represents the condition where the temperature goes above and below freezing three or more times vertically through the atmosphere while the surface temperature is higher than 0°C. These regions are located in areas where adjacent pixels change by more than one color when compared against the color scale.

Case 4 (Figure 5-36, Case 4) is relatively simple and represents the condition where the temperature at the surface is above freezing and the air generally cools with height crossing the 0°C line once.

5.3.4 Use

Freezing level graphics are used to assess the lowest freezing level heights and their values relative to flight paths. Clear, rime and mixed icing are found in layers with below-freezing (negative) temperatures and super-cooled water droplets. Users should be aware that official forecast freezing level information is specified within the AIRMET Zulu Bulletins (Contiguous U.S. and Hawaii) and the AIRMET "ICE AND FZLVL" information embedded within the Area Forecasts (Alaska only)

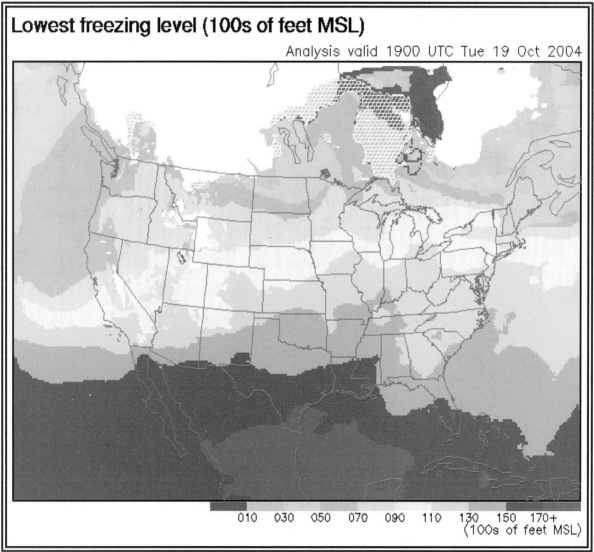

Figure 5-37. ADDS Freezing Level Graphic Example

5.4 Lifted Index Analysis Chart

The Lifted Index Analysis Chart (Figure 5-38) provides a data plot of observed lifted index (LI) and K index values for radiosonde sites and an analysis of LI for the contiguous U.S., southern Canada and northern Mexico.

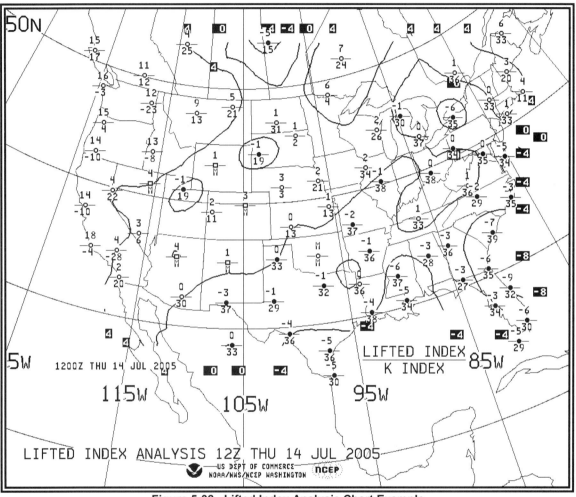

Figure 5-38. Lifted Index Analysis Chart Example

5.4.1 Issuance
The Lifted Index Analysis Chart is issued twice daily by the NWS and based solely on radiosonde observations from 00Z and 12Z. It is available at the NWS Fax Chart web site at: http://weather.noaa.gov/pub/fax/QXUA00.TIF.

5.4.2 Lifted Index (LI)
The lifted index (LI) is a common measure of atmospheric stability. The Lifted Index Analysis Chart depicts a number associated with the stability of a surface parcel of air lifted to 500 mb. For more complete information on the computation, refer to the Aviation Weather manual (AC 00-6A).

Lifted index values range from positive to negative. A positive lifted index indicates stable air. Larger positive numbers imply greater stability. A negative lifted index indicates unstable air.

Larger negative numbers imply greater instability. A zero lifted index indicates neutrally stable air.

5.4.2.1 Data Plot
Values of lifted index are plotted above the station circle for each available radiosonde station. Missing values are denoted with **M**. Station circles are blackened for LI values of zero or less.

5.4.2.2 Analysis
Isopleths, or lines of equal value, of lifted index are drawn for intervals of 4 units for index values of +4 and lower.

5.4.2.3 Use of Lifted Index
LI values on the Lifted Index Analysis Chart must be used with caution. The chart is only issued twice per day and significant changes can occur between chart times. LI values are typically lowest (least stable) during the afternoon due to daytime heating and highest (most stable) at sunrise due to nighttime cooling.

LI values can change rapidly due to moving fronts, drylines, outflow boundaries, and other boundaries which change surface airmass characteristics. LI (and thus stability) is particularly sensitive to changes of surface dew point. Temperature changes at 500 mb also affect LI, but these changes are usually much less dramatic than those which occur with temperature and dew point near the surface.

The Lifted Index Analysis chart only uses radiosonde data in its LI analysis. This means only large synoptic-scale stability patterns can be determined. Smaller, mesoscale LI variations will be missed.

An unstable airmass (denoted by negative LI values) only implies the potential for thunderstorms. A lifting mechanism such as a front, dryline, upslope flow, outflow boundary from prior storms, or frictional convergence around lows and troughs is still necessary to initiate a thunderstorm.

5.4.3 K Index
The K index (Figure 5-38) is a measure of thunderstorm potential based on vertical temperature lapse rate, moisture content of the lower atmosphere, and vertical extent of the moist layer. For more complete information on the computation, refer to the Aviation Weather (AC 00-6A).

5.4.3.1 Data Plot
Values of K will be plotted below the station circle for each available radiosonde station. Missing values are denoted with **M**. No analysis of the K-index is made on the Lifted Index Chart.

5.4.3.2 Use of K Index
With the K index, the higher the positive number, the likelihood of thunderstorm development is greater. The computation of the K-Index biases it in favor of general thunderstorms and it works better for non-severe convection. The K-index is also an index for forecasting heavy rain.

Although K-index values can be correlated to a probability of thunderstorm occurrence, these values will vary with seasons, locations, and synoptic settings. The values listed in Table 5-12 were empirically-derived and should be used with caution.

Table 5-12. K Index and Coverage of General Thunderstorms

K INDEX West of the Rockies	K INDEX East of the Rockies	Coverage of General Thunderstorms
less than 15	less than 20	None
15 to 20	20 to 25	Isolated thunderstorms
21 to 25	26 to 30	Widely scattered thunderstorms
26 to 30	31 to 35	Scattered thunderstorms
Above 30	Above 35	Numerous thunderstorms
Note: K value may not be representative of airmass if 850 mb level is near the surface.		

The chart only uses radiosonde data in its K index analysis. This means only large synoptic-scale stability patterns can be determined. Smaller, mesoscale K index variations will be missed.

5.5 Weather Depiction Chart

The Weather Depiction Chart (Figure 5-39) contains a plot of weather conditions at selected METAR stations and an analysis of weather flying category. It is designed primarily as a briefing tool to alert aviation interests to the location of critical or near-critical operational minimums at terminals in the conterminous US and surrounding land areas. The chart can be found at: http://weather.noaa.gov/pub/fax/QGUA00.TIF

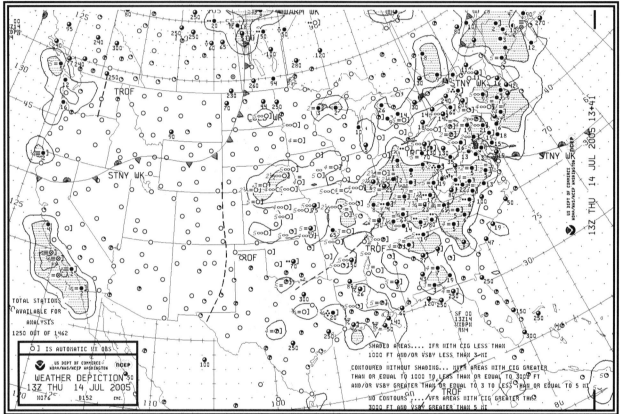

Figure 5-39. Weather Depiction Chart Example

5.5.1 Issuance
The Weather Depiction chart is issued eight times daily at the following times:

Table 5-13. Weather Depiction Charts Issuance Schedule

Valid Time (UTC)	01	04	07	10	13	16	19	22

5.5.2 Station Plot Model
METAR elements (Section 2.1) associated with weather flying category (visibility, present weather, sky cover, and ceiling) are plotted for each station on the chart (Figure 5-41). The station is located at the center of the sky cover symbol. Most stations are not plotted due to space limitations. However, all reporting stations are used in the weather flying category analysis.

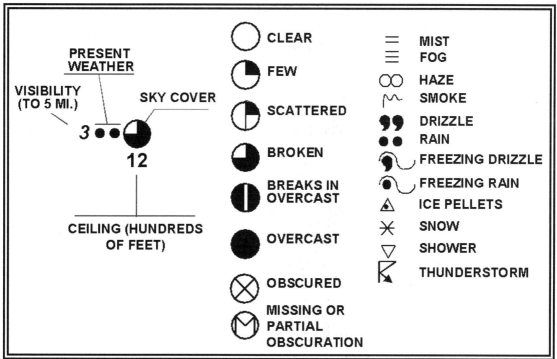

Figure 5-40. Weather Depiction Chart Station Plot Model

5.5.2.1 Visibility
When visibility is 5 miles or less, it is entered to the left of the station. Visibility is entered in statute miles and fractions of a mile.

5.5.2.2 Present Weather
Present weather symbols are entered to the left of the station. If the present weather information is obtained by an automated system, the right bracket symbol ( ]) is plotted to the right of the station.

When several types of weather and/or obstructions to visibility are reported, the most significant weather element is plotted. This is the first weather element coded in the METAR report (Section 2.1) and is usually the highest coded number in the Present Weather Symbols guide (Appendix I).

5.5.2.3 Sky Cover
Sky cover represents the summation total of the sky condition element from the METAR report. For example, if the METAR sky condition element was **SCT030 BKN060 OVC090**, the sky cover would be overcast. Sky cover symbols are listed in Figure 5-41.

5.5.2.4 Ceiling
Ceiling is the height from the base of the lowest layer aloft covering more than one-half the sky to the ground. Additionally, vertical visibility into a total surface-based obscuration is defined as a ceiling. For a METAR report, the first broken (BKN) or overcast (OVC) layer is the ceiling. For example, if the METAR sky condition element is **SCT030 BKN060 OVC090**, the ceiling is 6,000 feet.

For stations with broken to overcast layers, the ceiling height is plotted below the station. Ceilings are reported as hundreds of feet above ground level (AGL).

For a total surface-based obscuration, no ceiling is plotted and the METAR must be consulted.

Partial obscurations are not identified.

- For a partial obscuration <u>with no layer above</u>, the sky cover symbol will be plotted as missing (Figure 5-41).
- For a partial obscuration <u>with a layer above</u>, the sky cover and ceiling height will be plotted for the cloud layer only.

The METAR report should be consulted to identify the partial obscuration.

If the sky cover is clear, few, or scattered, no ceiling is plotted.

5.5.3 Weather Flying Category Analysis
Instrument Flight Rules (IFR) indicated on the Weather Depiction Chart represents ceilings less than 1,000 feet and/or visibility less than 3 statute miles and IFR operations must be in place. IFR areas are outlined on the chart with a solid line and are <u>shaded</u>. IFR areas are typically shaded red in colorized versions of the chart.

Marginal Visual Flight Rules (MVFR) indicated on the Weather Depiction Chart represents ceiling 1,000 to 3,000 feet and/or visibility 3 to 5 statute miles and VFR operations can take place. MVFR areas are outlined with a solid line, but the area is <u>not shaded</u>. MVFR areas are typically shaded blue in colorized versions of the chart.

Visual Flight Rules (VFR) indicated on the Weather Depiction Chart represents a ceiling greater than 3,000 feet or clear skies and visibility greater than 5 statute miles and VFR operations can take place. VFR conditions are not analyzed. This does not necessarily imply that the sky is clear.

5.5.4 Use
The Weather Depiction Chart is an ideal place to begin flight planning or to prepare for a weather briefing. This chart provides an overview of weather flying categories and other adverse weather conditions for the chart valid time. The chart, though, may not completely represent the en route conditions because of terrain variations and the possibility of weather occurring between reporting stations. This chart should be used in addition to the current METAR reports, pilot weather reports, and radar and satellite imagery for a complete look at the latest flying conditions.

5.6 Alaska Weather Depiction Charts

The Alaska Weather Depiction Charts (Figure 5-43) display color coded station plots which show: temperature, dew point, ceiling, visibility and wind direction/speed. A key to the station plots is found on each map.

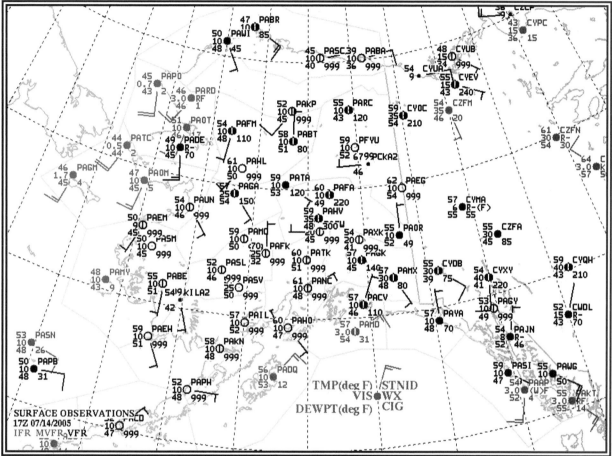

Figure 5-41. AAWU Alaska Weather Depiction Chart Example

Thirteen charts cover Alaska (except for the Aleutians) and adjacent areas of Canada.

Table 5-14. AAWU Alaska Weather Depiction Charts Coverage

Chart Coverage	Scale
Entire State of Alaska	(1:12 million)
All of Southeast Alaska	(1:5 million)
Southern Southeast Alaska	(1:3 million)
Northern Southeast Alaska	(1:3 million)
North Gulf Coast	(1:5 million)
South Central Alaska	(1:5 million)
Cook Inlet/Susitna Valley	(1:2 million)
Southwest Alaska	(1:6 million)
Western Interior	(1:5 million)
Central Interior	(1:5 million)
Northern Alaska	(1:6 million)
Southwest British Columbia	(1:7 million)
Yukon Territory/Northern British Columbia	(1:8 million)

5.6.1 Issuance

The charts are issued hourly and can be found on the Alaska Aviation Weather Unit (AAWU) web site at: http://aawu.arh.noaa.gov/Sigwx.php. The charts will first appear at about 10 minutes past the hour, with a second update at about 25 minutes past the hour.

5.6.2 Legends

The Alaska Weather Depiction Charts depict numerous parameters including the flying category, sky cover and wind.

5.6.2.1 Flying Category

Each station plot is color-coded according to the weather flying category reported (Table 5-16). Red indicates instrument flight rules (IFR), blue indicates marginal visual flight rules (MVFR), and black is plotted for stations reporting visual flight rules (VFR).

Table 5-15 AAWU Alaska Weather Flying Categories and Criteria

FLYING CATEGORY	CEILING (feet)	VISIBILITY (miles)
VFR (black)	Greater than 3,000 feet	Greater than 5 miles
MVFR (blue)	1,000 to 3,000 feet	3 to 5 miles
IFR (red)	Less than 1,000 feet	Less than 3 miles

5.6.2.2 Station Plot

METAR elements are plotted for each station on the chart (Figure 5-45). Some stations are not plotted due to space limitations, notably on the chart which covers the entire state of Alaska.

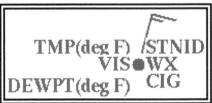

Figure 5-42. AAWU Alaska Weather Depiction Chart Station Plot Legend

5.6.2.3 Sky Cover

The sky cover symbol is plotted at the station location and is filled according to the summation total of the sky condition element from the METAR report. For example, if the METAR sky condition element was **SCT030 BKN060 OVC090**, the sky cover would be overcast. Sky cover symbols are listed in Figure 5-41.

5.6.2.4 Station Identifier (STNID)

The four-letter ICAO station identifier is entered to the upper right of the station.

5.6.2.5 Wind

Wind is plotted in increments of 5 knots (kts). The wind direction is referenced to "true" north and is depicted by a stem (line) pointed in the direction from which the wind is blowing. Wind speed is determined by adding the values of the flags (50 kts), barbs (10kts), and half barbs (5 kts) found on the stem.

A single circle over the station with no wind symbol indicates a calm wind.

Some sample wind symbols are shown on Figure 5-46.

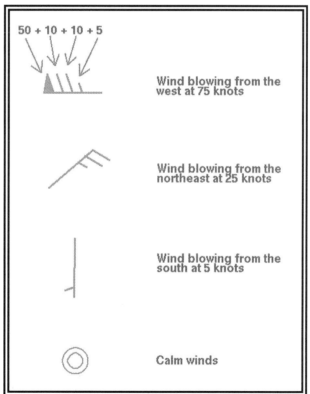

Figure 5-43. AAWU Alaska Weather Depiction Chart Wind Symbols

5.6.2.6 Temperature (TMP deg F)

Temperature in degrees Fahrenheit is plotted to the upper left of the sky cover symbol.

5.6.2.7 Visibility (VIS)

Visibility in statute miles is plotted to the left of the sky cover symbol. Decimals are used to represent tenths of miles when necessary.

5.6.2.8 Dew Point Temperature (DEWPT deg F)
Dew point temperature in degrees Fahrenheit is plotted to the lower left of the sky cover symbol.

5.6.2.9 Ceiling (CIG)
Ceiling is the height from the base of the lowest layer aloft covering more than one-half the sky. Additionally, vertical visibility into a total surface-based obscuration is defined as a ceiling. For a METAR report, the first broken (BKN) or overcast (OVC) layer is the ceiling. For example, if the METAR sky condition element is **SCT030 BKN060 OVC090**, the ceiling is 6,000 feet.

For a total surface-based obscuration, no ceiling is plotted and the METAR must be consulted.

If the sky cover is clear, few, or scattered, no ceiling is plotted.

The ceiling is plotted to the lower right of the station circle. Ceilings are reported as hundreds of feet above ground level (AGL). If no ceiling is present, the code **999** will be plotted.

5.6.2.10 Present Weather (WX)
Present weather symbols are entered to the left of the station. Note that the older Surface Aviation Observation (SAO) code is used instead Surface Analysis Chart symbols or the modern METAR code.

Table 5-16 Alaska Weather Depiction Charts Precipitation Symbols

Symbol	Meaning
T	Thunderstorm
R	Rain
RW	Rain Shower
L	Drizzle
ZR	Freezing Rain
ZL	Freezing Drizzle
A	Hail
IP	Ice Pellets
IPW	Ice Pellet Showers
S	Snow
SW	Snow Showers
SP	Snow Pellets
SG	Snow Grains
IC	Ice Crystals

Table 5-17 Alaska Weather Depiction Charts Obstruction to Visibility Symbols

Symbol	Meaning
BD	Blowing Dust
BN	Blowing Sand
BS	Blowing Snow
BY	Blowing Spray
D	Dust
F	Fog
GF	Ground Fog
H	Haze
IF	Ice Fog
K	Smoke

Table 5-18 Alaska Weather Depiction Charts Precipitation Intensity Symbols

Symbol	Meaning
-	Light
(No symbol)	Moderate
+	Heavy

5.6.3 Use

The Alaska Weather Depiction Charts provide an overview of weather flying categories and other adverse weather conditions for the chart valid time. The chart often does not completely represent the en route conditions because of terrain variations and the possibility of weather occurring between reporting stations. These charts should be used in addition to the latest METAR/SPECIs, pilot weather reports, and radar and satellite imagery for a complete look at the latest flying conditions.

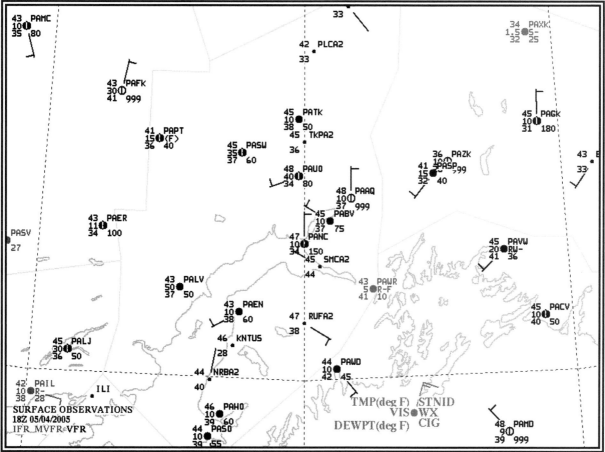

Figure 5-44. AAWU Alaska Weather Depiction Chart – South Central Alaska Example

5.7 Radar Summary Chart

The Radar Summary Chart (Figure 5-49) is a computer-generated mosaic of radar echo intensity contours based on Radar Weather Reports (Section 2.3) over the contiguous U.S. Possible precipitation types, cell movements, maximum tops, locations of line echoes, and remarks are plotted on this chart. Much of this information is often truncated due to space limitations. Severe thunderstorm and tornado watches are plotted if they are in effect when the chart is valid. The Radar Summary Chart is available on the National Weather Service (NWS) Fax Charts web site at: http://weather.noaa.gov/pub/fax/QAUA00.TIF

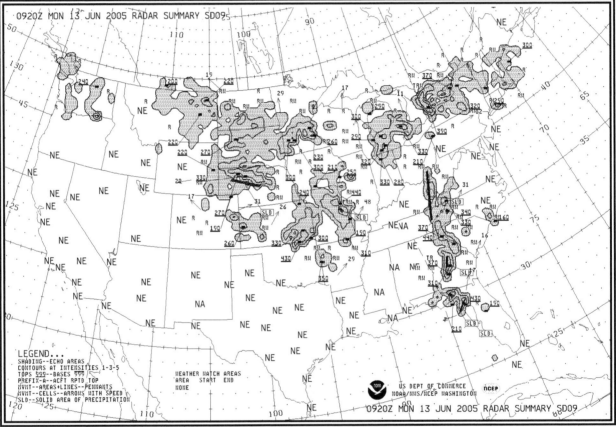

Figure 5-45. Radar Summary Chart Example

5.7.1 Issuance
The chart is issued hourly. Figure 5-50 depicts the WSR-88D weather radar network from which the chart is produced.

5.7.2 Format
The Radar Summary Chart depicts precipitation type, intensity, coverage, movement, echoes, and maximum tops.

5.7.2.1 Precipitation Type
The precipitation type, determined by a computer model, is indicated on the chart by symbols located adjacent to the precipitation areas. These symbols (Table 5-20) are <u>not</u> in METAR

format. Freezing precipitation is not reported in Radar Weather Reports and, thus, not plotted on the Radar Summary Chart.

Table 5-19. Radar Summary Chart Precipitation Type Symbols

SYMBOL	MEANING
R	Rain
RW	Rain shower
S	Snow
SW	Snow shower
T	Thunderstorms

5.7.2.2 Precipitation Intensity

The six precipitation intensity levels coded in the Radar Weather Report are consolidated into three contour intervals for the Radar Summary Chart (Figure 5-52). Precipitation intensity is correlated only for liquid precipitation, not solid precipitation (e.g., snow).

Figure 5-46. WSR-88D Weather Radar Network

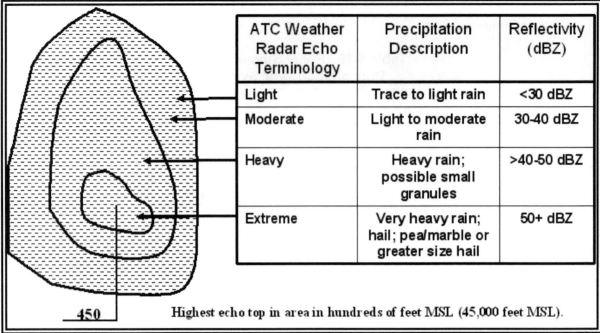

ATC Weather Radar Echo Terminology	Precipitation Description	Reflectivity (dBZ)
Light	Trace to light rain	<30 dBZ
Moderate	Light to moderate rain	30-40 dBZ
Heavy	Heavy rain; possible small granules	>40-50 dBZ
Extreme	Very heavy rain; hail; pea/marble or greater size hail	50+ dBZ

450   Highest echo top in area in hundreds of feet MSL (45,000 feet MSL).

Figure 5-47. Radar Summary Chart Precipitation Intensity

5.7.2.3 Echo Coverage

All of the shaded areas within the contours are assumed to contain precipitation. However, actual precipitation coverage is less. This is because only a fraction of a grid box needs to be covered with echoes for the entire grid box to be plotted as precipitation on the chart.

5.7.2.4 Line Echoes

When precipitation echoes are reported as a **LINE**, a line will be drawn through them on the chart (see Table 5-21). Where there is 8/10ths or more coverage, the line is labeled as solid (**SLD**) at both ends.

Table 5-20. Radar Summary Chart Echo Configuration Symbols

SYMBOL	MEANING
SLD	8/10ths or greater coverage in a line.
/	Line of echoes.
SLD TRW SLD	Solid line of thunderstorms with intense to extreme precipitation.

5.7.2.5 Cell Movement

Cell movement is the average motion of all cells within a configuration. An arrow indicates direction of cell movement. Speed in knots is entered near the arrowhead. **LM** identifies little movement. Movement of areas and lines can be significantly different from the motion of the individual cells that comprise these configurations.

Table 5-21. Radar Summary Chart Cell Movement Examples

SYMBOL	MEANING
↗ 35	Cell movement to the northeast at 35 knots
→ 24	Cell movement to the east at 24 knots
↓ 18	Cell movement to the south at 18 knots
↙ 12	Cell movement to the southwest at 12 knots
LM	Little cell movement

5.7.2.6 Maximum Top

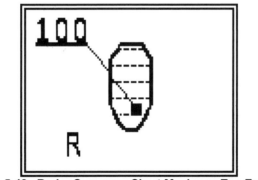

Figure 5-48. Radar Summary Chart Maximum Top Example

A maximum top is the altitude of the highest precipitation echo as coded on a Weather Radar Report (Section 2.3). Altitudes are sometimes augmented by satellite data. Individual Radar Weather Reports should be used to determine if satellite data was used for precipitation echo tops.

Tops are plotted in 3-digit groups representing height in hundreds of feet MSL and are underlined. Where it is necessary to offset a top for reasons of insufficient space, a line is drawn from one end of the underline to a small black square which represents the location of the top.

Maximum echo top does not equal maximum <u>cloud</u> top. The maximum echo top is the altitude of the highest light precipitation echo, not highest cloud top. Also, all radar heights are approximations due to radar wave propagation variations depending on atmosphere conditions.

5.7.2.7 Weather Watch Areas

Heavy dashed lines outline Tornado (**WT**) (Section 5.5.2) and Severe Thunderstorm (**WS**) Watch (Section 5.4.2) areas. The type of watch and the watch number are enclosed in a rectangle and positioned as closely as possible to the northeast corner of the watch. If there is no room at the northeast corner of the watch, the watch information is offset and connected to the watch by a thin line. The watch number is also printed at the bottom of the chart (in Mexico) together with the issuance time and expiration time under a label reading "**WEATHER WATCH AREAS**". In case no weather watch is in effect, "**NONE**" is printed at the bottom of the chart.

Table 5-22. Radar Summary Chart Weather Watch Area Examples

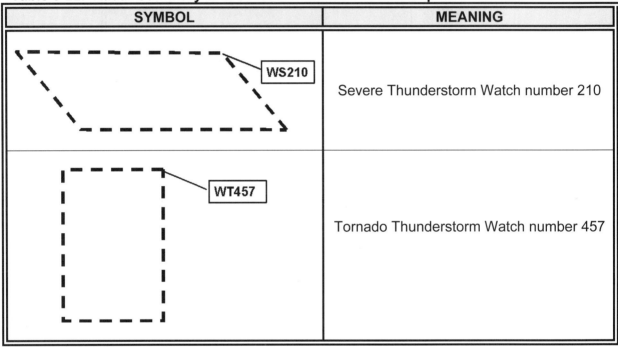

SYMBOL	MEANING
WS210	Severe Thunderstorm Watch number 210
WT457	Tornado Thunderstorm Watch number 457

5.7.2.8 Operational Contractions

Radar sites which report **PPINA**, **PPINE**, and **PPIOM** in their Weather Radar Reports (Section 2.3.1.10) are abbreviated to **NA**, **NE**, and **OM** respectively and plotted over the radar sites on the chart.

Table 5-23. Radar Summary Chart
Operational Contractions

SYMBOL	MEANING
NA	Not available
NE	No echoes
OM	Out for maintenance

5.7.3 Use

The Radar Summary Chart aids in preflight planning by identifying areas of precipitation and highlighting its characteristics. This chart displays precipitation only; it does <u>not</u> display clouds, fog, fronts, or other boundaries. Therefore, the absence of echoes does not equal clear weather. Cloud tops will be somewhat higher than precipitation tops detected by radar. The chart must be used in conjunction with other charts, reports, and forecasts.

The radar summary chart is for <u>preflight</u> planning only and should always be cross-checked and updated by current WSR-88D images. Once airborne, the pilot must evade individual storms by in-flight observations. This can be done by using visual sighting or airborne radar as well as by requesting weather radar information from En route Flight Advisory Service "Flight Watch" briefers at Automated Flight Service Station (AFSS). AFSS Flight Watch briefers have access to current weather radar imagery.

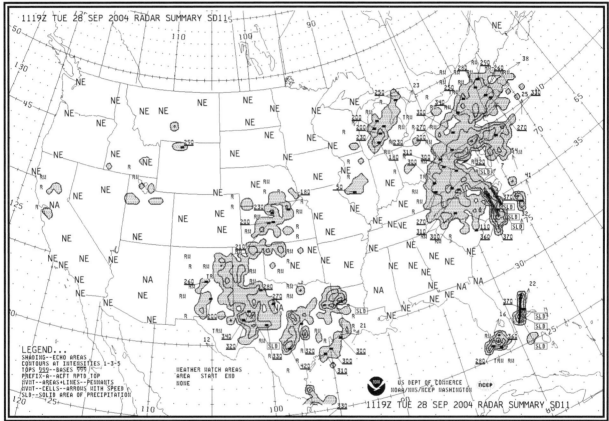

Figure 5-49. Radar Summary Chart Example

5.8 Alaska Initial Geopotential Heights and Winds Charts

The Alaska Initial Geopotential Heights and Winds Charts (Figure 5-54) display an analysis of the observed height contours and winds at selected constant pressure surfaces (flight levels).

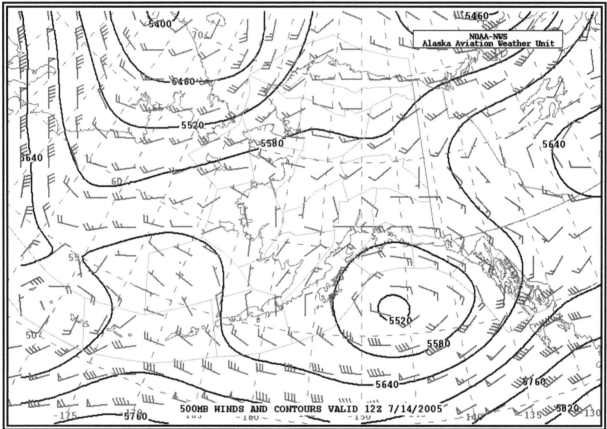

Figure 5-50. Alaskan Initial Geopotential Heights and Winds Chart Example

Table 5-24. Pressure Altitudes of Alaska Initial Geopotential Heights and Winds Charts

CHART	PRESSURE ALTITUDE (Feet, MSL)	PRESSURE ALTITUDE (Meters, MSL)
200 MB	39,000 ft	12,000 m
300 MB	30,000 ft	9,000 m
500 MB	18,000 ft	5,500 m
700 MB	10,000 ft	3,000 m
850 MB	5,000 ft	1,500 m

5.8.1 Issuance

The charts are issued twice daily with valid times of 00z and 12z and can be found on the Alaska Aviation Weather Unit (AAWU) web site at: http://aawu.arh.noaa.gov/upperwinds.php.

5.8.2 Analysis

The analysis of both height contours and winds are based on output from the North American Mesoscale (NAM) computer forecast model.

5.8.2.1 Height Contours

Height contours are lines of constant height referenced to MSL and are used to map the height variations of constant pressure surfaces. They identify and characterize pressure systems on constant pressure charts.

Contours are drawn as solid lines and labeled in meters. The intervals at which the contours are drawn are 60 meters on all of the charts.

5.8.2.2 Winds

Wind is plotted in increments of 5 knots (kts). The wind direction is referenced to "true" north and is depicted by a stem (line) pointed in the direction from which the wind is blowing. Wind speed is determined by adding the values of the flags (50 kts), barbs (10kts), and half barbs (5 kts) found on the stem.

A single circle over the station with no wind symbol indicates a calm wind.

Figure 5-55 contains some examples wind symbols.

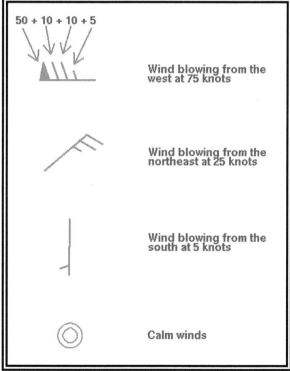

Figure 5-51. Alaskan Initial Geopotential Heights and Winds Chart Wind Plotting Model

5.8.3 Use

The Alaska Initial Geopotential Heights and Winds Charts are used to provide an overview of heights, pressure patterns and winds at specified pressure altitudes. Pressure patterns cause and characterize much of the weather. Typically, lows and troughs are associated with bad weather, clouds and precipitation, while highs and ridges are associated with good weather.

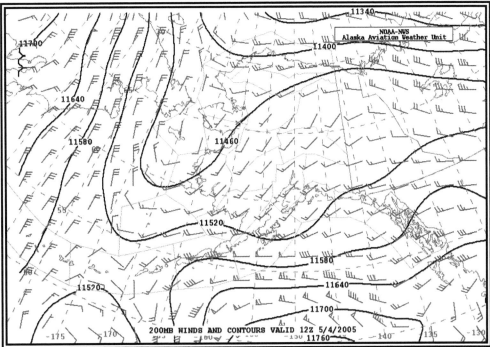

Figure 5-52. Alaskan Initial Geopotential Heights and Winds Chart - 200MB Winds and Contours Chart Example

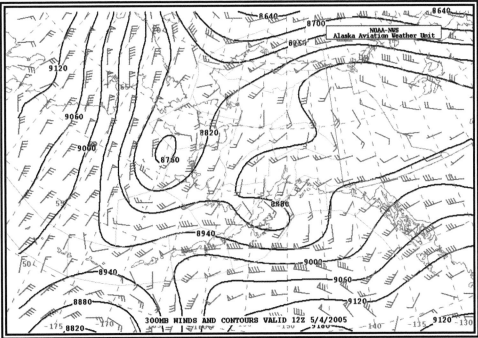

Figure 5-53. Alaskan Initial Geopotential Heights and Winds Chart - 300MB Winds and Contours Chart Example

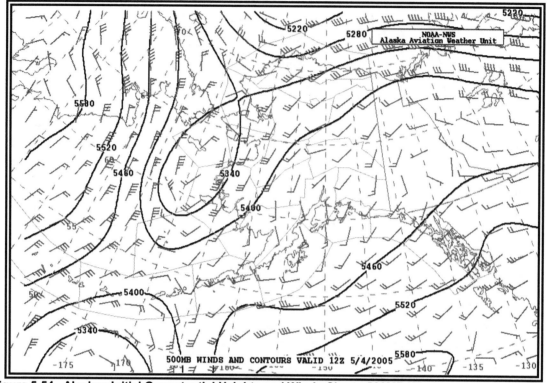

Figure 5-54. Alaskan Initial Geopotential Heights and Winds Chart - 500MB Winds and Contours Chart Example

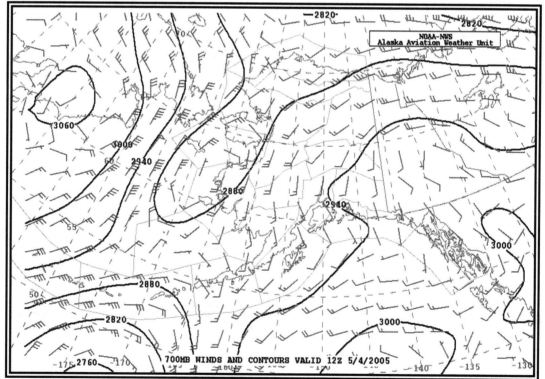

Figure 5-55. Alaskan Initial Geopotential Heights and Winds Chart - 700MB Winds and Contours Chart Example

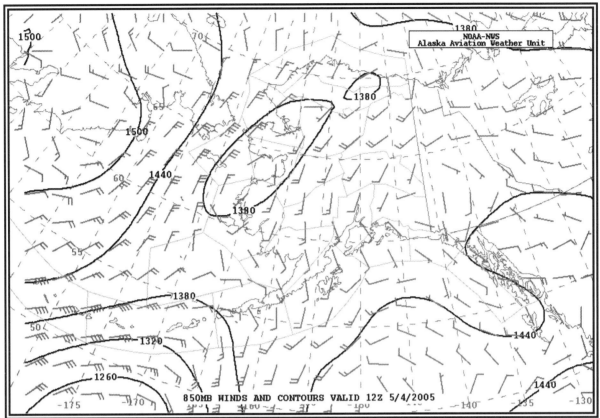

Figure 5-56. Alaskan Initial Geopotential Heights and Winds Chart - 850MB Winds and Contours Chart Example

6 PRODUCTS FOR AVIATION HAZARDS

6.1 Significant Meteorological Information (SIGMET)

Significant Meteorological Information (SIGMETs) provides aircraft operators and aircrews notice of potentially hazardous en route phenomena such as thunderstorms and hail, turbulence, icing, sand and dust storms, tropical cyclones, and volcanic ash.

- Domestic SIGMETs for the conterminous U.S. (CONUS) are available on the Aviation Digital Data Service (ADDS) web site at: http://adds.aviationweather.noaa.gov/airmet/

- International SIGMETs (including Alaska and Hawaii) are available on the Aviation Weather Center (AWC) website at: http://aviationweather.gov/products/sigmets/intl/

  o Alaska SIGMETs are also available on the Alaska Aviation Weather Unit (AAWU) web site at: http://aawu.arh.noaa.gov/

  o Hawaii SIGMETs are also available on the NWS WFO Honolulu web site at: http://www.prh.noaa.gov/hnl/pages/aviation.php

6.1.1 SIGMET Criteria (Non-Convective)
A SIGMET may be issued when any of the following conditions occur or is expected to occur in an area affecting at least 3,000 square miles or an area deemed to have a significant effect on the safety of aircraft operations.

- Thunderstorm* of the following type *except for the contiguous U.S.* (see Section 6.1.8)

  o Obscured (OBSCN TS)

  o Embedded (EMBD TS)

  o Widespread (WDSPR TS)

  o Squall line (SQL TS)

  o Isolated severe (ISOL SEV TS)

- Severe or greater Turbulence (SEV TURB)

- Severe Icing (SEV ICE)

- Widespread Duststorm (WDSPR DS)

- Widespread Sandstorm (WDSPR SS)

- Volcanic Ash (VA)

- Tropical Cyclone (TC)

NOTE: Obscured, embedded, or squall line thunderstorms, or mountain waves do not have to reach 3,000 square miles.

* Tornado (TDO), Funnel Cloud (FC), Waterspout (WTSPT), and Heavy Hail (HVYGR) may be used as a further description of the thunderstorm as necessary.

6.1.2 Standardization
SIGMETs follow these standards:

- All heights or altitudes are referenced to Mean Sea Level (MSL) and consist of three (3) digits depicting height in hundreds of feet. Flight Level (**FL**) is used for heights at or above 18,000 feet above. Examples: 100, FL190.

- References to latitude and longitude are in whole degrees and minutes following the model: Nnn[nn] or Snn[nn], Wnnn[nn] or Ennn[nn] with a space between latitude and longitude and a hyphen between successive points. Products issued by AWC have the N, S, W, or E behind the latitude/longitude numbers. Example: N3106 W07118

- Messages are prepared using approved ICAO contractions, abbreviations and numerical values of self-explanatory nature.

- Weather and obstructions to visibility are the same as weather abbreviations used for surface observations (METAR or SPECI) (Section 2.1).

- All amended (AMD) or corrected (COR) en route forecasts or advisory products follow the same format procedures. An AMD or COR is identified as a change in the first line after the time and date, and where appropriate, a comment - the reason for the change, etc. - is added as the last line of the product.

6.1.3 SIGMET Format
The WMO SIGMET header for non-convective SIGMETs is **WS**.

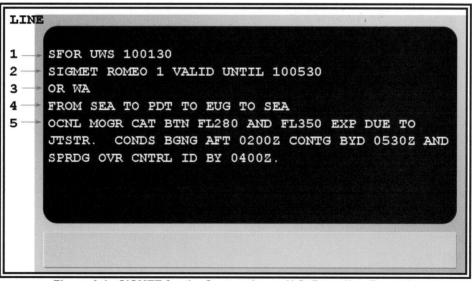

Figure 6-1. SIGMET for the Conterminous U.S. Decoding Example

Table 6-1. Decoding a Domestic SIGMET for the Conterminous U.S.

Line	Content	Description
1	SFO R WS 100130	SIGMET area identifier SIGMET series letter Product type Issuance UTC date/time
2	SIGMET ROMEO 1 VALID UNTIL 100530	Product type SIGMET series name Issuance number Ending valid UTC date/time
3	OR WA	Phenomenon location (states)
4	FROM SEA TO PDT TO EUG TO SEA	Phenomenon location (VOR coordinates)
5	OCNL MOGR CAT BTN FL280 AND FL350 EXP DUE TO JTSTR. CONDS BGNG AFT 0200Z CONTG BYD 0530Z AND SPRDG OVR CNTRL ID BY 0400Z.	Phenomenon description

The SIGMET in Figure 6-1 can be decoded as the following:

(Line 1) *SIGMET ROMEO series issued for the San Francisco Area at 0130 UTC on the 10th day of the month.*

(Line 2) *This is the first issuance of the SIGMET ROMEO series and is valid until the 10th day of the month at 0530 UTC.*

(Line 3) *The affected states within the SFO area are Oregon and Washington.*

(Line 4) *From Seattle, WA; to Pendleton, OR; to Eugene, OR; to Seattle, WA;*

(Line 5) *Occasional moderate or greater clear air turbulence between Flight Level 280 and Flight Level 350, expected due to jet stream. Conditions beginning after 0200Z continuing beyond 0530Z and spreading over central Idaho by 0400Z.*

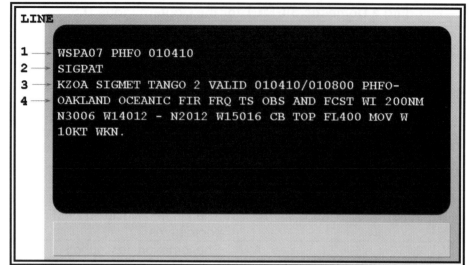

Figure 6-2. SIGMET Outside the Conterminous U.S. Decoding Example

Table 6-2. Decoding a SIGMET Outside of the Conterminous U.S.

Line	Content	Description
1	WSPA07 PHFO 010410	ICAO communication header Issuance MWO Issuance UTC date/time
2	SIGPAT	NWS AWIPS communication header
3	KZOA SIGMET TANGO 2 VALID 010410/010800 PHFO	Area Control Center Product type SIGMET series Issuance number Valid period UTC date/time Issuance office
4	OAKLAND OCEANIC FIR FRQ TS OBS AND FCST WI 200NM N3006 W14012 - N2012 W15016 CB TOP FL400 MOV W 10KT WKN.	Flight Information Region (FIR) Phenomenon description

The SIGMET in Figure 6-2 can be decoded as the following:

(Line 1) The WMO product header is WSPA07. Issued by the PHFO on the 1st day of the month at 0410 UTC.

(Line 2) The NWS AWIPS communication header is SIGPAT.

(Line 3) For the Oakland (KZOA) Area Control Center. This is the 2nd issuance of SIGMET Tango series, valid from the 1st day of the month at 0410 UTC until the 1st day of the month at 0800 UTC, issued by the Honolulu Meteorological Watch Office.

(Line 4) Concerning the Oakland Oceanic Flight Information Region (FIR), frequent thunderstorms observed and forecast within 200 nautical miles of 30 degrees and 6 minutes north; 140 degrees and 12 minutes west; to 20 degrees and 12 minutes north, 150 degrees and 16 minutes west, cumulonimbus tops to flight level 400 moving west at 10 knots, weakening.

6.1.3.1 SIGMET Phenomena Information

A SIGMET contains the following information related to the specific phenomena and in the order indicated:

- Phenomena and its description; e.g., **SEV TURB.**

- An indication whether the information is observed, using **OBS** or **FCST** with the time of observation will be given in UTC.

- Location (referring, when possible, to latitude and longitude and/or locations or geographic features which are well known internationally) and flight level (altitude).

- Movement towards or expected movement using sixteen points of the compass, with speed in knots, or stationary, if appropriate.

- Thunderstorm maximum height as FL.

- Changes in intensity; using as appropriate, the abbreviations Intensifying (**INTSF**), Weakening (**WKN**), or No Change (**NC**).

- On the last line, an outlook beyond the valid period for forecast trajectory of a volcanic ash cloud or tropical cyclone.

6.1.4 Issuance

SIGMETs are issued from Meteorological Watch Offices (MWO). The U.S. has three MWOs: the Aviation Weather Center (AWC), the Alaska Aviation Weather Unit (AAWU), and the Weather Forecast Office (WFO) in Honolulu. Their areas of responsibility are as follows:

- The AWC:

  o Forecast areas for the conterminous U.S. (CONUS) out to the domestic Flight Information Region (FIR) boundary (Figure 6-3).

  o The New York, Houston, Miami, and San Juan Oceanic FIRs (Figure 6-4).

  o The Oakland Oceanic FIR north of 30 north latitude, and the portion east of 140 west longitude which is between the equator and 30 north latitude (Figure 6-5).

- The AAWU is responsible for the Anchorage FIR (Figures 6-5 and 6-6).

- WFO Honolulu is responsible for the Oakland Oceanic FIR south of 30 north latitude to the equator and between 140 west and 160 east longitude (Figure 6-5).

Figure 6-3. AWC SIGMET Areas of Responsibility - Conterminous U.S.

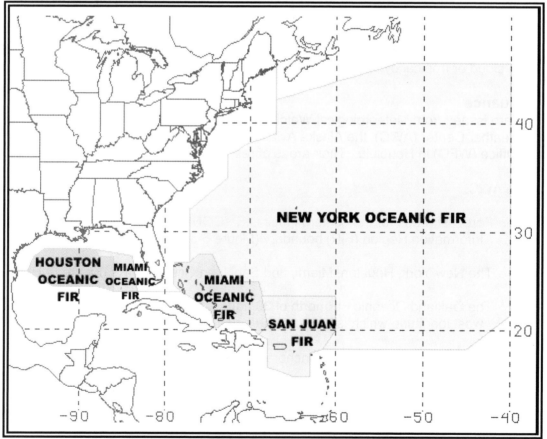

Figure 6-4. AWC SIGMET Areas of Responsibility – Atlantic Basin

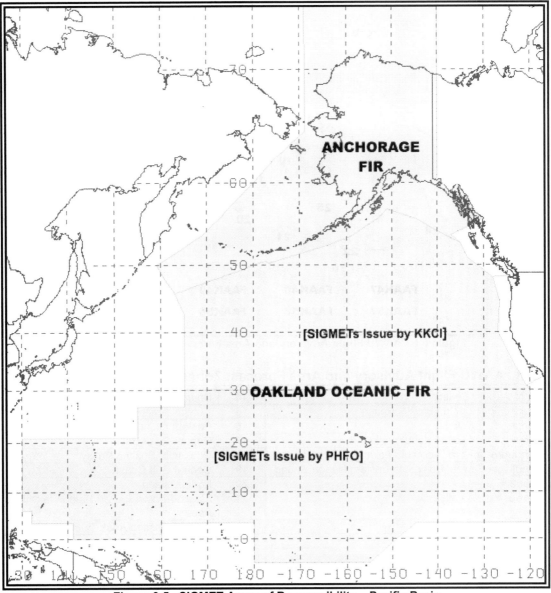

Figure 6-5. SIGMET Areas of Responsibility – Pacific Basin

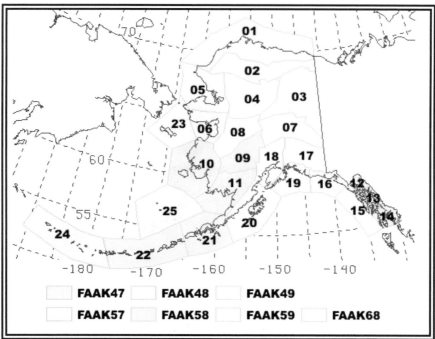

Figure 6-6. AAWU Flight Advisory and Area Forecast Zones – Alaska

Table 6-3. AAWU Flight Advisory and Area Forecast Zones – Alaska

1	Arctic Coast Coastal	14	Southern Southeast Alaska
2	North Slopes of the Brooks Range	15	Coastal Southeast Alaska
3	Upper Yukon Valley	16	Eastern Gulf Coast
4	Koyukuk and Upper Kobuk Valley	17	Copper River Basin
5	Northern Seward Peninsula-Lower Kobuk Valley	18	Cook Inlet-Susitna Valley
6	Southern Seward Peninsula-Eastern Norton Sound	19	Central Gulf Coast
7	Tanana Valley	20	Kodiak Island
8	Lower Yukon Valley	21	Alaska Peninsula-Port Heiden to Unimak Pass
9	Kuskowim Valley	22	Unimak Pass to Adak
10	Yukon-Kuskowim Delta	23	St. Lawrence Island-Bering Sea Coast
11	Bristol Bay	24	Adak to Attu
12	Lynn Canal and Glacier Bay	25	Pribilof Islands and Southeast Bering Sea
13	Central Southeast Alaska		

6.1.4.1 SIGMET Identification

When a SIGMET is issued, it is assigned a unique series identifier:

- AWC for CONUS
  - **NOVEMBER** through **YANKEE**, excluding **SIERRA** and **TANGO**
- AWC for Oakland Oceanic FIR
  - **ALFA** through **HOTEL**
- Honolulu MWO for Oakland Oceanic FIR
  - **NOVEMBER** through **ZULU**
- AAWU for Anchorage FIR
  - **INDIA** through **MIKE**

A number is assigned sequentially with each issuance until the phenomenon ends. At 0000 UTC each day, all continuing SIGMETs are renumbered to one (1) regardless of a continuation of the phenomena.

Examples: YANKEE 1, YANKEE 2, YANKEE 3, etc.

6.1.4.2 SIGMET Valid Period
SIGMETs for volcanic ash or tropical cyclones are valid up to six (6) hours with an outlook of up to twelve (12) hours beyond the valid period. They are reissued at least every six (6) hours while the volcanic ash or tropical cyclone exists or is forecast to exist. SIGMET messages for volcanic ash or tropical cyclones expected to affect a FIR are issued up to 12 hours before the start of the valid period or as soon as practicable if such advance warning of the existence of these phenomena is not available.

SIGMETs for all other phenomena (other than domestic Convective SIGMETs) have a valid period not to exceed four (4) hours. These SIGMETs are issued up to four (4) hours before the initial valid time. SIGMETs for continuing phenomena are reissued at least every four (4) hours as long as SIGMET criteria are met.

6.1.4.3 SIGMET Cancellation
SIGMETs are cancelled when the phenomena is no longer occurring or expected to occur in the area.

6.1.4.4 SIGMET Amendments
Updates to SIGMETs are issued as necessary. This is done by issuing a new SIGMET in the current series, which advances the SIGMET number and replaces the previous SIGMET. The valid time of the new SIGMET is updated to reflect a new four (4) hour period.

6.1.4.5 SIGMET Corrections
Corrections to SIGMETs are issued as necessary. This is done by issuing a new SIGMET in the series, which advances the SIGMET number and cancels the previous SIGMET. The start time of the new, corrected SIGMET is updated, but the end valid time remains the same as the original SIGMET. At the end of the SIGMET, the product states "CORRECTS SIGMET {SERIES} {#}", where {SERIES} is the SIGMET designator and {#} is the series number.

6.1.5 SIGMET Examples

```
BOSR WS 050600
SIGMET ROMEO 2 VALID UNTIL 051000
ME NH VT
FROM MLT TO YSJ TO CON TO MPV TO MLT
OCNL SEV TURB BLW 080 EXP DUE TO STG NWLY FLOW. CONDS CONTG BYD
1000Z.
```

SIGMET issued for the Boston Area Forecast region on the 5th day of the month at 0600 UTC. This is the second (2nd) issuance of SIGMET series Romeo and is valid until the 5th day of the month at 1000 UTC. The affected states are Maine (ME), New Hampshire (NH) and Vermont (VT). Within an area bounded from Millinocket, Maine; to St. Johns, New Brunswick; to Concord, New Hampshire; to Montpelier, Vermont; to Millinocket, Maine. Occasional severe turbulence below 8,000 feet due to strong northwesterly flow. Conditions are expected to continue beyond 1000 UTC.

```
KZOA SIGMET TANGO 1 VALID 010400/010800 PHFO-
OAKLAND OCEANIC FIR ACT TS OBS BY SATELLITE WITHIN 100 NM EITHER SIDE
OF LINE N3006 W14012 - N2012 W15016. CB TO TOPS FL400. MOV W 10 KT.
WKN.
```

SIGMET issued for the Honolulu area of the Oakland Oceanic FIR. This first (1) issuance of SIGMET Series Tango valid from the 1st day of the month at 0400 UTC to the 1st day of the month at 0800 UTC. Issued by the Honolulu Weather Forecast Office. Within the Oakland Oceanic FIR, active thunderstorms observed by satellite within 100 nautical miles either side of a line from 30 degrees, 6 minutes north to 140 degrees, 12 minutes west to 20 degrees 12 minutes north, 150 degrees, 16 minutes west. Cumulonimbus tops to 40,000 feet. The thunderstorms are moving west at 10 knots and weakening.

```
KZOA SIGMET DELTA 2 VALID 081530/081930 KKCI-
OAKLAND OCEANIC FIR FRQ TS WITHIN AREA BOUNDED BY N3935 W16920 - N3414
W17050 - N3010 W17325 TOPS FL470 MOV NNE 10KT NC.
```

SIGMET issued for the Aviation Weather Center's area of the Oakland FIR. This is the second (2) issuance of SIGMET series Delta valid from the 8th day of the month at 1530 UTC to the 8th day of the month at 1930 UTC. Within the Oakland Oceanic FIR, frequent thunderstorms within an area bounded by 39 degrees 35 minutes north, 169 degrees 20 minutes west to 34 degrees 14 minutes north, 170 degrees, 50 minutes west to 30 degrees 10 minutes north, 173 degrees 25 minutes west. Thunderstorm tops to 47,000 feet, thunderstorms are moving to the north-northeast at 10 knots with no change observed.

```
DFWY WS 121450 COR
SIGMET YANKEE 4 VALID UNTIL 121700
MO AR LA MS
FROM STL TO 30 N MEI TO BTR TO MLU TO STL
OCNL SVR MXD ICING 90 TO 130 EXPCD.
FRZLVL 80 E TO 120 W. CONDS CONTG BYD 1700Z
CORRECTS YANKEE 3
```

Corrected SIGMET issued for the Dallas/Fort Worth Area Forecast region on the 12th day of the month at 1450 UTC. This is the fourth (4) issuance of SIGMET series Yankee and is valid until the 12th day of the month at 1700 UTC. The affected states are Missouri (MO), Arkansas (AR), Louisiana (LA) and Mississippi (MS). Bounded within an area from Saint Louis, Missouri; to 30 nautical miles north of Meridian, Mississippi; to Baton Rouge, Louisiana; to Monroe, Louisiana; to Saint Louis, Missouri. Occasional severe mixed icing between 9,000 and 13,000 feet is expected. The freezing level is 8,000 feet over the eastern portion of the Area Forecast region to 12,000 feet over the western portion of the Area Forecast region. Conditions are expected to continue beyond 1700 UTC. This SIGMET corrects SIGMET YANKEE 3.

6.1.6 SIGMET for Volcanic Ash

A SIGMET for volcanic ash (VA) may be issued for all volcanic eruptions, regardless of the eruption's magnitude. Volcanic ash SIGMETs are issued until the ash cloud is no longer a threat to aviation. The forecast position information for the volcanic ash cloud is based on advisories provided by a Volcanic Ash Advisory Center (VAAC). Initial VA SIGMETs may be issued based on credible pilot or aircraft reports in the absence of a Volcanic Ash Advisory (VAA) but are updated once a VAA is issued.

6.1.6.1 Examples of SIGMETs for Volcanic Ash

```
ANCI UWS 190530

PAZA SIGMET INDIA 1 VALID 190530/190930 PANC-
SATELLITE IMAGERY SHOWS DEVELOPING VA FROM ANOTHER POSSIBLE
ERUPTION OF CHIKURACHKI VOLCANO AT 0500 UTC IN THE NORTHERN KURIL
ISLANDS. HEIGHT IS ESTIMATED AT FL300 MOVEMENT IS E AT 75KTS. FURTHER
UPDATES TO FOLLOW ASAP.
FCSTR APRIL 2003 AAWU

ANCI UWS 190930
PAZA SIGMET INDIA 2 VALID 190930-191330Z PANC-
AT 0830 UTC SATELLITE IMAGERY SHOWED THE PLUME FROM THE 0500 UTC
ERUPTION OF CHIKURACHKI VOLCANO IN THE NORTHERN KURIL ISLANDS BECOMING
VERY DIFFUSE IN AN APPROXIMATELY 60 NM WIDE BAND FROM N48/E167 AND
EXTENDING SE FOR 250 NM. HEIGHT IS ESTIMATED AT FL300. MOVEMENT IS E
AT 90 KTS. THE PLUME IS MOVING SE OF ALASKA AIRSPACE INTO THE
WASHINGTON VAAC AREA OF RESPONSIBILITY. SEE WASHINGTON VAAC VOLCANIC
ASH SIGMETS AND ADVISORIES FOR FURTHER FORECASTS.
RB APRIL 2003
```

6.1.7 SIGMET for Tropical Cyclone

A SIGMET for a tropical cyclone may be issued for non-frontal synoptic-scale cyclones over oceanic FIRs (Figure 6-4 and 6-5) meeting the following criteria:

- Originate over tropical or sub-tropical waters with organized convection and definite cyclonic surface wind circulation; and

- Wind speeds reach 34 knots or more, independent of the wind averaging time used by the Tropical Cyclone Advisory Center (TCAC).

SIGMETs for tropical cyclones do not include references to associated turbulence and icing.

6.1.7.1 Example of a SIGMET for Tropical Cyclone

```
KZNY SIGMET CHARLIE 4 VALID 081500/082100 KKCI-
NEW YORK OCEANIC FIR TC KYLE OBS N3106 W07118 AT 1500Z CB TOPS FL500
WI 80NM OF CENTER MOV SSW 5KT NC FCST 2100Z TC CENTER N2930 W07130
OTLK TC CENTER 090000 N3018 W07142 091200 UTC N2918 W07224
```

SIGMET issued by the Aviation Weather Center (AWC) for the New York Area Control Center. This is the fourth (4th) issuance of SIGMET series Charlie valid from the 8th day of the month at 1500 UTC to the 8th day of the month at 2100 UTC. Within the New York Oceanic FIR. Tropical Cyclone Kyle was observed at 31 degrees 6 minutes north, 71 degrees 18 minutes west at 1500 UTC. Cumulonimbus tops at 50,000 feet within 80 nautical miles of the center, moving south-southwest at 5 knots, with no change observed. At 2100 UTC, the tropical cyclone's center is forecast to be located at 29 degrees, 30 minutes north and 71 degrees, 30 minutes west. The outlook on the 9th day of the month at 0000 UTC is for the tropical cyclone center to be located at 30 degrees, 18 minutes north and 71 degrees, 42 minutes west. On the 9th day of the month at 1200 UTC, the center is forecast to be at 29 degrees, 18 minutes north and 72 degrees, 24 minutes west.

6.1.8 Convective SIGMET for CONUS

Convective SIGMETs (also known as SIGMETs for Convection) are issued for the contiguous U.S. instead of SIGMETs for convection. Each bulletin includes one or more Convective SIGMETs for a specific region of the CONUS (Figure 6-7). Convective SIGMETs are issued for thunderstorms and related phenomena and do not include references to all weather associated with thunderstorms such as turbulence, icing, low-level wind shear and IFR conditions.

6.1.8.1 Convective SIGMET Criteria

A Convective SIGMET may be issued when any of the following occurs and/or is forecast to occur:

- Severe thunderstorms and embedded thunderstorms occurring for more than 30 minutes of the valid period regardless of the size of the area.
  - A thunderstorm is classified as severe when it is accompanied by tornadoes, hail ¾-inch or greater, or wind gusts of 50 knots or greater
  - A thunderstorm is classified as embedded when it is obscured by haze, non-convective clouds or precipitation.
- A line of thunderstorms
  - A line of thunderstorms must be at least 50 miles long with thunderstorms affecting at least 40 percent of its length.
- An area of active thunderstorms affecting at least 3,000 square miles.
  - Thunderstorms are classified as active when they are heavy (>40 dBZ) or greater and affect at least 40 percent of the area. In the absence of radar, AWC meteorologists may identify active thunderstorms using satellite or lightning information.

Obscured, embedded, or squall line thunderstorms do not have to reach 3000 square miles to be included in Convective SIGMETs.

6.1.8.2 Special Convective SIGMET

A special Convective SIGMET may be issued when either of the following criteria is occurring or expected to occur for more than 30 minutes of the valid period of the current Convective SIGMET:

- Tornado, hail greater than or equal to 3/4 inch, or wind gusts greater than or equal to 50 knots is reported or indicated when the previous Convective SIGMET did not mention severe thunderstorms; and/or

- Indications of rapidly changing conditions, if, in the forecaster's judgment, they are not sufficiently described in existing Convective SIGMETs.

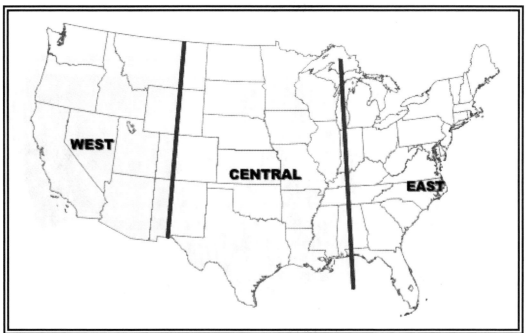

Figure 6-7. AWC Convective SIGMET Areas of Responsibility

6.1.8.3 Convective SIGMET Issuance

Three (3) Convective SIGMET bulletins describing conditions in the eastern, central and western regions of the CONUS are issued hourly at 55 minutes past the hour (Figure 6-7). Special Convective SIGMETs are issued as required. Each Convective SIGMET bulletin is made up of one or more individually numbered Convective SIGMETs for conditions within the region and are valid for up to two (2) hours or until superseded by the next hourly issuance. An outlook message is included which describes areas where Convective SIGMET issuances are expected between two (2) and six (6) hours after issuance time.

Since the Convective SIGMET bulletin is a scheduled product, a message must be transmitted each hour. If no Convective SIGMETs are expected within a region, a bulletin with **CONVECTIVE SIGMET...NONE** is transmitted.

Convective SIGMETs are not cancelled but expire as soon as the next bulletin is issued.

6.1.8.4 Format of a Convective SIGMET

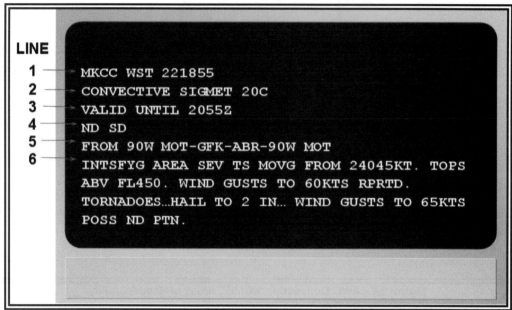

Figure 6-8. Convective SIGMET Decoding Example

Table 6-4. Decoding a Convective SIGMET

Line	Content	Description
1	MKC C WST 221855	Issuing Office W, C, or E Contraction WST Six-digit date/time group (DDHHMM)
2	CONVECTIVE SIGMET 20 C	Product type Issuance number Location (West, Central or East)
3	VALID UNTIL 2055Z	Ending valid UTC date/time
4	ND SD	Phenomenon location (states)
5	FROM 90W MOT-GFK-ABR-90W MOT	Phenomenon location (VOR coordinates)
6	INTSFYG AREA SEV TS MOVG FROM 24045KT. TOPS ABV FL450. WIND GUSTS TO 60KTS RPRTD. TORNADOES...HAIL TO 2 IN... WIND GUSTS TO 65KTS POSS ND PTN	Phenomenon description

The Convective SIGMET in Figure 6-8 is decoded as the following:

(Line 1) *Convective SIGMET issued for the central portion of the United States on the 22nd at 1855Z.*

(Line 2) *This is the 20th Convective SIGMET issued on the 22nd for the central United States as indicated by "20C."*

(Line 3) *Valid until 2055Z*

(Line 4) *The affected states are North and South Dakota.*

(Line 5) *From 90 nautical miles west of Minot, ND; to Grand Forks, ND; to Aberdeen, SD; to 90 nautical miles west of Minot, ND.*

(Line 6) *An intensifying area of severe thunderstorms moving from 240 degrees at 45 knots (to the northeast). Thunderstorm tops above Flight Level 450. Wind gusts to 60 knots reported. Tornadoes, hail to 2 inches in diameter, and wind gusts to 65 knots possible in the North Dakota portion.*

6.1.8.5 Convective SIGMET Bulletin Example

```
CONVECTIVE SIGMET 54C
VALID UNTIL 1855Z
WI IL
FROM 30E MSN-40ESE DBQ
DMSHG LINE TS 15 NM WIDE MOV FROM 30025KT. TOPS TO FL450. WIND GUSTS
TO 50 KT POSS.

CONVECTIVE SIGMET 55C
VALID UNTIL 1855Z
TX OK NM
FROM 70SE TBE-60NW AMA-40NW TCC-30ESE CIM-70SE TBE
AREA SEV TS MOV FROM 33025KT. TOPS TO FL400.
HAIL TO 2 IN...WIND GUSTS TO 70KT POSS.

OUTLOOK VALID 251855-252255
FROM 60NW ISN-INL-TVC-GIJ-UIN-FSD-BIL-60NW ISN
WST ISSUANCES EXPD. REFER TO MOST RECENT ACUS01 KWNS FROM STORM
PREDICTION CENTER FOR SYNOPSIS AND METEOROLOGICAL DETAILS.
```

Convective SIGMET 54C is the 54th Convective SIGMET issued for the central region of the US on the 25th day of the month. Valid until 1855Z. States affected include Wisconsin and Illinois. Bounded within an area from 30 east of Madison, WI; to 40 miles east-southeast of Dubuque, Iowa. A diminishing line of thunderstorms 15 nautical miles wide moving from 300 degrees (to the southeast) at 25 knots. Thunderstorms tops to FL450 (approximately 45,000 ft MSL). Wind gusts to 50 knots are possible.

Convective SIGMET 55C is the 55th Convective SIGMET issued for the central region of the US on the 25th day of the month. Valid until 1855 UTC. States affected include Texas, Oklahoma and New Mexico. Bounded within an area from 70 miles southeast of Tuba City, Arizona; to 60 miles northwest of Amarillo, Texas; to 40 northwest of Tucumcari, New Mexico; to 30 miles east-southeast of Cimarron, New Mexico; to 70 miles southeast of Tuba City, Arizona. An area of severe thunderstorms is moving from 330 degrees (to the southeast) at 25 knots.

Thunderstorms tops to Flight Level 400 (approximately 40,000 feet MSL). Hail up to 2 inches in diameter and wind gust to 70 knots are possible.

The outlook portion of the Convective SIGMET bulletin is valid from the 25th day of the month at 1855 UTC to the 25th day of the month at 2255 UTC. Within an area bounded by 60 miles northwest of Williston, North Dakota; to International Falls, Minnesota; to Traverse City, Michigan; to Niles, Michigan; to Quincy, Illinois; to Sioux Falls, South Dakota; to Billings, Montana; to 60 miles northwest of Williston, North Dakota. Convective SIGMET issuances are expected for this area. Refer to the most recent Day 1 Convective Outlook (ACUS01 KWNS) from the Storm Prediction Center for a synopsis and meteorological details.

6.2 Airmen's Meteorological Information (AIRMET)

An Airmen's Meteorological Information (AIRMET) is a concise description of the occurrence or expected occurrence in time and space of specified en route weather phenomena. The intensities are lower than those of a SIGMET although the phenomena can still affect the safety of aircraft operations. AIRMETs are intended for dissemination to all pilots in flight to enhance safety and are of particular concern to operators and pilots of aircraft sensitive to the phenomena described and to pilots without instrument ratings. Freezing level information is also included.

An AIRMET provides notice of significant weather phenomena, issued as scheduled products, for icing, turbulence, strong surface winds and low-level wind shear, and Instrument Flight Rules (IFR) and mountain obscuration, all at intensities that DO NOT meet SIGMET criteria.

- AIRMETs are available for the conterminous U.S. (CONUS) on the Aviation Digital Data Service (ADDS) web site at: http://adds.aviationweather.noaa.gov/airmets/

- AIRMETs are available for Alaska on the Alaska Aviation Weather Unit (AAWU) web site at: http://aawu.arh.noaa.gov/

- AIRMETs are available for Hawaii on the NWS WFO Honolulu web site at: http://www.prh.noaa.gov/hnl/pages/aviation.php

6.2.1 AIRMET Criteria
An AIRMET may be issued when any of the following weather phenomena are occurring or expected to occur over an area of at least 3,000 square miles:

- Sustained surface wind greater than 30 knots - **STG SFC WND**
  - Cause and direction will not be given

- Ceiling less than 1,000 feet (**IFR, CIG BLW 010**) or visibility less than 3 statue miles - **IFR, VIS BLW 3 SM BR**
  - The cause of the visibility restriction is included but limited to precipitation (PCPN), smoke (FU), haze (HZ), mist (BR), fog (FG), and blowing snow (BLSN)

- Widespread mountain obscuration - **MTN OBSCN**
  - The cause of the mountain obscuration is included but limited to clouds (CLDS) precipitation (PCPN), smoke (FU), haze (HZ), mist (BR), and fog (FG)

- Moderate turbulence - **MOD TURB**

- Moderate icing - **MOD ICE**
  - Will not reference the location of the icing with respect to either in clouds or in precipitation
  - The freezing level is defined as the lowest freezing level above ground level or the surface (SFC)
  - Freezing levels above the surface are delineated using high altitude VOR locations at intervals of 4,000 feet above MSL or the surface (SFC)
  - Areas with multiple freezing levels are delineated with high altitude VOR locations

- The range of freezing levels over the AIRMET area is included

- Non-convective LLWS potential below 2,000 ft - **LLWS POTENTIAL**
  - Will include a list of affected states and be bounded by high altitude VOR locations

6.2.2 Standardization
All in-flight advisories follow these standards:

- All heights or altitudes are referenced to Mean Sea Level (MSL) and consist of three (3) digits depicting height in hundreds of feet. Flight Level (**FL**) is used for heights at or above 18,000 feet above. Examples: 100, FL190.

- References to latitude and longitude are in whole degrees and minutes following the model: Nnn[nn] or Snn[nn], Wnnn[nn] or Ennn[nn] with a space between latitude and longitude and a hyphen between successive points. Products issued by AWC have the N, S, W, or E behind the latitude/longitude numbers.

- Messages are prepared using approved ICAO contractions, abbreviations and numerical values of self-explanatory nature.

- Weather and obstructions to visibility will be limited to clouds, (CLDS), precipitation (PCPN), smoke (FU), haze (HZ), mist (BR), fog, (FG), and blowing snow (BLSN).

- All amended (AMD) or corrected (COR) en route forecasts or advisory products follow the same format procedures. An AMD or COR is identified as a change in the first line after the time and date (Figure 6-9).

6.2.3 AIRMET Format

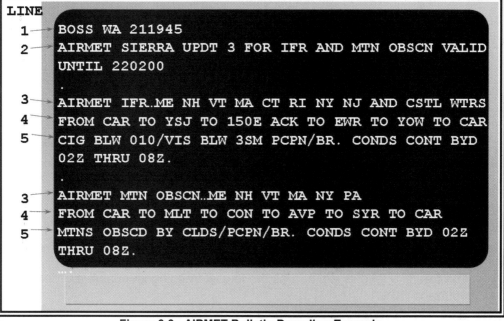

Figure 6-9. AIRMET Bulletin Decoding Example

Table 6-5. Decoding an AIRMET Bulletin

Line	Content	Description
1	BOS S WA 211945	AIRMET area identifier AIRMET series Product type Issuance UTC date/time
2	AIRMET SIERRA UPDT 3 FOR IFR AND MTN OBSCN VALID UNTIL 220200	Product type AIRMET series Update number Product description Ending UTC date/time
3	AIRMET IFR..ME NH VT MA CT RI NY NJ AND CSTL WTRS AIRMET MTN OBSCN..ME NH VT MA NY PA	Product type/series... Phenomenon location (states)
4	FROM CAR TO YSJ TO 150E ACK TO EWR TO YOW TO CAR FROM CAR TO MLT TO CON TO AVP TO SYR TO CAR	Phenomenon location (VOR locations)
5	CIG BLW 010/VIS BLW 3SM PCPN/BR. CONDS CONT BYD 02Z THRU 08Z. MTNS OBSCD BY CLDS/PCPN/BR. CONDS CONT BYD 02Z THRU 08Z.	Phenomenon description

The AIRMET bulletin in Figure 6-9 is decoded as follows:

(Line 1) *AIRMET SIERRA issued for the Boston area at 1945Z on the 21st day of the month. "SIERRA" contains information on Instrument Flight Rules (IFR) and/or mountain obscuration.*

(Line 2) *This is the third updated issuance of this Boston AIRMET series as indicated by "SIERRA UPDT 3" and is valid until 0200Z on the 22nd.*

(Line 3) *The affected states within the BOS area are: Maine, New Hampshire, Vermont, Massachusetts, Connecticut, Rhode Island, New York, New Jersey, and coastal waters.*

(Line 4) *From Caribou, ME; to Saint Johns, New Brunswick; to 150 nautical miles east of Nantucket, MA; to Newark, NJ; to Ottawa, Ontario; to Caribou, ME*

(Line 5) *Ceiling below 1,000 feet/visibility below 3 statute miles, precipitation/mist. Conditions continuing beyond 0200Z through 0800Z.*

6.2.3.1 AIRMET Series
The AIRMET series consists of Sierra, Tango, and Zulu.

- AIRMET Sierra describes IFR conditions and/or extensive mountain obscurations.

- AIRMET Tango describes moderate turbulence, sustained surface winds of 30 knots or greater, and non-convective low-level wind shear.

- AIRMET Zulu describes moderate icing and provides freezing level heights.

6.2.3.2 AIRMET Bulletins

AIRMETs are issued in AIRMET bulletins, each containing one or more AIRMET messages. The bulletins are issued on a scheduled basis every 6 hours and, except in Alaska, at 0300, 0900, 1500 and 2100 UTC. In Alaska, AIRMET bulletins are issued every six hours at the same time as the Area Forecast (Section 6.1). An AIRMET bulletin may be issued for each forecast area (Figures 6-10, 6-11, and 6-12).

6.2.3.3 SIGMET Information in AIRMET Bulletin

A reference to the appropriate SIGMET series is included in AIRMET bulletins which cover the affected area and for similar phenomena; for example, **SEE SIGMET BRAVO SERIES**.

6.2.3.4 AIRMET Phenomena Information

An AIRMET message contains the following information as necessary and in the order indicated relating to the phenomena that caused the AIRMET to be issued:

- Location (using locations or geographic features well known nationally if possible)

- Phenomena and its description from Section 6.2.1.1; e.g., **MOD TURB**

- If appropriate, level (altitude), or vertical extent

- Expected beginning and ending time of phenomena, if different from the AIRMET bulletin's valid time

- Remarks

6.2.3.5 AIRMET Remarks

A remark is included at the end of each AIRMET regarding whether the condition is expected to continue after the valid time of the AIRMET.

6.2.3.6 AIRMET Outlook (Except Alaska)

If AIRMET conditions are expected to develop during the 6-hour period after the ending valid time of the AIRMET bulletin, the information is included in an outlook section.

6.2.3.7 AIRMET Outlook (Alaska Only)

If AIRMET conditions are expected to develop during the 6-hour period after the ending valid time of the AIRMET bulletin, the information is included in the appropriate Area Forecast zone.

6.2.4 AIRMET Issuance

AIRMETs are issued from the three Meteorological Watch Offices (MWO) located at the: the Aviation Weather Center (AWC), the Alaska Aviation Weather Unit (AAWU), and the Weather Forecast Office (WFO) in Honolulu. Their areas of responsibility are:

- AWC: The conterminous U.S. and adjacent coastal waters (CONUS) (Figure 6-10).

- AAWU: Alaska and adjacent coastal waters (Figure 6-11).

- WFO Honolulu: Hawaii and adjacent waters (Figure 6-12).

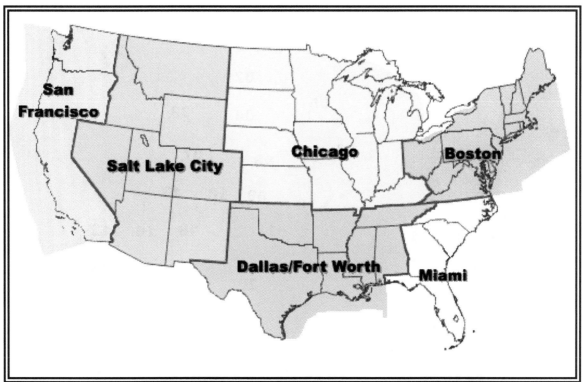

Figure 6-10 AWC AIRMET Areas of Responsibility – Conterminous U.S.

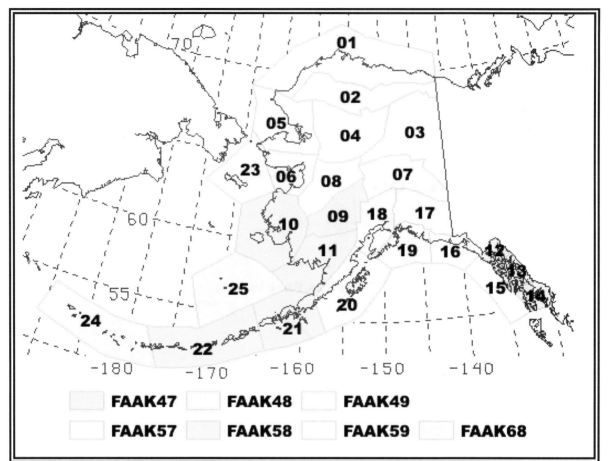

Figure 6-11. AAWU Flight Advisory and Area Forecast Zones - Alaska

Table 6-6. AAWU Flight Advisory and Area Forecast Zones – Alaska

1	Arctic Coast Coastal	14	Southern Southeast Alaska
2	North Slopes of the Brooks Range	15	Coastal Southeast Alaska
3	Upper Yukon Valley	16	Eastern Gulf Coast
4	Koyukuk and Upper Kobuk Valley	17	Copper River Basin
5	Northern Seward Peninsula-Lower Kobuk Valley	18	Cook Inlet-Susitna Valley
6	Southern Seward Peninsula-Eastern Norton Sound	19	Central Gulf Coast
7	Tanana Valley	20	Kodiak Island
8	Lower Yukon Valley	21	Alaska Peninsula-Port Heiden to Unimak Pass
9	Kuskowim Valley	22	Unimak Pass to Adak
10	Yukon-Kuskowim Delta	23	St. Lawrence Island-Bering Sea Coast
11	Bristol Bay	24	Adak to Attu
12	Lynn Canal and Glacier Bay	25	Pribilof Islands and Southeast Bering Sea
13	Central Southeast Alaska		

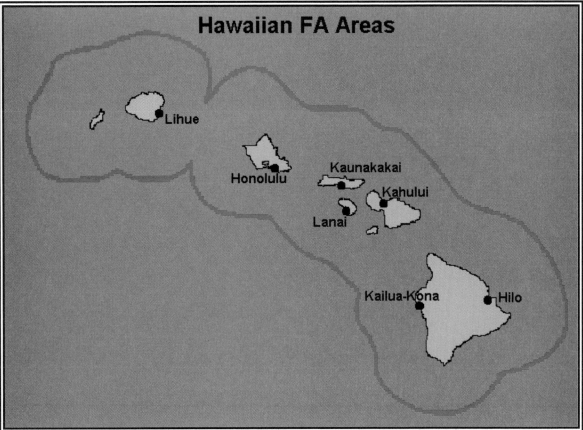

Figure 6-12. WFO Honolulu AIRMET Areas of Responsibility - Hawaii

6.2.4.1 AIRMET Valid Time
AIRMETs are issued for conditions occurring or expected to develop within the six-hour (6) valid time of the AIRMET Bulletin. An AIRMET's valid period is the same as the AIRMET bulletin's valid time unless otherwise noted.

6.2.4.2 AIRMET Updates and Amendments
Unscheduled updates to AIRMET bulletins are issued as necessary. If an AIRMET is amended, **AMD** is added after the date/time group on the FAA product line (Figure 6-9, Line1). **UPDT** is added to the end of the line containing the list of affected states. This issue time of the AIRMET bulletin is updated to reflect the time beginning valid time with the ending valid time remaining unchanged.

6.2.4.3 AIRMET Corrections
AIRMETs containing errors are corrected by adding **COR** after the date/time group on the FAA product line (Figure 6-9, Line 1). The issuance time is updated with the ending valid time unchanged. UPDT is added to the third line after the list of affected states.

6.2.4.3.1 Example of a Corrected/Amended AIRMET

```
SLCZ WA 222000 COR/AMD
AIRMET ZULU UPDATE 3 FOR ICE AND FZ LVL VALID UNTIL 230200
.
AIRMET ICE...ID MT WY CO... UPDT
```

```
FROM GTF TO 80NW RAP TO BFF TO GLD TO DEN TO OCS TO TWF TO BOI TO GTF
MOD ICE BTW 080 AND FL200. FZ LVLS SFC-080 OVR MOST OF AREA RSG TO
080-100 OVR SWRN PORTIONS AREA. CONDS CONTG BYD 02Z THRU 08Z.
.
FRZLVL...RANGING FROM SFC-130
     040...GEG-DBS-OCS-CYS
     080...FMG-DTA-DBL-AKO
     120...LAS-ABQ-TCC
```

6.2.5 AIRMET Examples

```
CHIS WA 171345
AIRMET SIERRA UPDT 3 FOR IFR AND MTN OBSCN VALID UNTIL 172000
.
AIRMET IFR...NE KS
FROM 60NE BFF TO ANW TO 40NE OBH TO OSW TO 40N GAG TO 40N GLD TO 60NE
BFF
CIG BLW 010/VIS BLW 3SM PCPN BR. CONDS ENDG 18Z.
.
AIRMET IFR...MO IL IN KY
FROM CVG TO HNN TO HMV TO BWG TO RZC TO SGF TO 60S DEC TO CVG
CIG BLW 010/VIS BLW 3SM BR. CONTG BYD 20Z ENDG 23Z.
```

AIRMET Sierra bulletin for the Chicago Area Forecast region issued on the 17th day of the month at 1345 UTC and contains two AIRMETS. This is the 3rd AIRMET Sierra bulletin update issued for IFR and Mountain Obscurations and is valid until the 17th day of the month at 2000 UTC.

The first AIRMET is for IFR conditions affecting Nebraska and Kansas. Bounded within an area from 60 nautical miles (NM) northeast of Scottsbluff, Nebraska; to Ainsworth, Nebraska; to 40 NM northeast of Wolbach, Nebraska; to Oswego, Kansas; to 40 NM miles north of Gage, Oklahoma; to 60 NM miles northeast of Scottsbluff, Nebraska. Ceilings below 1,000 feet and visibilities less than 3 statute miles (SM) with precipitation and mist. Conditions are forecast to end by 1800 UTC.

The second AIRMET is for IFR conditions affecting Missouri, Illinois, Indiana and Kentucky. Bounded within an area from Covington, Kentucky; to Henderson, West Virginia; to Holston Mountain, Tennessee; to Bowling Green, Kentucky; to Razorback, Arkansas; to Springfield, Missouri; to 60 NM miles south of Decatur, Illinois; to Covington, Kentucky. Ceilings below 1,000 feet and visibilities less than 3 SM due to mist. Conditions are forecast to continue beyond 2000 UTC and end by 2300 UTC.

```
HNLS WA 080945
AIRMET SIERRA UPDATE 1 FOR IFR VALID UNTIL 081600.
.
NO SGFNT IFR EXP.
```

AIRMET Sierra bulletin for the Hawaii forecast region issued on the 8th day of the month at 0945 UTC. This is the 1st AIRMET Sierra bulletin update issued for IFR and Mountain Obscurations and is valid until the 8th day of the month at 1600 UTC.

No conditions meeting IFR criteria are expected.

```
HNLT WA 080945
AIRMET TANGO UPDT 1 FOR TURB VALID UNTIL 081600
.
AIRMET TURB...HI
OVR AND IMT S THRU W OF MT OF ALL ISLANDS.
MOD TURB BLW 060. COND CONT BYD 1600Z THRU 22Z.
```

AIRMET Tango bulletin for the Hawaii Area Forecast region issued on the 8th day of the month at 0945 UTC. This is the 1st AIRMET Tango bulletin update issued for turbulence and is valid until the 8th day of the month at 1600 UTC.

AIRMET for turbulence for Hawaii. Over and immediately south through west of the mountains for all islands. Moderate turbulence is expected below 6,000 feet above MSL. Conditions are expected to continue beyond 1600 UTC through 2200 UTC.

```
SFOZ WA 171345
AIRMET ZULU UPDT 2 FOR ICE AND FRZLVL VALID UNTIL 172000
.
AIRMET ICE...WA OR ID MT
FROM 60SSW YXH TO 70ESE MLP TO PDT TO PDX TO SWA TO BLI TO 60SSW YXH
MOD ICE BTN 120 AND FL200. CONDS CONTG BYD 20Z THRU 02Z.
.
FRZLVL...RANGING FROM 090-160
   120...ONP-LKV-REO
   160...SNS-40SW BTY
```

AIRMET Zulu bulletin for the San Francisco Area Forecast area issued on the 17th day of the month at 1345 UTC. This is the 2nd AIRMET Zulu bulletin update issued for icing and freezing levels and is valid until the 17th day of the month at 2000 UTC.

AIRMET for icing for Washington, Oregon, Idaho, and Montana. Bounded within an area from 60 miles south-southwest of Medicine Hat, Alberta; to 70 NM miles east-southeast of Mullan Pass, Idaho; to Pendleton, Oregon; to Portland, Oregon; to Seattle, Washington; to Billings, Montana; to 60 NM miles south-southwest of Medicine Hat, Alberta. Moderate icing between 12,000 and 20,000 feet above MSL. Conditions continuing beyond 2000 UTC through 0200 UTC.

Freezing levels over the area ranging between 9,000 and 16,000 feet above MSL, Freezing level at 12,000 feet above MSL along a line from the Newport, Oregon VOR to the Lake County, Oregon VOR to the Rome, Oregon VOR. Freezing level at 16,000 feet above MSL along a line from the Salinas, California VOR to 40 NM southwest of the Beatty, Nevada VOR.

6.3 Center Weather Advisory (CWA)

A Center Weather Advisory (CWA) is an aviation weather warning for conditions meeting or approaching national in-flight advisory (AIRMET, SIGMET or SIGMET for convection) criteria. The CWA is primarily used by aircrews to anticipate and avoid adverse weather conditions in the en route and terminal environments. CWAs are available on the Aviation Weather Center (AWC) web site at: http://aviationweather.gov/products/cwsu/.

6.3.1 CWA Issuance

CWAs are issued by the NWS Center Weather Service Units (CWSUs). CWSU areas of responsibility in the contiguous U.S. are depicted on Figure 6-13. CWSU Anchorage area of responsibility for Alaska is depicted on Figure 6-14.

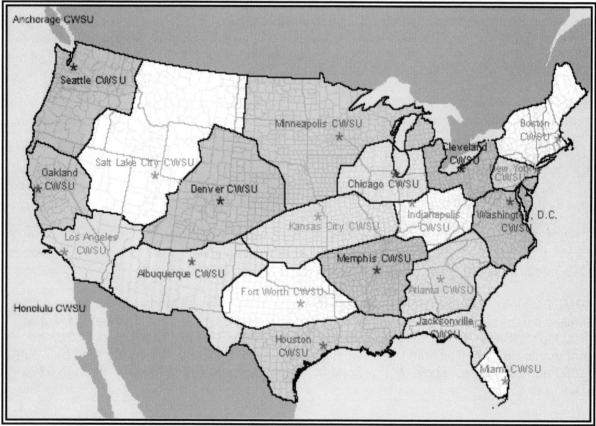

Figure 6-13. Center Weather Service Unit (CWSU) Areas of Responsibility, Contiguous U.S.

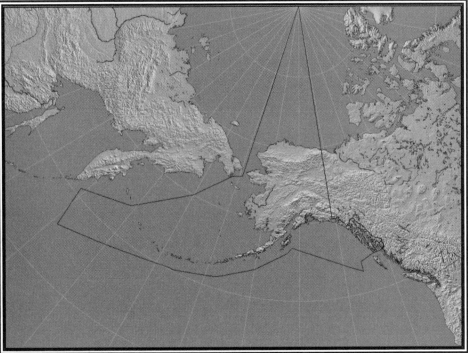

Figure 6-14. CWSU Anchorage, AK (PAZA) Area of Responsibility

CWAs are valid for up to two (2) hours and may include forecasts of conditions expected to begin within two (2) hours of issuance. If conditions are expected to persist after the advisory's valid period, a statement to that effect is included in the last line of the text. Additional CWAs will subsequently be issued as appropriate. Notice of significant changes in the phenomenon described in a CWA is provided by a new CWA issuance for that phenomenon. If the forecaster deems it necessary, CWAs may be issued hourly for convective activity.

6.3.2 CWA Communications Headers (UCWA / CWA)
The Urgent CWA (**UCWA**) communications header is intended for those situations where weather conditions have an immediate effect on the safe flow of air traffic within the ARTCC area of responsibility. It is only used when the CWSU meteorologist believes any delay in dissemination to FAA facilities would impact aviation safety. The routine CWA header is used for subsequent issuances of the same phenomenon.

6.3.3 CWA Criteria
CWAs are used in the four (4) following situations:

- **Precede an Advisory**
  - When the AWC has not yet issued an advisory, but conditions meet or will soon meet advisory criteria.
  - In the case of an impending advisory, the CWA can be issued as an Urgent CWA (UCWA) to allow the fastest possible dissemination.

- **Refine an existing Advisory**
  - To supplement an existing AWC advisory for the purpose of refining or updating the location, movement, extent, or intensity of the weather event relevant to the ARTCC's area of responsibility.

- **Highlight significant conditions not meeting Advisory criteria**
  - When conditions do not meet advisory criteria, but conditions, in the judgment of the CWSU meteorologist, will adversely impact air traffic within the ARTCC area of responsibility.

- **To cancel a CWA when the phenomenon described in the CWA is no longer expected.**

6.3.4 CWA Format

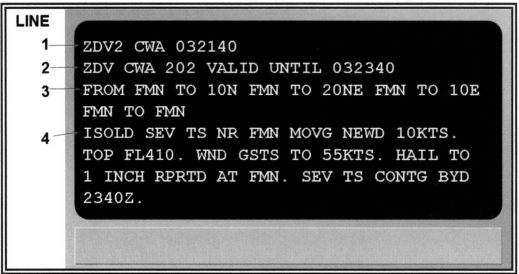

LINE

```
1 ── ZDV2 CWA 032140
2 ── ZDV CWA 202 VALID UNTIL 032340
3 ── FROM FMN TO 10N FMN TO 20NE FMN TO 10E
     FMN TO FMN
4 ── ISOLD SEV TS NR FMN MOVG NEWD 10KTS.
     TOP FL410. WND GSTS TO 55KTS. HAIL TO
     1 INCH RPRTD AT FMN. SEV TS CONTG BYD
     2340Z.
```

Figure 6-15. Center Weather Advisory (CWA) Decoding Example

Table 6-7. Decoding a Center Weather Advisory (CWA)

Line	Content	Description
1	ZDV 2 CWA 032140	ARTCC Identification Phenomenon Number (single digit, 1-6) Product Type (UCWA/CWA) Beginning and/or issuance UTC date/time
2	ZDV CWA 2 02 VALID TIL 032340Z	ARTCC Identification Product Type Phenomenon Number (single digit, 1-6) Issuance Number (issued sequentially for each Phenomenon Number) Ending valid UTC date/time
3	FROM FMN TO 10N FMN TO 20NE FMN TO 10E FMN TO FMN	Phenomenon Location
4	ISOLD SEV TS NR FMN MOVG NEWD 10KTS. TOP FL410. WND GSTS TO 55KTS. HAIL TO 1 INCH RPRTD AT FMN. SEV TS CONTG BYD 2340Z	Phenomenon Description

Time permitting, any CWA overlapping into another center's airspace is coordinated and a statement is included in the text, e.g., **SEE ZOB CWA 201 FOR TS CONDS IN ZOB CTA** (CTA is control area). If issuance prior to coordination is necessary, a statement regarding the area(s) affected is included in the text, e.g., **LINE TS EXTDS NW INTO ZOB CTA**.

AIRMETs/SIGMETs being augmented by the CWA will be referenced in a text remark, e.g. **SEE CONVECTIVE SIGMET 8W**.

The CWA in Figure 6-15 is decoded as follows:

(Line 1) Center Weather Advisory issued for the Denver ARTCC (ZDV) CWSU. The "2" after ZDV in the first line denotes this is the second meteorological event of the local calendar day. This CWA was issued/begins on the 3rd day of the month at 2140 UTC.

(Line 2) The Denver ARTCC (ZDV) is identified again. The "202" in the second line denotes the phenomena number again (2) and the issuance number (02) for this phenomenon. This CWA is the valid until the 3rd day of the at 2340 UTC.

(Line 3) From Farmington, New Mexico to 10 nautical miles north of Farmington, New Mexico to 20 nautical miles northeast of Farmington, NM to 10 nautical mile east of Farmington, New Mexico to Farmington, New Mexico.

(Line 4) Isolated severe thunderstorms near Farmington moving northeastward at 10 knots. Tops to Flight Level 410. Wind gusts to 55 knots. Hail to one inch reported at Farmington. Severe thunderstorms continuing beyond 2340 UTC.

6.3.5 Examples

```
ZME1 CWA 081300

ZME CWA 101 VALID TIL 081500
FROM MEM TO JAN TO LIT TO MEM
OCNL TS MOV FM 26025KT. TOPS TO FL450.
```

Center Weather Advisory issued for the Memphis, Tennessee ARTCC on the 8th day of the month at 1300 UTC. The 1 after the ZME in the first line denotes this CWA has been issued for the first weather phenomenon to occur for the local calendar day. The 101 in the second line denotes the phenomenon number again (1) and the issuance number (01) for this phenomenon. The CWA is valid until the 8th of the month at 1500 UTC. From Memphis, Tennessee to Jackson, Mississippi to Little Rock, Arkansas to Memphis, Tennessee. Occasional thunderstorms moving from 260 degrees at 25 knots. Tops to flight level 450.

```
ZLC3 CWA 271645

ZLC CWA 303 VALID TIL 271745
CNL CWA 302.
SEE CONVECTIVE SIGMET 8W.
```

Center Weather Advisory issued for the Salt Lake City, Utah ARTCC on the 27th day of the month at 1645 UTC. The 3 after the ZLC in the first line denotes this CWA has been issued for the third weather phenomenon to occur for the local calendar day. The 303 in the second line denotes the phenomenon number again (3) and the issuance number (03) for this phenomenon. The CWA is valid until the 27th day of the month at 1745 UTC. CWA number 302 has been cancelled. See Convective SIGMET 8W.

```
ZME1 CWA 040100

ZME CWA 101 VALID TIL 040300
VCY MEM
SEV CLR ICE BLW 020 DUE TO FZRA. NUMEROUS ACFT REP RAPID
ACCUMULATION OF ICE DRG DES TO MEM. NO ICE REPS ABV 020. CONDS CONTG
AFT 03Z. NO UPDATES AFT 040200Z.
```

Center Weather Advisory issued for the Memphis, Tennessee ARTCC on the 4th day of the month at 0100 UTC. The 1 after the ZLC in the first line denotes this CWA has been issued for the first weather phenomenon to occur for the local calendar day. The 101 in the second line denotes the phenomenon number again (1) and the issuance number (01) for this phenomenon. The CWA is valid until the 4th day of the month at 0300 UTC. For the Memphis, Tennessee vicinity. Severe clear icing below 2,000 feet MSL due to freezing rain. Numerous aircraft report rapid accumulation of icing during descent to Memphis. No icing reports above 2,000 feet MSL. Conditions continuing after 0300 UTC. No updates after 4th day of the month at 0200 UTC.

```
ZNY5 UCWA 021400

ZNY CWA 502 VALID TIL 021600
FROM BGM TO 18WNW JFK TO HAR TO SLT TO BGM
NUMEROUS ACFT REP SEV TURB AND WS BLW 020.
CONDS EXTD NE INTO ZBW CTA. CONDS EXP TO CONT AFT 16Z.
```

Center Weather Advisory issued for the New York ARTCC on the 2nd day of the month at 1400 UTC. The 5 after the ZNY in the first line denotes this CWA has been issued for the fifth weather phenomenon to occur for the local calendar day. The 502 in the second line denotes the phenomenon number again (5) and the issuance number (02) for this phenomenon. The CWA is valid until the 2nd day of the month at 1600 UTC. From Binghamton, New York; to 18 nautical miles west-northwest of New York (JFK Airport), New York; to Harrisburg, Pennsylvania; to Slate Run, Pennsylvania; to Binghamton, New York. Numerous aircraft report severe turbulence and wind shear below 2,000 feet MSL. Conditions extending northeast into Nashua, New Hampshire control area. Conditions expected to continue after 1600 UTC.

```
ZNY4 UCWA 041500

ZNY CWA 401 VALID TIL 041700
40N SLT TO 18WNW JFK
DEVELOPING LINE TS 25 NM WIDE MOV 24020KT. TOPS ABV FL350.
LINE TS EXTDS NW INTO ZOB CTA.
```

Urgent Center Weather Advisory issued for the New York ARTCC on the 4th day of the month at 1500 UTC. The 4 after the ZNY in the first line denotes this CWA has been issued for the fourth weather phenomenon to occur for the local calendar day. The 401 in the second line denotes the phenomenon number again (4) and the issuance number (01) for this phenomenon. The

CWA is valid until the 4th day of the month at 1700 UTC. From 40 nautical miles north of Slate Run, Pennsylvania; to 18 nautical miles west-northwest of New York (JFK Airport), New York. Developing line of thunderstorms 25 nautical miles wide moving from 240 degrees at 20 knots. Tops above flight level 350. The line of thunderstorms extends northwest into the Oberlin, Ohio control area.

6.4 Additional Products for Convection

The National Weather Service (NWS) in addition to the SIGMETs (Section 6.1), Convective SIGMETs (Section 6.1.8), and CWAs (Section 6.3) already discussed, offers a few more products informing the aviation community about the potential for convective weather.

6.4.1 Convective Outlooks (AC)

The NWS Storm Prediction Center (SPC) issues narrative and graphical convective outlooks to provide the contiguous U.S. NWS Weather Forecast Offices (WFOs), the public, media and emergency managers with the potential for severe (tornado, wind gusts 50 knots or greater, or hail 3/4 inch diameter size or greater) and non-severe (general) convection and specific severe weather threats during the following three days. The Convective Outlook defines areas of slight risk (**SLGT**), moderate risk (**MDT**) or high risk (**HIGH**) of severe thunderstorms for a 24-hour period beginning at 1200 UTC (Figure 6-16). The Day 1 and Day 2 Convective Outlooks also depict areas of general thunderstorms (**GEN TSTMS**), while the Day 1, Day 2, and Day 3 Convective Outlooks may use **SEE TEXT** for areas where convection may approach or slightly exceed severe criteria. The outlooks are available on the SPC web site at: http://www.spc.noaa.gov/products/outlook/.

6.4.1.1 Issuance

Convective Outlooks are scheduled products issued at the following times:

Table 6-8. Convective Outlook Issuance Schedule

Convective Outlook	Issuance Time (UTC)	Valid Period (UTC)
Day 1	0600	1200 – 1200
	1300	1300 – 1200
	1630	1630 – 1200
	2000	2000 – 1200
	0100	0100 – 1200
Day 2	0730 (Daylight Savings Time) 0830 (Standard Time)	Day 2/1200 – 1200
	1730	Day 2/1200 – 1200
Day 3	1100	Day 3/1200 – 1200

SPC corrects outlooks for format and grammatical errors and amends outlooks when the current forecast does not or will not reflect the ongoing or future convective development.

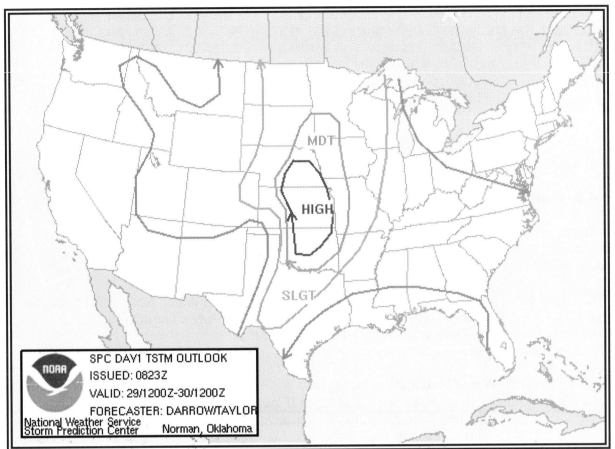

Figure 6-16. Day 1 Categorical Convective Outlook Graphic Example

6.4.1.2 Format of the Categorical Convective Outlook Narrative

SPC AC ddhhmm [SPC - issuing office, AC – product type, ddhhmm – date and time the product was issued]

```
DAY (ONE, TWO OR THREE) CONVECTIVE OUTLOOK
NWS STORM PREDICTION CENTER NORMAN OK
time am/pm time_zone day mon dd yyyy
```

VALID DDHHMM – DDHHMMZ

THERE IS A (SLIGHT, MODERATE, HIGH) RISK OF SEVERE THUNDERSTORMS TO THE RIGHT OF LINE (LIST OF ANCHOR POINTS AND DIRECTION AND DISTANCE IN STATUTE MILES FROM THE LINE). THE LINE WILL ENCLOSE THE AREA OF RISK. THERE MAY BE ONE OR MORE AREAS OF RISK AT THE APPROPRIATE LEVEL OF RISK. WHEN A MODERATE OR HIGH RISK IS FORECAST, THE INDIVIDUAL STATES ARE ALSO LISTED WITH THE TWO LETTER POSTAL STATE IDENTIFIERS.

GEN TSTMS ARE FCST TO THE RIGHT OF A LINE FROM (LIST OF ANCHOR POINTS AND DIRECTION AND DISTANCE IN STATUTE MILES FROM THE LINE). THERE MAY BE ONE OR MORE AREAS OF GEN TSTMS LISTED.

...AREA OF CONCERN #1...

```
AREAS OF HIGHEST RISK ARE DISCUSSED FIRST (HIGH SEVERE RISK,
MODERATE SEVERE RISK, SLIGHT SEVERE RISK, APPROACHING SEVERE
LIMITS). THE FORECAST PROVIDES A NARRATIVE TECHNICAL DISCUSSION.

...AREA OF CONCERN #2...
NARRATIVE TECHNICAL DISCUSSION

$$

...FORECASTER NAME... MM/DD/YY
```

6.4.2 Watch Notification Messages

The NWS Storm Prediction Center (SPC) issues Watch Notification Messages to alert the aviation community, NWS offices (WFOs), the public, media and emergency managers to organized thunderstorms forecast to produce tornadic and/or severe weather in the conterminous U.S.

SPC issues three types of Watch Notification Messages: Aviation Watch Notification Message, Public Severe Thunderstorm Watch Notification Message and Public Tornado Watch Notification Message. They are available on the SPC web site at: http://www.spc.noaa.gov/products/watch/.

6.4.2.1   Aviation Watch Notification Message

SPC issues Aviation Watch Notification Messages (Figure 6-17) to alert the aviation community to organized thunderstorms forecast to produce tornadic and/or severe weather as indicated in Public Watch Notification Messages.

6.4.2.1.1 Format of an Aviation Watch Notification Message

```
SPC AWW ddhhmm
WWnnnn SEVERE TSTM ST LO DDHHMMZ - DDHHMMZ
AXIS...XX STATUTE MILES EITHER SIDE OF A LINE
XXDIR CCC/LOCATION ST/ - XXDIR CCC/LOCATION ST
..AVIATION COORD.. XX NM EITHER SIDE /XXDIR CCC - XXDIR CCC
HAIL SURFACE AND ALOFT..X X/X INCHES. WIND GUSTS..XX KNOTS.
MAX TOPS TO XXX. MEAN STORM MOTION VECTOR DIR/SPEED
```

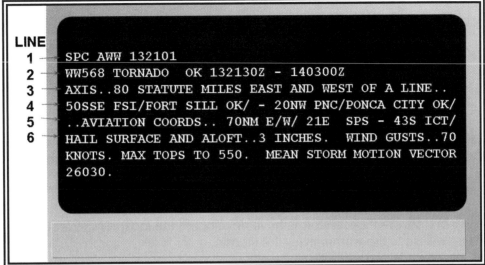

Figure 6-17. Aviation Watch Notification Message Decoding Example

Table 6-9. Decoding a Severe Weather Watch Bulletin

Line	Content	Description
1	SPC AWW 132101	Issuing office Product Type Issuance date/time
2	WW568 TORNADO OK 132130Z – 140300Z	Watch number Watch Type States affected Valid date/time period
3	AXIS..80 STATUTE MILES EAST AND WEST OF A LINE...	Watch axis
4	50SSE FSI/FORT SILL OK/ – 20NW PNC/PONCA CITY OK/	Anchor points
5	...AVIATION COORDS.. 70NM E/W/ 21E SPS – 43S ICT/	Aviation coordinates
6	HAIL SURFACE AND ALOFT...3 INCHES. WIND GUSTS..70 KNOTS. MAX TOPS TO 550. MEAN STORM MOTION VECTOR 26030.	Type, intensity, max tops, and mean storm motion using standard contractions.

The Severe Weather Watch Bulletin in Figure 6-17 is decoded as follows:

(Line 1) *Alert Severe Weather Watch Bulletin (AWW), issued by the Storm Prediction Center on the 13th at 2101Z,*

(Line 2) *for Tornado Watch number 568 (WW568) for Oklahoma, valid from the 13th at 2130Z until the 14th at 0300Z.*

(Line 3) *The Tornado Watch area is along and 80 statute miles east and west of a line from*

(Line 4) *50 statute miles south southeast of Fort Sill (Lawton), OK to 20 statute miles northwest of Ponca City, OK.*

(Line 5) *Aviation coordinates for this Tornado Watch are 70 nautical miles east and west of a line from 21 nautical miles east of Sheppard AFB (Wichita Falls), TX to 43 nautical miles south of Wichita, KS.*

(Line 6) *Hail surface and aloft to 3 inches in diameter, wind gusts to 70 knots, max tops to Flight Level 550, mean storm motion from 260 degrees at 30 knots*

6.4.2.1.2 Issuance

Watch Notification Messages are non-scheduled, event driven products valid from the time of issuance to expiration or cancellation time. Valid times are in UTC. SPC will correct watches for format and grammatical errors.

When tornadoes or severe thunderstorms have developed, the local NWS Weather Forecast Offices (WFOs) will issue the warnings for the storms.

SPC forecasters may define the watch area as a rectangle (some number of miles either side of line from point A to point B) or as a parallelogram (some number of miles north and south or east and west of line from point A to point B). The axis coordinates are measured in statute miles. The aviation coordinates are measured in nautical miles and referenced to VHF Omni-Directional Range (VOR) navigational aid locations. The watch half-width is in statute miles. The Aviation Watch Notification Message contains hail size in inches or half inches at the surface and aloft, surface convective wind gusts in knots, maximum tops, and the Mean Storm Motion Vector. Forecasters have discretion in including hail size for tornado watches associated with hurricanes.

6.4.3 Public Severe Thunderstorm Watch Notification Message

SPC issues a Public Severe Thunderstorm Watch Notification Message when forecasting six or more hail events of 3/4 inch (penny) diameter or greater or damaging winds of 50 knots (58 mph) or greater. The forecast event minimum threshold is at least 2 hours over an area at least 8,000 square miles. Below these thresholds, SPC, in collaboration with affected NWS offices may issue convective watches along coastlines, near the Canadian and Mexican borders, and for any ongoing organized severe convection.

A Public Severe Thunderstorm Watch Notification Message contains the area description and axis, watch expiration time, a description of hail size and thunderstorm wind gusts expected, the definition of the watch, a call to action statement, a list of other valid watches, a brief discussion of meteorological reasoning, and technical information for the aviation community.

SPC includes the term "adjacent coastal waters" when the watch affects coastal waters adjacent to the Pacific/Atlantic coast, Gulf of Mexico, or Great Lakes. Adjacent coastal waters refers to a WFO's near-shore responsibility (out to 20 miles for oceans), except for convective watches which include portions of the Great Lakes.

SPC issues a watch cancellation message when **no** counties, parishes, independent cities and/or marine zones remaining are in the watch area prior to the expiration time. The text of the message will specify the number and area of the cancelled watch.

6.4.3.1 Format of Public Severe Thunderstorm Watch Notification Message

WWUS20 KWNS ddhhmm (*ICAO communication header*)

```
URGENT - IMMEDIATE BROADCAST REQUESTED
SEVERE THUNDERSTORM WATCH NUMBER nnnn
NWS STORM PREDICTION CENTER NORMAN OK
time am/pm time zone day mon dd yyyy

THE STORM PREDICTION CENTER HAS ISSUED A
SEVERE THUNDERSTORM WATCH FOR PORTIONS OF

              PORTION OF STATE
              PORTION OF STATE
              AND ADJACENT COASTAL WATERS (IF REQUIRED)

EFFECTIVE (TIME PERIOD) UNTIL hhmm am/pm time zone.

...THIS IS A PARTICULARLY DANGEROUS SITUATION (IF FORECAST)...

HAIL TO X INCHES IN DIAMETER...THUNDERSTORM WIND GUSTS TO XX MPH...AND
DANGEROUS LIGHTNING ARE POSSIBLE IN THESE AREAS.

NARRATIVE DESCRIPTION OF WATCH AREA USING A LINE AND ANCHOR
POINTS. DISTANCES TO EITHER SIDE OF THE LINE WILL BE IN STATUTE MILES.

CALL TO ACTION STATEMENTS

OTHER WATCH INFORMATION...OTHER WATCHES IN EFFECT AND IF THIS
WATCH REPLACES A PREVIOUS WATCH.

NARRATIVE DISCUSSION OF REASON FOR THE WATCH.

AVIATION...BRIEF DESCRIPTION OF SEVERE WEATHER THREAT TO AVIATORS.
HAIL SIZE WILL BE GIVEN IN INCHES AND WIND GUSTS IN KNOTS. MAXIMUM
STORM TOPS AND A MEAN STORM VECTOR WILL ALSO BE GIVEN.

$$

..FORECASTER NAME.. MM/DD/YY
```

6.4.3.2 Example of a Public Severe Thunderstorm Watch Notification Message

```
WWUS20 KWNS 161711 (ICAO communication header)
SPC WW 161710

URGENT - IMMEDIATE BROADCAST REQUESTED
SEVERE THUNDERSTORM WATCH NUMBER 647
NWS STORM PREDICTION CENTER NORMAN OK
1210 PM CDT FRI JUL 16 2004

THE NWS STORM PREDICTION CENTER HAS ISSUED A
SEVERE THUNDERSTORM WATCH FOR PORTIONS OF

        EASTERN IOWA
        NORTHERN ILLINOIS
```

```
NORTHWEST INDIANA
LAKE MICHIGAN

EFFECTIVE THIS FRIDAY AFTERNOON FROM 1210 PM UNTIL 500 PM CDT.

HAIL TO 2 INCHES IN DIAMETER...THUNDERSTORM WIND GUSTS TO 70
MPH...AND DANGEROUS LIGHTNING ARE POSSIBLE IN THESE AREAS.

THE SEVERE THUNDERSTORM WATCH AREA IS ALONG AND 75 STATUTE MILES
EITHER SIDE OF A LINE FROM 40 MILES SOUTHEAST OF SOUTH BEND
INDIANA TO 35 MILES SOUTHWEST OF CEDAR RAPIDS IOWA.

REMEMBER...A SEVERE THUNDERSTORM WATCH MEANS CONDITIONS ARE
FAVORABLE FOR SEVERE THUNDERSTORMS IN AND CLOSE TO THE WATCH AREA.
PERSONS IN THESE AREAS SHOULD BE ON THE LOOKOUT FOR THREATENING
WEATHER CONDITIONS AND LISTEN FOR LATER STATEMENTS AND POSSIBLE
WARNINGS. SEVERE THUNDERSTORMS CAN AND OCCASIONALLY DO PRODUCE
TORNADOES.

OTHER WATCH INFORMATION...CONTINUE...WW 646...

DISCUSSION...THUNDERSTORMS WILL CONTINUE TO INCREASE ACROSS WATCH
AREA WHERE AIR MASS HAS BECOME STRONGLY UNSTABLE AND UNCAPPED.
VEERING SHEAR PROFILE SUPPORT STORMS EVOLVING INTO SHORT LINE
SEGMENTS ENHANCING WIND DAMAGE POTENTIAL

AVIATION...A FEW SEVERE THUNDERSTORMS WITH HAIL SURFACE AND ALOFT
TO 2 INCHES. EXTREME TURBULENCE AND SURFACE WIND GUSTS TO 60
KNOTS. A FEW CUMULONIMBI WITH MAXIMUM TOPS TO 500. MEAN STORM
MOTION VECTOR 33025.

...HALES
```

6.4.4 Public Tornado Watch Notification Message

SPC issues a Public Tornado Watch Notification Message when forecasting three or more tornadoes or any tornado which could produce F2 or greater damage. The forecast event minimum thresholds are at least 2 hours over an area at least 8,000 square miles. Below these thresholds, SPC, in collaboration with affected NWS offices, may issue convective watches along coastlines, near the Canadian and Mexican borders and for any ongoing organized severe convection.

A Public Tornado Watch Notification Message contains the area description and axis, watch expiration time, the term "damaging tornadoes", a description of the largest hail size and strongest thunderstorm wind gusts expected, the definition of the watch, a call to action statement, a list of other valid watches, a brief discussion of meteorological reasoning, and technical information for the aviation community.

SPC includes the term "adjacent coastal waters" when the watch affects coastal waters adjacent to the Pacific/Atlantic coast, Gulf of Mexico, or Great Lakes. Adjacent coastal waters refers to a WFO's near shore responsibility (out to 20 nautical miles for oceans), except for convective watches which include portions of the Great Lakes.

SPC issues a watch cancellation message whenever it cancels a watch prior to the expiration time. The text of the message will specify the number and area of the cancelled watch.

6.4.4.1 Format of a Public Tornado Watch Notification Message

WWUS20 KWNS ddhhmm (*ICAO communication header*)

```
URGENT - IMMEDIATE BROADCAST REQUESTED
TORNADO WATCH NUMBER nnnn
NWS STORM PREDICTION CENTER NORMAN OK
time am/pm time zone day mon dd yyyy

THE STORM PREDICTION CENTER HAS ISSUED A
TORNADO WATCH FOR PORTIONS OF

        PORTION OF STATE
        PORTION OF STATE
        AND ADJACENT COASTAL WATERS (IF REQUIRED)

EFFECTIVE (TIME PERIOD) UNTIL hhmm am/pm time zone.

...THIS IS A PARTICULARLY DANGEROUS SITUATION (IF FORECAST)...

DESTRUCTIVE TORNADOES...HAIL TO X INCHES IN DIAMETER...THUNDERSTORM
WIND GUSTS TO XX MPH...AND DANGEROUS LIGHTNING ARE POSSIBLE IN THESE
AREAS.

NARRATIVE DESCRIPTION OF WATCH AREA USING A LINE AND ANCHOR POINTS.
DISTANCES TO EITHER SIDE OF THE LINE WILL BE IN STATUTE MILES.

CALL TO ACTION STATEMENTS

OTHER WATCH INFORMATION...OTHER WATCHES IN EFFECT AND IF THIS WATCH
REPLACES A PREVIOUS WATCH.

NARRATIVE DISCUSSION OF REASON FOR THE WATCH.

AVIATION...BRIEF DESCRIPTION OF SEVERE WEATHER THREAT TO AVIATORS.
HAIL SIZE WILL BE GIVEN IN INCHES AND WIND GUSTS IN KNOTS. MAXIMUM
STORM TOPS AND A MEAN STORM VECTOR WILL ALSO BE GIVEN.

$$

..FORECASTER NAME.. MM/DD/YY
```

6.4.4.2 Example of a Public Tornado Watch Notification Message

WWUS20 KWNS 050550 (*ICAO communication header*)

```
URGENT - IMMEDIATE BROADCAST REQUESTED
TORNADO WATCH NUMBER 243
NWS STORM PREDICTION CENTER NORMAN OK
```

```
1250 AM CDT MON MAY 5 2003

THE NWS STORM PREDICTION CENTER HAS ISSUED A
TORNADO WATCH FOR PORTIONS OF

        WESTERN AND CENTRAL ARKANSAS
        SOUTHERN MISSOURI
        FAR EASTERN OKLAHOMA

EFFECTIVE THIS MONDAY MORNING FROM 1250 AM UNTIL 600 AM CDT.

...THIS IS A PARTICULARLY DANGEROUS SITUATION...

DESTRUCTIVE TORNADOES...LARGE HAIL TO 2 INCHES IN DIAMETER...
THUNDERSTORM WIND GUSTS TO 70 MPH...AND DANGEROUS LIGHTNING ARE
POSSIBLE IN THESE AREAS.

THE TORNADO WATCH AREA IS ALONG AND 100 STATUTE MILES EAST AND
WEST OF A LINE FROM 15 MILES WEST NORTHWEST OF FORT LEONARD WOOD
MISSOURI TO 45 MILES SOUTHWEST OF HOT SPRINGS ARKANSAS.

REMEMBER...A TORNADO WATCH MEANS CONDITIONS ARE FAVORABLE FOR
TORNADOES AND SEVERE THUNDERSTORMS IN AND CLOSE TO THE WATCH AREA.
PERSONS IN THESE AREAS SHOULD BE ON THE LOOKOUT FOR THREATENING
WEATHER CONDITIONS AND LISTEN FOR LATER STATEMENTS AND POSSIBLE
WARNINGS.

OTHER WATCH INFORMATION...THIS TORNADO WATCH REPLACES TORNADO
WATCH NUMBER 237. WATCH NUMBER 237 WILL NOT BE IN EFFECT AFTER
1250 AM CDT. CONTINUE...WW 239...WW 240...WW 241...WW 242...

DISCUSSION...SRN MO SQUALL LINE EXPECTED TO CONTINUE EWD...WHERE
LONG/HOOKED HODOGRAPHS SUGGEST THREAT FOR EMBEDDED
SUPERCELLS/POSSIBLE TORNADOES. FARTHER S...MORE WIDELY SCATTERED
SUPERCELLS WITH A THREAT FOR TORNADOES WILL PERSIST IN VERY STRONGLY
DEEP SHEARED/LCL ENVIRONMENT IN AR.

AVIATION...TORNADOES AND A FEW SEVERE THUNDERSTORMS WITH HAIL
SURFACE AND ALOFT TO 2 INCHES. EXTREME TURBULENCE AND SURFACE WIND
GUSTS TO 60 KNOTS. A FEW CUMULONIMBI WITH MAXIMUM TOPS TO 500. MEAN
STORM MOTION VECTOR 26045.
.
..CORFIDI
```

6.5 Products for Tropical Cyclones

The NWS issues SIGMETs (Section 6.1), Convective SIGMETs (Section 6.1.8) and CWAs (Section 6.3) to inform the aviation community about the potential or existence of tropical cyclones and the adverse conditions associated with them. These above listed products are the primary source of information. The NWS also issues other products pertaining to convection. These additional products are defined in this section.

6.5.1 Aviation Tropical Cyclone Advisory (TCA)

The Aviation Tropical Cyclone Advisory (TCA) is intended to provide short-term tropical cyclone forecast guidance for international aviation safety and routing purposes. It is prepared by the National Hurricane Center (NHC) and the Central Pacific Hurricane Center (CPHC) in Honolulu, Hawaii, for all on-going tropical cyclone activity in their respective areas of responsibility. This requirement is stated in the World Meteorological Organization Region IV hurricane plan. Any valid TCA in the Atlantic or eastern Pacific is available on the NHC web site at: http://www.nhc.noaa.gov. Any valid TCA for the central Pacific is available on the CPHC web site at: http://www.prh.noaa.gov/hnl/cphc/

6.5.1.1 Issuance

TCAs are issued at 0300, 0900, 1500, and 2100 UTC and are valid from the time of issuance until the next scheduled issuance or update.

6.5.1.2 Content

TCAs list the current tropical cyclone position, motion and intensity, and 12-, 18- and 24-hour forecast positions and intensities. It is an alphanumeric text product produced by hurricane forecasters and consists of information extracted from the official forecasts. This forecast is produced from subjective evaluation of current meteorological and oceanographic data as well as output from numerical weather prediction models, and is coordinated with affected NWS offices, the NWS National Centers, and the Department of Defense.

6.5.1.3 Format

The format of the Aviation Tropical Cyclone Advisory is as follows:

```
FKaa2i CCCC DDHHMM (ICAO communication header)

(TROPICAL CYCLONE TYPE) ICAO ADVISORY NUMBER ##
ISSUING OFFICE CITY STATE
time am/pm time.zone day mon DD YYYY

TEXT

$$
```

6.5.1.4 Example of an Aviation Tropical Cyclone Advisory:

FKPZ21 KNHC 260215 (*ICAO communication header*)

TROPICAL DEPRESSION PATRICIA ICAO ADVISORY NUMBER 23
NWS TPC/NATIONAL HURRICANE CENTER MIAMI FL
0300Z SUN OCT 26 2003

```
TC ADVISORY
DTG:                      20031026/0300Z
TCAC:                     KNHC
TC:                       PATRICIA
NR:                       023
PSN:                      N1612 W11454
MOV:                      NW 05KT
C:                        1008HPA
MAX WIND:                 025KT
FCST PSN + 12 HR:         261200 N1636 W11500
FCST MAX WIND +           12 HR: 020KT
FCST PSN + 18 HR:         261800 N1654 W11506
FCST MAX WIND + 18 HR:    020KT
FCST PSN + 24 HR:         270000 N1712 W11512
FCST MAX WIND + 24 HR:    020KT
NXT MSG:                  NO MSG EXP
```

6.5.2 Tropical Cyclone Public Advisory (TCP)

A Tropical Cyclone Public Advisory (TCP) is the primary tropical cyclone information product issued to the public. The TCP provides critical tropical cyclone watch, warning, and forecast information for the protection of life and property.

6.5.2.1 Responsibility

The National Hurricane Center (NHC), as a part of the Tropical Prediction Center (TPC); the Central Pacific Hurricane Center (CPHC); and Weather Forecast Office (WFO) Tiyan, Guam, issue TCPs. In the Atlantic and central Pacific, NHC and CPHC issue TCPs for all tropical cyclones respectively. In the eastern Pacific, NHC will issue public advisories when watches or warnings are required, or the tropical cyclone is otherwise expected to impact nearby land areas. In the western Pacific, WFO Guam will issue public advisories for all tropical cyclones expected to affect land within 48 hours.

Valid TCP in the Atlantic or eastern Pacific is available on the NHC web site at: http://www.nhc.noaa.gov.

Valid TCP for the central Pacific is available on the CPHC web site at: http://www.prh.noaa.gov/hnl/cphc/.

TCPs issued by WFO Guam for the western Pacific are available at: http://www.prh.noaa.gov/pr/guam/cyclone.php.

6.5.2.2 Issuance

The initial advisory may be issued when data confirm a tropical cyclone has developed. The title of the advisory will depend upon the intensity of the tropical cyclone as listed below.

- A tropical depression advisory refers to a tropical cyclone with 1-minute sustained winds up to 33 knots (38 mph).

- A tropical storm advisory will refer to tropical cyclones with 1-minute sustained surface winds 34 to 63 knots (39 to 73 mph).

- A hurricane/typhoon advisory will refer to tropical cyclones with winds 64 knots (74 mph) or greater.

Public advisories are discontinued when the tropical cyclone:

- Becomes extra-tropical which is indicated by the center of the storm becoming colder than the surrounding air, fronts appear, and the strongest winds move to the upper atmosphere;

- Drops below tropical depression advisory criteria by dissipating or becoming a remnant low); or

- Moves inland and watches and warnings are no longer required.

Tropical Cyclone Public Advisories are issued according to the schedule below and are valid from the time of issuance until the next scheduled issuance or update. Valid position times correspond to the advisory time.

Table 6-10. Tropical Cyclone Public Advisory Issuance Schedule

TPC/CPHC ISSUANCE TIME (UTC)	WFO GUAM ISSUANCE TIME (UTC)
0300	0400
0900	1000
1500	1600
2100	2200

Times in advisories are local time of the affected area; however, local time and UTC are used when noting the storm's location. All advisories use statute miles and statute miles per hour. The Tropical Cyclone Center (TPC and CPHC) and WFO Guam, at their discretion, may use nautical miles/knots in parentheses immediately following statute miles/mph. Advisories include the metric units of kilometers and kilometers per hour following the equivalent English units except when the United States is the only country threatened.

NHC, CPHC and WFO Guam issue tropical storm/hurricane/typhoon watches if tropical storm/hurricane/typhoon conditions are possible over land areas within 36 hours, except 48 hours in the western north Pacific. Tropical storm watches are not issued if the tropical cyclone is forecast to reach hurricane/typhoon intensity within the watch period.

Tropical storm/hurricane/typhoon warnings are issued when tropical storm/hurricane/typhoon conditions along the coast are expected within 24 hours. Tropical storm warnings are issued at the discretion of the hurricane specialist when gale warnings, not related to the pending tropical storm, are already in place. Tropical storm warnings may be issued on either side of a hurricane/typhoon warning area.

6.5.2.2.1 Intermediate Issuances

Intermediate Public Advisories are issued on a 2- to 3-hourly interval between scheduled advisories (see times of issuance below). 3-hourly intermediate advisories are issued whenever a tropical storm or hurricane watch/warning is in effect. 2-hourly intermediates are issued whenever tropical storm or hurricane warnings are in effect and coastal radars are able to provide responsible Tropical Cyclone Centers with a reliable hourly center position. For clarity, when intermediate public advisories are issued, a statement is included at the end of the scheduled public advisory informing users when an intermediate advisory may be issued, i.e., "AN INTERMEDIATE ADVISORY WILL BE ISSUED BY THE CENTRAL PACIFIC HURRICANE CENTER AT 2 PM HST FOLLOWED BY THE NEXT COMPLETE ADVISORY ISSUANCE AT 5 PM HST."

Table 6-11. Intermediate Tropical Cyclone Public Advisory Issuance Schedule

	TPC/CPHC ISSUANCE TIME (UTC)	WFO GUAM ISSUANCE TIME (UTC)
3-Hourly Issuances	0000	0100
	0600	0700
	1200	1300
	1800	1900
2-Hourly Issuances	2300	0000
	0100	0200
	0500	0600
	0700	0800
	1100	1200
	1300	1400
	1700	1800
	1900	2000

Intermediate advisories are not used to issue tropical cyclone watches or warnings. They can be used to clear all, or parts of, a watch or warning area. Content is similar to the scheduled advisory.

6.5.2.3 Content

Advisories list all tropical cyclone watches and warnings in effect. The first advisory in which watches or warnings are mentioned will give the effective time of the watch or warning, except when it is being issued by other countries and the time is not known. Except for tropical storms and hurricanes/typhoons forming close to land, a watch will precede a warning. Once a watch is in effect, it will either be replaced by a warning or remain in effect until the threat of the tropical cyclone conditions has passed. A hurricane/typhoon watch and a tropical storm warning can be in effect for the same section of coast at the same time.

All advisories include the location of the center of the tropical cyclone by its latitude and longitude, and distance and direction from a well known point, preferably downstream from the tropical cyclone. If the forecaster is unsure of the exact location of a depression, the position may be given as within 50, 75, etc., miles of a map coordinate. When the center of the tropical cyclone is over land, its position is given referencing the state or country in which it is located and in respect to some well known city, if appropriate.

Movement forecasts apply to the tropical cyclone's center. The present movement is given to 16 points of the compass when possible. A 24-hour forecast of movement in terms of a continuance or departure from the present movement and speed is also included. This can be reduced to a 12-hour forecast. Uncertainties in either the tropical cyclone's location or movement will be explained in the advisory. An outlook beyond 24 hours (out to 72 hours when appropriate) may be included in the text of the advisory.

Maximum observed 1-minute sustained surface wind speed rounded to the nearest 5 mph is given. During landfall threats, specific gust values and phrases like "briefly higher in squalls" may be used. The area (or radius) of both tropical and hurricane/typhoon force winds is given. The storm may also be compared to some memorable hurricane or referred to by relative intensity. Where appropriate, the Saffir/Simpson Hurricane Scale (SSHS) is used in public releases.

Central pressure values in millibars and inches are provided as determined by available data.

The inland impacts of tropical cyclones will be highlighted in advisories. This includes the threat of strong winds, heavy rainfall, flooding, and tornadoes. The extent and magnitude of the inland winds is included as well as anticipated rainfall amounts and the potential for flooding and tornadoes. Tornado and flood watches will be mentioned as appropriate.

6.5.2.4 Format
The format of the Tropical Cyclone Public Advisory is as follows:

```
(TROPICAL CYCLONE TYPE) (NAME) ADVISORY NUMBER XX.
(ISSUING OFFICE CITY STATE)
time am/pm time zone day month DD YYYY

...HEADLINE...

TEXT

$$

FORECASTER NAME
```

6.5.2.5 Example of a Tropical Storm Public Advisory

```
BULLETIN
TROPICAL STORM FLOYD ADVISORY NUMBER 4
NWS TPC/NATIONAL HURRICANE CENTER MIAMI FL
11 AM AST WED SEP 08 1999

...FLOYD MOVING WEST-NORTHWESTWARD IN THE TROPICAL ATLANTIC...

AT 11 AM AST...1500Z...THE CENTER OF TROPICAL STORM FLOYD WAS
LOCATED NEAR LATITUDE 15.8 NORTH...LONGITUDE 50.0 WEST OR ABOUT
755 MILES...1210 KM...EAST OF THE LEEWARD ISLANDS.
```

FLOYD IS MOVING TOWARD THE WEST NORTHWEST NEAR 15 MPH ...24
KM/HR...AND THIS MOTION IS EXPECTED TO CONTINUE THROUGH TONIGHT.

MAXIMUM SUSTAINED WINDS ARE NEAR 45 MPH... 75 KM/HR...WITH HIGHER
GUSTS...AND SOME SLOW STRENGTHENING IS EXPECTED DURING THE NEXT
24 HOURS.

TROPICAL STORM FORCE WINDS EXTEND OUTWARD UP TO 85 MILES...140
KM FROM THE CENTER.

ESTIMATED MINIMUM CENTRAL PRESSURE IS 1003 MB...29.62 INCHES.

REPEATING THE 11 AM AST POSITION...15.8 N... 50.0 W. MOVEMENT
TOWARD...WEST NORTHWEST NEAR 15 MPH. MAXIMUM SUSTAINED
WINDS... 45 MPH. MINIMUM CENTRAL PRESSURE...1003 MB.

THE NEXT ADVISORY WILL BE ISSUED BY THE NATIONAL HURRICANE
CENTER AT 5 PM AST.

FORECASTER FRANKLIN

6.5.2.6 Example of a Hurricane/Typhoon Public Advisory

BULLETIN
HURRICANE FLOYD ADVISORY NUMBER 32
NWS TPC/NATIONAL HURRICANE CENTER MIAMI FL
11 AM EDT WED SEP 15 1999

...FRINGES OF HURRICANE CONTINUE TO IMPACT COAST OF NORTH FLORIDA
AND GEORGIA...BUT FLOYD IS HEADING FOR THE CAROLINAS...

AT 11 AM EDT...A TROPICAL STORM WATCH IS EXTENDED NORTHWARD AND
IS NOW IN EFFECT FROM NORTH OF CHINCOTEAGUE VIRGINIA TO
SANDYHOOK NEW JERSEY...INCLUDING DELAWARE BAY.

A HURRICANE WARNING REMAINS IN EFFECT FROM TITUSVILLE FLORIDA
TO THE NORTH CAROLINA/VIRGINIA BORDER...INCLUDING PAMLICO AND
ALBEMARLE SOUNDS. AT 11 AM EDT...HURRICANE WARNINGS ARE
DISCONTINUED SOUTH OF TITUSVILLE.

A HURRICANE WATCH CONTINUES IN EFFECT FROM THE NORTH
CAROLINA/VIRGINIA BORDER TO CHINCOTEAGUE VIRGINIA...INCLUDING
CHESAPEAKE BAY SOUTH OF SMITH POINT.

INTERESTS ALONG THE FLORIDA EAST COAST SOUTH OF TITUSVILLE
SHOULD EXERCISE CAUTION UNTIL WINDS AND SEAS SUBSIDE.

AT 11 AM EDT...1500Z...THE CENTER OF HURRICANE FLOYD WAS LOCATED
NEAR LATITUDE 29.9 NORTH...LONGITUDE 79.0 WEST OR ABOUT 165 MILES
EAST-SOUTHEAST OF JACKSONVILLE FLORIDA. THIS POSITION IS ALSO

ABOUT 260 MILES SOUTH OF MYRTLE BEACH SOUTH CAROLINA.

FLOYD IS MOVING TOWARD THE NORTH NORTHWEST NEAR 14 MPH AND A
GRADUAL TURN TOWARD THE NORTH IS EXPECTED TODAY.

MAXIMUM SUSTAINED WINDS ARE NEAR 125 MPH...205 KM/HR...WITH
HIGHER GUSTS. LITTLE CHANGE IN STRENGTH IS FORECAST BEFORE
LANDFALL...WHICH IS EXPECTED TONIGHT NEAR THE BORDER OF SOUTH
AND NORTH CAROLINA. ALL PREPARATIONS SHOULD BE RUSHED TO
COMPLETION.

HURRICANE FORCE WINDS EXTEND OUTWARD UP TO 140 MILES...220
KM...FROM THE CENTER...AND TROPICAL STORM FORCE WINDS EXTEND
OUTWARD UP TO 230 MILES...370 KM.

THE LATEST MINIMUM CENTRAL PRESSURE REPORTED BY U.S. AIR FORCE
HURRICANE HUNTER AIRCRAFT IS 943 MB...27.85 INCHES.

STORM SURGE FLOODING OF 10 TO 13 FEET ABOVE NORMAL TIDE
LEVELS...ALONG WITH LARGE AND DANGEROUS BATTERING WAVES...ARE
EXPECTED NEAR AND TO THE EAST OF WHERE THE CENTER CROSSES THE
COAST. HEAVY SURF ADVISORIES ARE IN EFFECT FOR THE U.S. EAST COAST
NORTHWARD TO CHATHAM MASSACHUSETTS. REFER TO STATEMENTS
ISSUED BY LOCAL NATIONAL WEATHER SERVICE OFFICES FOR ADDITIONAL
INFORMATION.

RAINFALL TOTALS OF 5 TO 10 INCHES ARE EXPECTED ALONG THE PATH OF
THE HURRICANE.

ISOLATED TORNADOES ARE POSSIBLE OVER THE COASTAL COUNTIES OF
SOUTH AND NORTH CAROLINA.

REPEATING THE 11 AM EDT POSITION...29.9 N... 79.0 W. MOVEMENT
TOWARD...NORTH NORTHWEST NEAR 14 MPH. MAXIMUM SUSTAINED
WINDS...125MPH. MINIMUM CENTRAL PRESSURE... 943 MB.

FOR STORM INFORMATION SPECIFIC TO YOUR AREA...PLEASE MONITOR
PRODUCTS ISSUED BY YOUR LOCAL WEATHER OFFICE.

INTERMEDIATE ADVISORIES WILL BE ISSUED BY THE NATIONAL HURRICANE
CENTER AT 1 PM EDT AND 3 PM EDT FOLLOWED BY THE NEXT COMPLETE
ADVISORY AT 5 PM EDT.

FORECASTER LAWRENCE

6.5.2.7 Example of an Intermediate Public Advisory

BULLETIN
HURRICANE FLOYD INTERMEDIATE ADVISORY NUMBER 32B
NWS TPC/NATIONAL HURRICANE CENTER MIAMI FL
3 PM EDT WED SEP 15 1999

...FRINGES OF HURRICANE CONTINUE TO IMPACT COAST OF NORTH FLORIDA
AND GEORGIA...BUT FLOYD IS HEADING FOR THE CAROLINAS...

A HURRICANE WARNING REMAINS IN EFFECT FROM NORTH OF FERNANDINA
BEACH FLORIDA TO THE NORTH CAROLINA/VIRGINIA BORDER...INCLUDING
PAMLICO AND ALBEMARLE SOUNDS. AT 3 PM EDT...WARNINGS ARE
DISCONTINUED FROM FERNANDINA BEACH SOUTHWARD. WARNINGS WILL
LIKELY BE DISCONTINUED FOR PORTIONS OF GEORGIA LATER TODAY.

A HURRICANE WATCH REMAINS IN EFFECT FROM THE NORTH
CAROLINA/VIRGINIA BORDER TO CHINCOTEAGUE VIRGINIA...INCLUDING
CHESAPEAKE BAY SOUTH OF SMITH POINT.

A TROPICAL STORM WATCH REMAINS IN EFFECT FROM NORTH OF
CHINCOTEAGUE VIRGINIA TO MONTAUK POINT LONG ISLAND...INCLUDING
DELAWARE BAY AND LONG ISLAND SOUND.

INTERESTS ALONG THE FLORIDA EAST COAST SHOULD EXERCISE CAUTION
UNTIL WINDS AND SEAS SUBSIDE.

AT 3 PM EDT...1900Z...THE CENTER OF HURRICANE FLOYD WAS LOCATED
NEAR LATITUDE 30.8 NORTH...LONGITUDE 79.1 WEST OR ABOUT 200 MILES
SOUTH OF MYRTLE BEACH SOUTH CAROLINA.

FLOYD IS MOVING ALMOST DUE NORTHWARD AT 15 MPH AND THIS MOTION
IS EXPECTED TO CONTINUE TODAY WITH A GRADUAL TURN TOWARD THE
NORTH-NORTHEAST ON THURSDAY.

MAXIMUM SUSTAINED WINDS HAVE DECREASED TO NEAR 120 MPH...WITH
HIGHER GUSTS. ALTHOUGH THE HURRICANE HAS BEEN SLOWLY
WEAKENING...IT IS OVER THE WARM WATERS OF THE GULF STREAM COULD
MAINTAIN ITS PRESENT STRENGTH UNTIL LANDFALL TONIGHT. ALL
PREPARATIONS IN THE WARNING AREA SHOULD BE RUSHED TO
COMPLETION.

HURRICANE FORCE WINDS EXTEND OUTWARD UP TO 140 MILES...220 KM...
FROM THE CENTER...AND TROPICAL STORM FORCE WINDS EXTEND
OUTWARD UP TO 230 MILES...370 KM.

THE LATEST MINIMUM CENTRAL PRESSURE REPORTED BY U.S. AIR FORCE
HURRICANE HUNTER AIRCRAFT IS 947 MB...27.96 INCHES.

STORM SURGE FLOODING OF 10 TO 13 FEET ABOVE NORMAL TIDE
LEVELS...ALONG WITH LARGE AND DANGEROUS BATTERING WAVES...ARE
EXPECTED NEAR AND TO THE EAST OF WHERE THE CENTER CROSSES THE
COAST.

HEAVY SURF ADVISORIES ARE IN EFFECT FOR THE U.S. EAST COAST
NORTHWARD TO CHATHAM MASSACHUSETTS. REFER TO STATEMENTS
ISSUED BY LOCAL NATIONAL WEATHER SERVICE OFFICES FOR ADDITIONAL
INFORMATION.

RAINFALL TOTALS OF 5 TO 10 INCHES ARE EXPECTED ALONG THE PATH OF
THE HURRICANE.

ISOLATED TORNADOES ARE POSSIBLE OVER THE COASTAL COUNTIES OF
SOUTH AND NORTH CAROLINA.

FOR STORM INFORMATION SPECIFIC TO YOUR AREA...PLEASE MONITOR
PRODUCTS ISSUED BY YOUR LOCAL WEATHER OFFICE.

REPEATING THE 3 PM EDT POSITION...30.8 N... 79.1 W. MOVEMENT
TOWARD...NORTH NEAR 15 MPH. MAXIMUM SUSTAINED WINDS...120 MPH.
MINIMUM CENTRAL PRESSURE... 947 MB.

THE NEXT ADVISORY WILL BE ISSUED BY THE NATIONAL HURRICANE
CENTER AT 5 PM EDT.

FORECASTER LAWRENCE

6.5.2.8 Example of a Special Public Advisory

BULLETIN
HURRICANE ANDREW SPECIAL ADVISORY NUMBER 25
NWS TPC/NATIONAL HURRICANE CENTER MIAMI FL
900 AM EDT MON AUG 24 1992

...HURRICANE ANDREW MOVING INTO THE GULF OF MEXICO...

HURRICANE WARNINGS REMAIN POSTED FOR THE FLORIDA WEST COAST
SOUTH OF VENICE TO FLAMINGO AND FOR LAKE OKEECHOBEE. AT 9 AM
EDT A HURRICANE WATCH WILL GO INTO EFFECT FOR THE NORTHERN GULF
COAST FROM MOBILE ALABAMA TO SABINE PASS TEXAS. ALL OTHER
POSTED WATCHES AND WARNINGS ARE DISCONTINUED.

WIND GUSTS TO HURRICANE FORCE CONTINUE TO OCCUR ALONG THE
SOUTHEAST FLORIDA COAST BUT WILL GRADUALLY DIMINISH DURING THE
DAY. SMALL CRAFT ADVISORIES REMAIN IN EFFECT. RESIDENTS IN THESE
AREAS SHOULD MONITOR LOCAL NWS OFFICES FOR THE LATEST
FORECASTS AND CONDITIONS IN THEIR AREA.

AT 9 AM EDT THE CENTER OF HURRICANE ANDREW WAS LOCATED NEAR
LATITUDE 25.6 NORTH AND LONGITUDE 81.8 WEST OR APPROXIMATELY 45
MILES SOUTH OF NAPLES FLORIDA.

HURRICANE ANDREW IS MOVING TOWARD THE WEST AT 18 MPH. THIS
MOTION IS EXPECTED TO CONTINUE THIS MORNING WITH A GRADUAL TURN
TO THE WEST NORTHWEST LATER TODAY.

MAXIMUM SUSTAINED WINDS ARE NEAR 140 MPH. LITTLE CHANGE IN
STRENGTH IS LIKELY DURING THE NEXT 24 HOURS.

HURRICANE FORCE WINDS EXTEND OUTWARD TO 30 MILES...50 KM FROM
THE CENTER WITH TROPICAL STORM FORCE WINDS EXTENDING OUTWARD
TO 140 MILES. ESTIMATED MINIMUM CENTRAL PRESSURE IS 945 MB...27.91
INCHES.

STORM SURGES OF 5 TO 8 FEET ARE POSSIBLE ON THE FLORIDA WEST
COAST NEAR AND TO THE SOUTH OF THE CENTER FOLLOWING PASSAGE OF
THE HURRICANE. ALONG THE SOUTHEAST COAST OF FLORIDA STORM
SURGE TIDES ARE DECREASING. PRELIMINARY REPORTS FROM THE SOUTH
FLORIDA WATER MANAGEMENT DISTRICT INDICATE A STORM SURGE OF 8
FEET ABOVE NORMAL WAS RECORDED IN BISCAYNE BAY NEAR
HOMESTEAD FLORIDA.

RAINFALL AMOUNTS OF 5 TO 8 INCHES AND ISOLATED TORNADOES ARE
POSSIBLE ACROSS SOUTHERN AND CENTRAL FLORIDA TODAY.

FOR STORM INFORMATION SPECIFIC TO YOUR AREA...PLEASE MONITOR
PRODUCTS ISSUED BY YOUR LOCAL WEATHER OFFICE.

REPEATING THE 9 AM EDT POSITION...LATITUDE 25.6 NORTH AND
LONGITUDE 81.8 WEST AND MOVING TOWARD THE WEST AT 18 MPH.
MAXIMUM SUSTAINED WINDS NEAR 140 MPH. MINIMUM CENTRAL
PRESSURE OF 945 MB...27.91 INCHES.

THE NEXT SCHEDULED ADVISORY WILL BE ISSUED BY THE NATIONAL
HURRICANE CENTER AT 11 AM EDT MON.

6.5.2.9 Example of a Public Advisory Correction

HURRICANE ANDREW ADVISORY NUMBER 25...**CORRECTED**
NWS TPC/NATIONAL HURRICANE CENTER MIAMI FL
500 AM EDT MON AUG 24 1992

CORRECTED FOR CENTRAL PRESSURE...

BODY OF TEXT

6.6 Volcanic Ash Advisory Products

In addition to SIGMETs (Section 6.1), the NWS issues the following products to notify the aviation community of volcanic ash.

6.6.1 Volcanic Ash Advisory Statement (VAAS)
A Volcanic Ash Advisory Statement (VAAS) provides information on hazards to aircraft flight operations caused by a volcanic eruption.

6.6.1.1 Issuance
Volcanic Ash Advisory Centers (VAACs) are responsible for providing ash movement and dispersion guidance to Meteorological Watch Offices (MWOs) and neighboring VAACs. There are nine VAACs worldwide, two of which are located in the US (Figure 6-18).

Each VAAC issues Volcanic Ash Advisory Statements and provide guidance to Meteorological Watch Offices (MWOs) for SIGMETs involving volcanic ash.

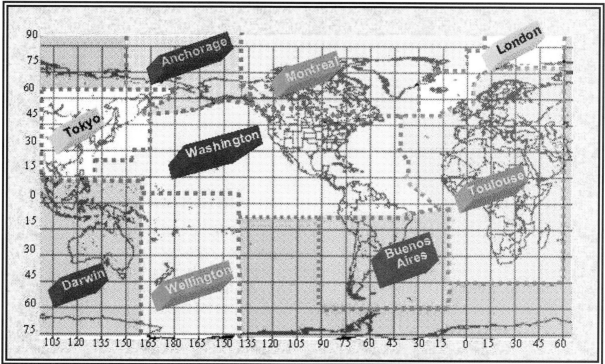

Figure 6-18. Volcanic Ash Advisory Centers (VAACs) Area of Responsibility

6.6.1.2 Format
A VAAS may be issued within 6 hours of an eruption and every 6 hours thereafter. However, it can be issued more frequently if new information about the eruption is received.

A VAAS summarizes the known information about an eruption. It may include the location of the volcano, height of the volcano summit, height of the ash plume, a latitude/longitude box of the ash dispersion cloud, and a forecast of ash dispersion. The height of the ash cloud is estimated by meteorologists analyzing satellite imagery and satellite cloud drift winds combined with any pilot reports, volcano observatory reports, and upper-air wind reports.

6.6.1.3 Example of a VAAS Issued by the Washington VAAC

```
VOLCANIC ASH ADVISORY
ISSUED: 2003JUL10/1300Z      VAAC: WASHINGTON

VOLCANO: ANATAHAN 0804-20
LOCATION: N1621E14540    AREA: MARIANA ISLANDS

SUMMIT ELEVATION: 2585 FT (788 M)

ADVISORY NUMBER: 2003/251

INFORMATION SOURCE: GOES 9 IMAGERY. GFS MODEL WINDS FORECAST

ERUPTION DETAILS: ASH AND GAS EMISSIONS SINCE MAY 10.

OBS ASH DATE/TIME: 09/1202Z.

OBS ASH CLOUD: ASH NOT IDENTIFIABLE FROM SATELLITE DATA.

WINDS SFC/FL080 MOVING SW 10-15 KNOTS.

FCST ASH CLOUD +6H: SEE SIGMETS.

REMARKS: THE ASH PLUME OBSERVED IN VISIBLE IMAGERY IS TOO THIN AND
DIFFUSE TO BE SEEN IN INFRARED AND MULTISPECTRAL IMAGAERY. ANY ASH UP
TO FL080 SHOULD MOVE TOWARDS THE SW AT 10-15 KNOTS.

NEXT ADVISORY: WILL BE ISSUED BY 2003JUL10/1900Z.
```

6.6.2 Volcanic Ash Advisory (VAA)

Volcanic Ash Advisory Centers (VAACs) issue Volcanic Ash Advisories (VAAs) when airborne volcanic ash is observed or reported which may affect the atmosphere in the VAAC's area of responsibility. The VAA is intended as guidance to support MWOs in meeting their responsibility to issue the volcanic ash SIGMET. The VAA also may be issued as a watch for an imminent eruption expected to produce airborne ash.

6.6.2.1 VAA Responsibility.

The U.S. has two VAACs with responsibilities defined in ICAO Annex 3. The Washington VAAC is jointly managed by the National Environmental Satellite Data and Information Service (NESDIS) Satellite Analysis Branch (SAB) and the NWS National Centers for Environmental Prediction (NCEP) Central Operations (NCO). The Anchorage VAAC is managed by the AAWU. The areas of responsibility for each VAAC are:

- Washington VAAC

    o FIRs in CONUS and adjacent coastal waters (Figures 6-3 and 6-18)
    o The Oakland Oceanic FIR over the Pacific Ocean (Figures 6-5 and 6-18)
    o The New York FIR over the western Atlantic Ocean (Figures 6-4 and 6-18)

      o FIRs over and adjacent to the Caribbean, and Central and South America north of 10 degrees south latitude (Figure 6-4 and 6-18)

- Anchorage VAAC
  - o The Anchorage FIR (Figures 6-5 and 6-18).
  - o Russian FIRs north of 60 degrees north latitude and east of 150 degrees east longitude (Figure 6-18).

6.6.2.2 VAA Issuance and Update Times
The VAA may be issued as soon as possible after credible information is received on the presence of airborne volcanic ash in the VAAC's area of responsibility or when responsibility for an existing VAA is transferred between VAACs. The VAA contains information on an ash cloud up to 18 hours. It may be issued any time to account for changing or new information. Any necessary updates are issued at a minimum of every 6 hours.

6.6.2.3 VAA Content
The VAA follows international recommendations contained in ICAO Annex 3, chapter 3.6.2 and contains the name of the erupting volcano and number, if known; its location (latitude and longitude) and summit height (in meters or feet); the information source; the volcano aviation color code if applicable; eruption details; the date and time of the observed ash; information about the observed ash cloud; the forecast area and height of the ash cloud at 6, 12, and 18 hours after the issuance of the VAA; any pertinent remarks on the eruption/ash event; and the next VAA issuance time.

A VAA watch is not an official WMO/ICAO product. However, if it is issued, it contains all information **except** for the eruption details, and observed and forecast ash clouds. Information on the direction the ash likely will spread in the event of an eruption will be included in remarks. In Alaska, a VAA watch may be issued for a non-erupting seismically monitored volcano in color code orange or red. A one-time VAA Watch may be issued when a monitored Alaska volcano goes from color code green to yellow.

6.6.2.4 VAA Cancellation
The VAA will be canceled when it is determined airborne volcanic ash is no longer a threat to aircraft or has moved out of the VAAC's area of responsibility.

6.6.2.5 Interchange of VAAs among Volcanic Ash Advisory Centers (VAAC)
When an ash cloud is forecast to move from one VAAC's area of responsibility into another VAAC's area of responsibility, the two VAACs will coordinate by telephone or telephone fax on handoff procedures. The VAAC passing off responsibility will include in remarks of its last VAA the name of the VAAC assuming responsibility for issuing subsequent VAAs for the event, the new WMO header, and the date/time of next expected issuance. The accepting VAAC will include in remarks the name of the VAAC from which it is accepting responsibility and the WMO header of the current VAA it will be updating. Generally, only one (1) VAAC will issue VAAs for a particular ash event. If the ash area affects more than one VAAC area of responsibility, the VAAC issuing the VAA will include the entire ash area in the advisory. In the rare situation of large or persistent ash emissions, adjacent responsible VAACs, upon coordination, may agree to divide operational responsibilities.

6.6.2.6 VAA Dissemination
VAAs will be disseminated to Meteorological Watch Offices (MWOs), Area (Traffic) Control Centers, World Area Forecast Centers (WAFCs), relevant Regional Area Forecast Centers

(RAFCs), international operational meteorological data banks, and other government and commercial meteorological offices, in accordance with regional air navigation agreements.

6.6.2.7 Example of a Volcanic Ash Advisory (VAA)

```
VOLCANIC ASH ADVISORY - ALERT

ISSUED 2003 APR 19/0615Z

VAAC: ANCHORAGE

VOLCANO: CHIKURACHKI, 900-36

LOCATION: N5019 E15527

AREA: KAMCHATKA NORTHERN KURIL ISLANDS

SUMMIT ELEVATION: 7674 FT (2339 M)

ADVISORY NUMBER: 2003-02

INFORMATION SOURCE: SATELLITE

AVIATION COLOR CODE: NOT GIVEN

ERUPTION DETAILS: A NEW ERUPTION OCCURRED AT APPROXIMATELY 190500
UTC. HEIGHT IS ESTIMATED AT FL300. ESTIMATE IS BASED ON OBSERVED
AND MODEL WINDS. MOVEMENT APPEARS TO BE E AT 75 KTS.

OBS ASH DATA/TIME: 19/0500Z

OBS ASH CLOUD: VA EXTENDS FM NEAR VOLCANO EWD TO N50 E160.
FCST ASH CLOUD +6HR: 30NM EITHER SIDE OF LN FM NIPPI N49 E159 -
N50 E175.

FCST ASH CLOUD +12HR: 30NM EITHER SIDE OF LN FM N50 E168 -
N50 E180.

FCST ASH CLOUD +18HR: 30NM EITHER SIDE OF LN FM N51 E175 -
N50 E185.

NEXT ADVISORY: 20030419/1500Z

REMARKS: UPDATES AS SOON AS INFO BECOMES AVAILABLE.
```

7 FORECAST TEXT PRODUCTS

7.1 Area Forecasts (FA)

The NWS issues Area Forecasts (FA) to provide an overview of regional weather conditions that could impact aviation operations in the U.S. and adjacent coastal waters. Area forecasts are issued by the following offices for the following areas:

- The Aviation Weather Center (AWC)
    - Conterminous U.S. and adjacent coastal waters (CONUS)
    - Gulf of Mexico
    - Caribbean Sea and north Atlantic Ocean

- The Alaskan Aviation Weather Unit (AAWU)
    - Alaska and adjacent coastal waters

- WFO Honolulu, Hawaii
    - Hawaii and adjacent coastal waters

They are all available on the Aviation Weather Center (AWC) web site at: http://aviationweather.gov/products/fa/.

7.1.1 CONUS (FAUS) and Hawaii (FAHW) Area Forecasts

A CONUS and Hawaii Area Forecast (FA) describe, in abbreviated language, specified en route weather phenomena below FL450. To understand the complete weather picture, the FA **must** be used in conjunction with the AIRMETs (Section 6.2) and SIGMETs (Section 6.1). Together, they are used to determine forecast en route weather and to interpolate conditions at airports for which no Terminal Aerodrome Forecasts (TAFs) are issued.

The CONUS and Hawaii FAs are available on the Aviation Weather Center (AWC) web site at: http://aviationweather.gov/products/fa/.

The Hawaii Area Forecast can also be found on the NWS WFO Honolulu web site at: http://www.prh.noaa.gov/hnl/pages/aviation.php.

The FA contains forecast information for VFR/MVFR clouds and weather for a 12-hour period with a 12- to 18-hour categorical outlook forecast for IFR, MVFR, and/or VFR. The following weather elements are included in the 12-hour forecast:

- Thunderstorms and precipitation;
- Sky condition (cloud height, amount, and tops) if bases are at or below (AOB) FL180 MSL. (Tops will only be forecast for broken (BKN) or overcast (OVC) clouds);
- Obstructions to visibility (fog, mist, haze, blowing dust, etc.) if surface visibilities are three (3) to six (6) miles; and
- Sustained surface wind speed of 20 knots or greater.

Hazardous weather (e.g., IFR, icing, turbulence, etc.) meeting AIRMET or SIGMET criteria is <u>not</u> forecast in the CONUS or Hawaii FA. Valid AIRMETs and SIGMETs must be used in conjunction with the FA to determine hazardous weather information for the flight.

The Aviation Weather Center (AWC) issues the following CONUS FAs for six (6) geographical areas (Figure 7-1). The Weather Forecast Office (WFO) Honolulu issues FAs for the main Hawaiian Islands and coastal waters extending out to 40 NM of the coastlines (Figure 7-2).

Figure 7-1. AWC Area Forecast Regions- Contiguous U.S.

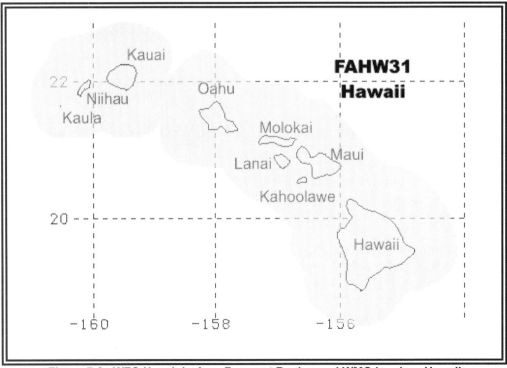

Figure 7-2. WFO Honolulu Area Forecast Region and WMO header - Hawaii

An Area Forecast (FA) provides an overview of regional weather conditions that could impact aviation operations in the U.S. and adjacent coastal waters. The Area Forecast **does not** include forecast for IFR conditions so the Area Forecast **must** be used in conjunction with SIGMETs and AIRMETs. Each FA contains a precautionary statement, prior to the synopsis, saying **SEE AIRMET SIERRA** followed by a reminder of what thunderstorm activity implies and a reference to how heights not reported in MSL are denoted. This is not a reference to a specific AIRMET but a reminder the FA does not include forecasted IFR conditions.

7.1.1.1 Standardization
The CONUS FA follows these standards:

- All referenced heights or altitudes are referenced above mean sea level (AMSL), unless otherwise noted, and annotated using the height in hundreds of feet, consisting of three (3) digits (e.g., 040). For heights at or above 18,000 feet, the level will be preceded by FL (e.g., FL180).

- Messages are prepared in abbreviated plain language using contractions from the Federal Aviation Administration (FAA) Order 7340.1Y for domestic products and International Civil Aviation Organization (ICAO) document 8400 for international products issued for Oceanic FIRs. A limited number of non-abbreviated words, geographic names and numerical values of a self-explanatory nature may also be used.

- Weather and obstructions to visibility are described using the weather and abbreviations for surface airways observations (METAR or SPECI).

7.1.1.1.1 Height Reference
All heights are referenced to Mean Sea Level (MSL) except when prefaced by **AGL**, **CIG** or **CEILING**. Tops are always referenced to MSL.

Examples:

`SCT030 BKN100`
　　Scattered at 3,000 feet MSL, broken at 10,000 feet MSL

`AGL SCT030 CIG BKN050`
　　Scattered at 3,000 feet AGL, broken at 5,000 feet AGL

`AGL SCT-BKN015-025. TOPS 070-090`
　　Scattered to broken at 1,500 to 2,500 feet AGL. Tops 7,000 to 9,000 feet MSL.

7.1.1.2 CONUS and Hawaii Area Forecast Format
The FA is an 18 hour forecast composed of the following 4 sections: communication and product header, precautionary statements, synopsis and visual flight rules (VFR) clouds and weather forecast.

7.1.1.2.1 Communication and Product header
The Communication and Product header section (Figure 7-3) contains descriptive information about the product.

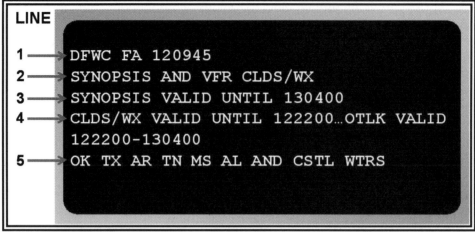

Figure 7-3. Area Forecast - Communication and Product Header Example

Table 7-1. Decoding the Communications and Product Header

Line	Content	Description
1	DFW C FA 120945	Area Forecast region identifier Indicates VFR clouds and weather forecast Product type Issuance and beginning of valid date/time (UTC)
2	SYNOPSIS AND VFR CLDS/WX	Statement of weather information contained in this forecast message
3	SYNOPSIS VALID UNTIL 130400	Synopsis valid date and time
4	CLDS/WX VALID UNTIL 122200...OTLK VALID 122200-130400	The main forecast for VFR clouds and weather valid time. The valid date and time of the outlook.
5	OK TX AR TN MS AL AND CSTL WTRS	Description of the area for which the FA is valid.

7.1.1.2.2 Precautionary Statements

The Precautionary Statements section (Figure 7-4) consists of three lines.

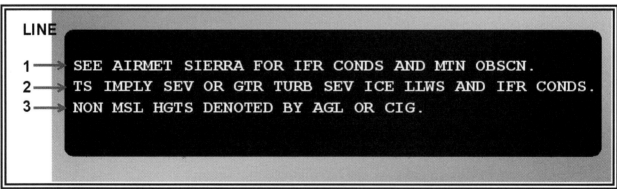

Figure 7-4. Area Forecast- Precautionary Statements Example

Line 1 is included to alert users that IFR conditions and/or mountain obscurations may be occurring or are forecast to occur and are not included in the product.

Line 2 is included as a reminder of all hazards associated with thunderstorms. These hazards are not spelled out in the body of the FA.

Line 3 indicates height references are Mean Sea Level (MSL) unless they are preceded by **AGL** or **CIG**.

7.1.1.2.3 Synopsis

The Synopsis section (Figure 7-5) contains a brief summary of the location and movement of fronts, pressure systems, and other circulation features for the entire 18 hour (FA) valid period.

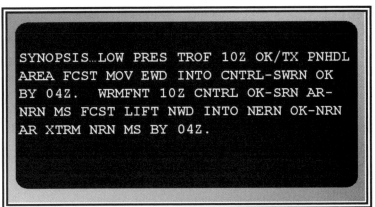

```
SYNOPSIS...LOW PRES TROF 10Z OK/TX PNHDL
AREA FCST MOV EWD INTO CNTRL-SWRN OK
BY 04Z.  WRMFNT 10Z CNTRL OK-SRN AR-
NRN MS FCST LIFT NWD INTO NERN OK-NRN
AR XTRM NRN MS BY 04Z.
```

Figure 7-5. Area Forecast - Synopsis Example

The Synopsis is decoded as follows:

Synopsis...low pressure through at 1000 UTC over the Oklahoma and Texas Panhandle area forecast to move eastward into central/southwestern Oklahoma by 0400 UTC. A warm front at 1000 UTC from central Oklahoma to southern Arkansas to northern Mississippi is forecast to lift northward into northeastern Oklahoma to northern Arkansas to extreme northern Mississippi by 0400 UTC.

7.1.1.2.4 VFR clouds and Weather (CLDS/WX)

The VFR CLDS/WX section (Figure 7-6) describes conditions consisting of MVFR cloud ceilings (1,000 to 3,000 feet AGL), MVFR obstructions to visibility (3-5 statute miles), and any other significant VFR clouds (bases at or below FL180) or VFR precipitation. The CLDS/WX section also includes widespread sustained surface winds of 20 knots or greater. Occasionally, IFR conditions may be forecast in the Hawaii FA as IFR conditions may not reach AIRMET geographical coverage criteria.

This section contains a 12-hour forecast, followed by a 6-hour categorical outlook of IFR, MVFR and/or VFR, giving a total forecast period of 18 hours. In the CONUS, the CLDS/WX section is divided into regions with generally uniform weather conditions. These divisions may be by geographical regions (e.g., LM – Lake Michigan) or states using their 2-letter designators (e.g. ND – North Dakota). See Appendix H for geographical regions.

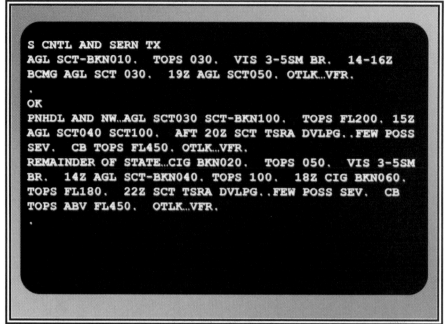

Figure 7-6. Area Forecast - VFR Clouds and Weather Example

The VFR CLDS/WX section is decoded as follows:

South central and southeast Texas.
Scattered to broken at 1,000 feet above ground level (AGL). Tops at 3,000 feet above mean sea level (MSL). Visibility 3 to 5 statute miles in mist. Between 1400 and 1600 UTC…clouds becoming scattered at 3,000 feet AGL. 1900 UTC…scattered at 5,000 feet AGL. Outlook…VFR.

Oklahoma
Panhandle and northwest…scattered at 3,000 feet AGL, scattered to broken at 10,000 feet AGL. Tops at flight level 20,000 feet MSL. 1500 UTC…scattered at 4,000 feet AGL, scattered at 10,000 feet AGL. After 2000 UTC…scattered thunderstorms with rain showers developing..a few possible severe. Cumulonimbus tops to flight level 45,000 feet MSL. Outlook…VFR.
Remainder of the state…Ceilings broken at 2,000 feet AGL. Tops at 5,000 feet MSL. Visibilities 3 to 5 statute miles in mist. 1400 UTC…scattered to broken at 4,000 feet AGL. Tops at 10,000 feet MSL. 1800 UTC…ceilings broken 6,000 feet AGL. Tops to flight level 18,000 feet MSL. 2200 UTC…scattered thunderstorm with rain showers developing…a few possibly severe. Cumulonimbus tops above flight level 45,000 feet MSL. Outlook…VFR.

7.1.1.3 CONUS and Hawaii Area Forecast Issuance
The CONUS FAUSs are issued three times daily for each of six areas (see following table).

Table 7-2. Area Forecast (FAUS) Issuance Schedule - CONUS

Area Forecast (FAUS)	Boston and Miami (UTC)	Chicago and Fort Worth (UTC)	San Francisco and Salt Lake City (UTC)
1st Issuance	0845 DT 0945 ST	0945 DT 1045 ST	1045 DT 1145 ST
2nd Issuance	1745 DT 1845 ST	1845 DT 1945 ST	1945 DT 2045 ST
3rd Issuance	0045 DT 0145 ST	0145 DT 0245 ST	0245 DT 0345 ST

The Hawaii Area Forecast is issued four times daily at 0340, 0940, 1540, and 2140 UTC.

7.1.1.3.1 FA Amendments

An amended FA may be issued to notify pilots and briefers that a weather phenomena and/or condition that was not forecast is now expected or a forecast phenomenon or condition has improved or did not develop as expected. The new condition is expected to exceed half of the time of a regular issuance and is expected to no longer affect low-level flights. An amended FA is denoted by an **AMD** after the date/time group on the FAA product line (line 1 in Table 7-1) and will contain an **UPDT** contraction following the affected geographical area in the CLDS/WX section.

Example

```
CHIC FA 231345 AMD
SYNOPSIS AND VFR CLDS/WX
SYNOPSIS VALID UNTIL 240400
CLDS/WX VALID UNTIL 232200...OTLK VALID 232200-240400
ND SD NE KS MN IA MO WI LM LS MI LH IL IN KY

.
SD...UPDT
EXTRM SWRN/EXTRM S CNTRL...CIG BKN-OVC010 TOP 120. 18Z AGL SCT015
SCT-BKN035. OTLK...VFR.
RMNDR WRN/CNTRL...SCT CI. OTLK...VFR TSRA.
ERN...AGL SCT-BKN035 TOP 120. OTLK...VFR.
```

7.1.1.3.2 FA Corrections

FAs containing errors are corrected. This is identified by **COR** after the date/time group on the FAA product line. The first time indicated is the issuance time with the ending valid time unchanged. A corrected FA contains **UPDT** following the affected geographical area in the CLDS/WX section.

Example:

```
CHIC FA 231015 COR
SYNOPSIS AND VFR CLDS/WX
SYNOPSIS VALID UNTIL 240400
CLDS/WX VALID UNTIL 232200...OTLK VALID 232200-240400
ND SD NE KS MN IA MO WI LM LS MI LH IL IN KY

.
```

```
SD...UPDT
EXTRM SWRN/EXTRM S CNTRL...CIG BKN-OVC010 TOP 120. 18Z AGL SCT015
SCT-BKN035. OTLK...VFR.
RMNDR WRN/CNTRL...SCT CI. OTLK...VFR TSRA.
ERN...AGL SCT-BKN035 TOP 120. OTLK...VFR.
```

7.1.1.4 Example of a CONUS Area Forecast

```
FAUS5 KDFW 030953  (ICAO Communication Header)
FA4W
DFWC FA 030945  (AMD or COR if needed)
SYNOPSIS AND VFR CLDS/WX
SYNOPSIS VALID UNTIL 040400
CLDS/WX VALID UNTIL 032200...OTLK VALID 032200-040400
OK TX AR TN LA MS AL
.
SEE AIRMET SIERRA FOR IFR COND AND MT OBSCN.
TS IMPLY SEV OR GTR TURB SEV ICE LLWS AND IFR CONDS.
NON MSL HGT DENOTED BY AGL OR CIG.
.
SYNOPSIS...HURCN LILI MOVG ONSHORE OVER CENTRAL LA COASTLINE. SEE
LATEST ADVISORY FM NHC. QUASI-STNR FRONTAL SYSTEM EXTENDS FM N OH
AND CENTRAL IN ACROSS S IL..SW MO..SW OK INTO SE CORNER OF NM. BY
04Z...COLD FRONT WILL EXTEND FM A LOW OVER SERN NE ACROSS CENTRAL KS
AND W OK INTO BIG BEND AREA OF SW TX.
.
OK
PANHANDLE/W OK...CIG OVC010. CLDS LYR TO FL240.
VIS 3-5SM BR. BECMG 1618 CIG OVC015-025. WIDELY SCT -SHRA/ISOL EMBD
-TSRA. CB TOP FL350. OTLK...MVFR CIG TSRA BR.
ERN OK...AGL SCT-BKN015-025. TOPS 030-050. VIS 3-5SM BR. BECMG
1417 AGL SCT030-050. OTLK...VFR.
.
NW TX
CIG010. CLDS LYR TO FL240. VIS 3-5SM BR.  BECMG 1618 CIG OVC015-025.
WIDELY SCT -SHRA/ISOL EMBD -TSRA. CB TOP FL350. OTLK...MVFR CIG TSRA
BR.
.
SW TX
AGL SCT040-060. OTLK...VFR.
.
CENTRAL TX
CIG BKN015-025. TOP 030-050. VIS 3-5SM BR. BECMG 1417 AGL SCT030-050.
OTLK...VFR.
.
E TX
SKC. BECMG 1316 AGL SCT030-050. OTLK...VFR.
.
AR
AGL SCT030-050. SCT-BKN100. TOP FL200. BKN CI. OTLK...MVFR CIG TSRA
BR.
.
```

LA
N LA...AGL SCT-BKN030-050. BKN100. TOPS FL240. ISOL -SHRA. BECMG 1618
CIG BKN030-050. WIDELY SCT TSRA/SHRA DEVELOPING. CB TOP FL400.
OTLK...MVFR CIG TSRA WIND.
S LA...CIG OVC010-020. CLDS LYR TO FL280. OCNL RA/+RA...SCT
+TSRA...POSS SEV. CB TOPS FL450. SFC WND 14030G50KT. E SECTIONS...WND
30025G40KT. WND DIMINISHING TO 20G30KT 19-22Z.
OTLK...MVFR CIG SHRA WND.

.
TN
BKN CI. OCNL VIS 3-5SM BR TIL 14Z. OTLK...VFR.

.
MS AL
N AND CENTRAL MS-AL/SE AL..SCT-BKN100. BKN150. TOPS FL280. BECMG 1618
AGL SCT-BKN050 BKN100 OVC150. OTLK...MVFR CIG TSRA.
S MS/SW AL...AGL SCT-BKN050 BKN100 OVC150. TOPS FL280. BECMG 1316 CIG
OVC015-025. OCNL RA/SCT EMBD TSRA. CB TOP FL410. OTLK...MVFR CIG TSRA.

7.1.1.5 Example of a Hawaii Area Forecast
FAHW31 PHFO 080940
FA0HI
HNLC FA 080940 *(AMD or COR, if necessary)*
SYNOPSIS AND VFR CLDS/WX
SYNOPSIS VALID UNTIL 090400
CLDS/WX VALID UNTIL 082200...OTLK VALID 082200-090400
HI

.
SEE AIRMET SIERRA FOR IFR CONDS AND MTN OBSCN.
TS IMPLY SEV OR GRTR TURB SEV ICE LLWS AND IFR CONDS.
NON MSL HGT DENOTED BY AGL OR CIG.

.
SYNOPSIS...SFC HIGH FAR N PHNL NEARLY STNR.

.
BIG ISLAND ABV 060.
SKC. 20Z SCT090. OTLK...VFR.

.
BIG ISLAND LOWER SLOPES...CSTL AND ADJ WTRS FROM UPOLU POINT TO
CAPE KUMUKAHI TO APUA POINT.
SCT030 BKN050 TOPS 080 ISOL BKN030 VIS 3-5SM -SHRA BR TIL 20Z ISOL
BKN010 VIS BELOW 3SM SHRA BR. 21Z SCT030 SCT-BKN050 TOPS 080 ISOL
BKN030 5SM -SHRA. OTLK...VFR.

.
BIG ISLAND LOWER SLOPES FROM APUA POINT TO SOUTH CAPE TO UPOLU POINT.
SKC. 21Z SCT-BKN060 TOPS 080. 23Z SCT030 SCT-BKN060 TOPS 080 ISOL
BKN030 -SHRA. OTLK...VFR.

.
BIG ISLAND CSTL AND ADJ WTRS FROM SOUTH CAPE TO PHKO TO UPOLU
POINT.
SCT050 ISOL BKN050 TOPS 080. 18Z FEW050. 23Z SCT-BKN050 TOPS080.
OTLK...VFR.

.

```
N AND E FACING SLOPES...CSTL AND ADJ WTRS OF THE REMAINING
ISLANDS.
SCT020 BKN045 TOPS070 TEMPO BKN020 VIS 3-5SM -SHRA...FM OAHU EASTWARD
ISOL CIG BLW 010 AND VIS BELOW 3SM SHRA BR WITH TOPS 120. 22Z SCT025
SCTBKN050 TOPS 070 ISOL BKN025 3-5SM -SHRA. OTLK...VFR.
.
REST OF AREA.
SCT035 SCT-BKN050 TOPS 070 ISOL BKN030 -SHRA. 20Z SCT050 ISOL SCT030
BKN045 TOPS 070 -SHRA. OTLK...VFR.
```

7.1.2 Gulf of Mexico Area Forecast (FAGX)

The Gulf of Mexico FA is an overview of weather conditions that could impact aviation operations over the northern Gulf of Mexico (Figure 7-7). It serves as a flight-planning and weather briefing aid and describes weather of significance to general aviation (GA), military and helicopter operations. The FAGX is a 24 hour forecast product with the synopsis valid the entire 24 hour period, the forecast section valid the first 12 hours, and the outlook section is valid the last 12 hours.

Each FA contains a statement before the synopsis indicating heights not reported in MSL are denoted and a reminder of what thunderstorm activity implies.

The Aviation Weather Center (AWC) produces this forecast and it can be found at: http://www.aviationweather.gov/products/fa/?area=gulf

Figure 7-7. AWC Area Forecast Region and WMO Header - Gulf of Mexico

7.1.2.1 Standardization

All forecasts follow these standards:

- All referenced heights or altitudes consist of three (3) digits depicting height in hundreds of feet Mean Sea Level (MSL).

- Messages are prepared using approved ICAO contractions, abbreviations and numerical values of self-explanatory nature.

- Weather and obstructions to visibility are the same as weather abbreviations used for surface airways observations (METAR or SPECI) (Section 2).

7.1.2.1.1 Height Reference

All heights are referenced to Mean Sea Level (MSL) except when prefaced by AGL or CIG. Tops are always referenced to MSL.

Examples:

SCT030 BKN100
> Scattered at 3,000 feet MSL, broken at 10,000 feet MSL

AGL SCT030 BKN100
> Scattered at 3,000 feet AGL, broken at 10,000 feet MSL

CIG BKN006 BKN070. TOP 100.
> Broken at 600 feet AGL, broken at 7,000 feet MSL. Top 10,000 feet MSL.

7.1.2.2 Gulf of Mexico Area Forecast Content

The Gulf of Mexico FA (FAGX) is a single product combining information contained in FAs prepared for the conterminous U.S. and the in-flight advisories -- AIRMET/SIGMET. Each section describes the phenomena impacting the respective areas and will always have an entry even if it is negative.

The FAGX contains a synopsis and a weather forecast section. The weather section includes:

- Flight precautions at or below 12,000 feet MSL for thunderstorms which are at least scattered or meet Convective SIGMET criteria;
- Moderate or greater turbulence; moderate or greater icing;
- Wind speeds greater than or equal to 25 knots below 1,000 feet;
- Ceilings and/or visibilities less than 1,000 feet and/or three (3) miles;
- Significant Clouds and Weather;
- Icing and freezing level; and
- Turbulence.

Table 7-3. Area Forecast Sections – Gulf of Mexico

SECTION	DESCRIPTION
SYNOPSIS...	A brief description of the location and movement of fronts, pressures systems, and other circulations as well as the weather associated with them.
FLIGHT PRECAUTIONS...	Flight precautions (at or below 12,000 feet MSL) include adverse weather. Examples may include thunderstorms, IFR, turbulence, strong winds, icing, etc...
SIGNIFICANT CLD/WX...	A 12-hour forecast of clouds and weather at or below 12,000 feet MSL. This forecast can include IFR conditions. It also includes a 12-hour categorical outlook (IFR/MVFR/VFR/WND) and the clouds or weather causing the categorical forecast.
ICE AND FZ LVL BLW 120	The location and altitudes of moderate or greater icing and the associated freezing levels at or below 12,000 feet MSL.

7.1.2.3 Example of a Gulf of Mexico Area Forecast
The following is an example of a Gulf of Mexico Area Forecast.

```
FAGX20 KKCI 091812
OFAGX
SYNOPSIS VALID TIL 101900Z
FCST...091900Z-100700Z
OTLK...100700Z-101900Z
.
 INTERNATIONAL OPERATIONS BRANCH
AVIATION WEATHER CENTER KANSAS CITY MISSOURI
.
CSTL WATERS FROM COASTLINE OUT TO HOUSTON OCEANIC FIR AND GLFMEX MIAMI
OCEANIC FIR AND W OF 85W. HOUSTON OCEANIC FIR AND GLFMEX MIAMI OCEANIC
FIR.
.
TS IMPLY SEV OR GTR TURB SEV ICE LLWS AND IFR CONDS. HGTS MSL.
.
01 SYNOPSIS...HIGH PRES OVR NRN GLFMEX.
.
02 SIGNIFICANT CLD/WX...
.
CSTL WATERS...
SCT020. OTLK...VFR.
.
HOUSTON OCEANIC FIR...SCT020. OTLK...VFR.
.
GLFMEX MIAMI OCEANIC FIR... SCT020. OTLK...VFR.
.
03 ICE AND FRZLVL...
CSTL WATERS...SEE AIRMETS ZULU WAUS44 KKCI AND WAUS42 KKCI.
HOUSTON OCEANIC FIR... NO SGFNT ICE EXP OUTSIDE CNVTV ACT.
```

```
GLFMEX MIAMI OCEANIC FIR...NO SGFNT ICE EXP OUTSIDE CNVTV ACT.
FRZLVL...140 THRUT.
.
04 TURB... CSTL WATERS...SEE AIRMETS TANGO WAUS44 KKCI AND WAUS42
KKCI.
HOUSTON OCEANIC FIR... NO SGFNT TURB EXP OUTSIDE CNVTV ACT.
GLFMEX MIAMI OCEANIC FIR...NO SGFNT TURB EXP OUTSIDE CNVTV ACT.
```

7.1.2.4 Gulf of Mexico Area Forecast Issuance

The FAGX, valid for 12 hours with a 12-hour extended outlook, is issued twice daily at 1030 and 1830 UTC.

7.1.2.4.1 Gulf of Mexico FA Amendments

Gulf of Mexico FAs are amended at the forecaster's discretion.

If any phenomena or conditions depicted in FA improve and are no longer expected to affect low-level flights (including VFR) and the new conditions will exceed half the period between regular issuances, a FA **AMD** message is sent indicating which section has been amended by adding **AMD**. The first time indicated is the issuance time with the ending valid time unchanged.

The product will not be amended between 0200 and 1100 UTC.

7.1.2.4.2 FA Corrections

FAs containing errors are corrected. A FA correction is sent indicating which section has changed by adding **COR**. The first time indicated is the issuance time with the ending valid time unchanged.

7.1.3 Caribbean Area Forecast (FACA)

The Caribbean FA is an overview of weather conditions that could impact aviation operations over the Caribbean Sea and adjacent landmasses and islands and the southwestern portions of the New York Oceanic FIR (Figure 7-8). Specifically, it covers the Atlantic south of 32N and W of 57W, the Caribbean from surface to FL240 (approximately 400 millibars).

The synopsis and forecast sections are valid for 12 hours each, with the outlook valid for 12 hours beyond the synopsis and forecast section valid period. In this form, it serves as a flight planning and weather briefing aid for general aviation pilots, and civil and military aviation operations.

The clouds/weather forecast section includes the following areas:

- Atlantic (Southwestern NY and MIA Oceanic FIRs)
- Caribbean Sea (San Juan FIR; Western Piarco FIR; Santo Domingo, Port au Prince and Habana FIRs; Northern Maiquetia, Curacao, and Northern Barranquilla FIRs; Kingston and Northern Central America FIRs; Northern Merida FIR; and Eastern Monterrey FIR)

The Caribbean Area Forecast is issued by the AWC and can be found at:
http://www.aviationweather.gov/products/fa/?area=carib

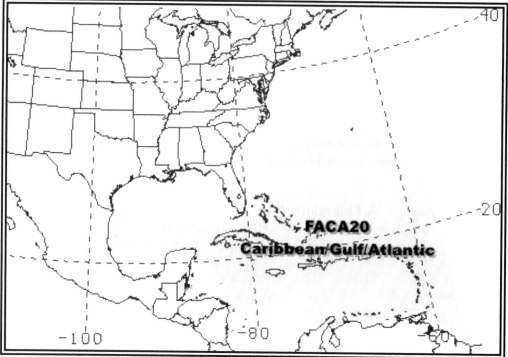

Figure 7-8. AWC Area Forecast Region and WMO Header - Caribbean

7.1.3.1 Standardization
All forecasts follow these standards:

- All Mean Sea Level (MSL) referenced heights or altitudes are annotated as FL for heights at or above 18,000 and consist of three (3) digits depicting height in hundreds of feet.

- Messages are prepared using approved ICAO contractions, abbreviations and numerical values of self-explanatory nature.

- Weather and obstructions to visibility are the same as weather abbreviations used for surface airways observations (METAR or SPECI) (Section 2).

7.1.3.1.1 Height Reference
All heights are referenced to Mean Sea Level (MSL) except when prefaced by AGL or CIG. Tops are always referenced to MSL.

Examples:

`SCT030 BKN100`
    Scattered at 3,000 feet MSL, broken at 10,000 feet MSL

`AGL SCT030 BKN100`
    Scattered at 3,000 feet AGL, broken at 10,000 feet MSL

`CIG BKN006 BKN070.  TOP 100.`
    Broken at 600 feet AGL, broken at 7,000 feet MSL. Top 10,000 feet MSL.

7.1.3.2 Caribbean Area Forecast Content

Table 7-4. Area Forecast Sections - Caribbean

Section	Description
SYNOPSIS...	A brief description of the location and movement of fronts, pressure systems, and other circulations, as well as the weather associated with them.
SIGNIFICANT CLD/WX...	A 12-hour forecast of clouds and weather (including IFR conditions) plus a 12-hour categorical outlook (IFR/MVFR/VFR/WND). The cause of IFR/MVFR conditions is specified. Wind is 25 knots or greater.
ICE AND FZ LVL...	The location and altitudes of moderate or greater icing and the associated freezing levels.
TURB...	The location and altitudes of moderate or greater turbulence.

7.1.3.3 Example of a Caribbean Area Forecast

```
FACA20 KKCI 121530
OFAMKC
INTERNATIONAL OPERATIONS BRANCH
AVIATION WEATHER CENTER KANSAS CITY MISSOURI
VALID 121600-130400
OUTLOOK...130400-131600
.
ATLANTIC S OF 32N W OF 57W...CARIBBEAN...GULF OF MEXICO BTN 22N AND
24N.
.
TS IMPLY SEV OR GTR TURB SEV ICE LLWS AND IFR CONDS. SFC TO 400 MB.
.
SYNOPSIS...WK CDFNT EXTDS FM NR 28N60W TO 23N63W TO THE MONA PASSAGE.
CDFNT WL MOV EWD AND WKN TODAY. EXP NARROW BAND OF CLDS WITH ISOL SHRA
INVOF CDFNT.
.
SIGNIFICANT CLD/WX...
ERN MONTERREY FIR...NRN MERIDA FIR
SCT025 SCT060. OTLK...VFR.
.
ATLC SWRN NEW YORK FIR...SAN JUAN FIR
NW OF CDFNT...SCT025 SCT060. LYR OCNL BKN. TOP 120. ISOL SHRA.
OTLK...VFR.
VCNTY CDFNT...SCT025 BKN060. OCNL BKN025. TOP 120. WDLY SCT SHRA. ISOL
TSRA TIL 20Z. OTLK...VFR SHRA.
SE OF CDFNT...SCT025 SCT060. ISOL SHRA. OTLK...VFR.
.
ATLC MIAMI FIR
SCT025 SCT060. LYR OCNL BKN. TOP 120. ISOL SHRA. OTLK...VFR.
.
WRN PIARCO FIR...NRN MAIQUETIA FIR...CURACAO FIR
BTN 61W-63W...SCT025 BKN060. OCNL BKN025. TOP 120. WDLY SCT SHRA.
OTLK...VFR SHRA.
RMNDR...SCT025 SCT060. ISOL SHRA. OTLK...VFR.
```

```
.
SANTO DOMINGO FIR...PORT-AU-PRINCE FIR
SCT025 SCT060. LYR OCNL BKN. TOP 120. ISOL SHRA. OTLK...VFR.
.
NRN BARRANQUILLA FIR...NRN PANAMA FIR
SCT025 SCT060. ISOL SHRA. SFC WND NE 20-25KT. OTLK...VFR.
.
KINGSTON FIR...NERN CNTRL AMERICAN FIR...HABANA FIR
SCT025 SCT060. ISOL SHRA. OTLK...VFR.
.
ICE AND FRZLVL...
NO SGFNT ICE EXP OUTSIDE CNVTV ACT.
FRZLVL... 145-170.
.
TURB...
NO SGFNT TURB EXP OUTSIDE CNVTV ACT.
```

7.1.3.4 Caribbean Area Forecast Issuance

The FACA is issued four times daily at 0330, 0930, 1530, and 2130 UTC.

7.1.3.4.1 FA Amendments

If any phenomena or condition included in the FA is no longer expected to affect flight operations (including VFR), and the new condition is expected to exceed half the period between regular issuances, a FA **AMD** message is sent indicating which section has been amended by adding **AMD**. The first time indicated is the issuance time with the ending valid time unchanged.

7.1.3.4.2 FA Corrections

FAs containing errors are corrected. A FA correction is sent indicating which section has changed by adding **COR**. The first time indicated is the issuance time with the ending valid time unchanged.

7.1.3.4.3 Routine Delayed (RTD) FAs

For FAs delayed in transmission, **RTD** is added after the date/time group on the FAA product line (section 7.1.3.3 line1). The first time indicated is the issuance time with the ending valid time unchanged.

7.1.4 Alaska Area Forecast

The Alaskan FAs contain an overview of weather conditions that could impact aviation operations over Alaska and it coastlines. The Alaskan FAs contain a short synopsis for the entire area and a forecast for each of a specified number of aviation zones (Figure 7-9). The valid period of the synopsis and flight precautions section is 12 hours. The outlook section is for eighteen (18) hours beyond the forecast valid period.

Each FA contains AIRMETs and references to SIGMETs. In addition, a statement about conditions implied by a forecast of thunderstorms and a reference to how heights not reported in MSL are denoted is included.

The Alaska Area Forecast is issued by the Alaska Aviation Weather Unit (AAWU) and can be found at:

http://aawu.arh.noaa.gov/areaforecasts.php and on the Aviation Weather Center (AWC) web site at http://aviationweather.gov/products/fa/?area=alaska

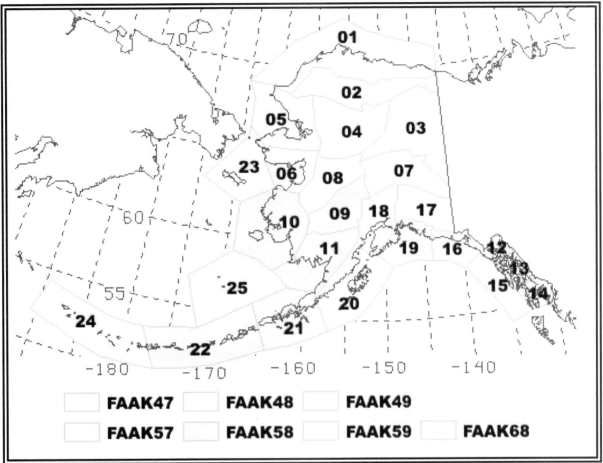

Figure 7-9. AAWU Flight Advisory and Area Forecast Zones - Alaska

Table 7-5. AAWU Flight Advisory and Area Forecast Zones – Alaska

1	Arctic Coast Coastal	14	Southern Southeast Alaska
2	North Slopes of the Brooks Range	15	Coastal Southeast Alaska
3	Upper Yukon Valley	16	Eastern Gulf Coast
4	Koyukuk and Upper Kobuk Valley	17	Copper River Basin
5	Northern Seward Peninsula-Lower Kobuk Valley	18	Cook Inlet-Susitna Valley
6	Southern Seward Peninsula-Eastern Norton Sound	19	Central Gulf Coast
7	Tanana Valley	20	Kodiak Island
8	Lower Yukon Valley	21	Alaska Peninsula-Port Heiden to Unimak Pass
9	Kuskowim Valley	22	Unimak Pass to Adak
10	Yukon-Kuskowim Delta	23	St. Lawrence Island-Bering Sea Coast
11	Bristol Bay	24	Adak to Attu
12	Lynn Canal and Glacier Bay	25	Pribilof Islands and Southeast Bering Sea
13	Central Southeast Alaska		

7.1.4.1 Standardization
All forecasts follow these standards:

- All referenced heights or altitudes are annotated as FL for heights at or above 18,000 and consist of three (3) digits depicting height in hundreds of feet Mean Sea Level (MSL).

- Messages are prepared using approved ICAO contractions, abbreviations and numerical values of self-explanatory nature.

- Weather and obstructions to visibility are the same as weather abbreviations used for surface airways observations (METAR or SPECI) (Section 2).

7.1.4.1.1 Height Reference
All heights are referenced to Mean Sea Level (MSL) except when prefaced by AGL or CIG. Tops are always referenced to MSL.

Examples:

SCT030 BKN100
> Scattered at 3,000 feet MSL, broken at 10,000 feet MSL

AGL SCT030 BKN100
> Scattered at 3,000 feet AGL, broken at 10,000 feet AGL

AGL SCT-BKN015-025. TOPS 030-050.
> Scattered to broken at 1,500 to 2,500 feet AGL. Tops 3,000 to 5,000 feet MSL.

7.1.4.2 Alaska Area Forecast Content
The Alaskan Area Forecast zones contain sections on Clouds and Weather, Turbulence, and Icing and Freezing Levels.

The Clouds and Weather section includes:
- SIGMETs for Thunderstorms and Volcanic Ash;
- AIRMETs for IFR ceiling and visibility, mountain obscuration, and strong surface winds;
- Bases and tops of significant cloud layers;
- Visibilities of six (6) miles or less and restricting phenomena;
- Precipitation and thunderstorms;
- Surface winds of 20 KTS or greater;
- Outlook using categorical terms (i.e., VFR CIG, MVFR BR, IFR SN WND); and
- Mountain-pass conditions using categorical terms (for selected zones only).

The Turbulence section includes:
- SIGMETs for Turbulence;
- AIRMETs for Turbulence and/or Low Level Wind Shear (LLWS);
- Forecast of significant turbulence not meeting SIGMET or AIRMET criteria or that is forecast for the period 6 to 12 hours after issuance; and
- If no significant turbulence is forecast, NIL SIG will be entered.

Icing section includes:
- SIGMETs for Icing;
- AIRMETs for Icing and freezing precipitation;

- Forecast of significant icing not meeting SIGMET or AIRMET criteria or which is forecast for the period 6 to 12 hours after issuance;
- Freezing Level; and
- If no significant icing is forecast, NIL SIG will be entered followed by the freezing level.

7.1.4.3 Example of an Alaska Area Forecast

```
FAAK01 PANC 251345 (AMD, COR, RTD if necessary)
FA8H
ANCH FA 251345
AK SRN HLF EXC SE AK...
.
AIRMETS VALID UNTIL 252000
TS IMPLY POSSIBLE SEV OR GREATER TURB SEV ICE LLWS AND IFR CONDS.
NON MSL HEIGHTS NOTED BY AGL OR CIG.
.
SYNOPSIS VALID UNTIL 260800
972 MB BRISTOL BAY LOW WL MOV N TO 50 S PAOM AT 987 MB BY END OF PD.
ASSOCIATED OCCLUDED FRONT FM PALJ..KENNEDY ENTRANCE..SE WL MOV NE
TO PAMH..PACV..SE
BY 08Z.
.
COOK INLET AND SUSITNA VALLEY AB...VALID UNTIL 260200
...CLOUDS/WX...
***AIRMET IFR/MT OBSC***AK RANGE/W SIDE COOK INLET..OCNL CIGS BLW 10
VIS BLW 3SM -RA BR. NC...
OTHERWISE..AK RANGE/W SIDE INLET..SCT005 OVC020 VIS 3-5SM -RA BR.
ELSEWHERE..SCT025 BKN045 OVC080 LYR ABV TO FL250. OCNL BKN025 OVC045
-RA.
COOK INLET..SFC WND NE 20G30 KTS. THRU TERRAIN GAPS..ERN MTS/AK
RANGE..SFC WND E 30G60 KTS.
OTLK VALID 260200-262000...MVFR CIG RA WND.
PASSES...LAKE CLARK..MERRILL..RAINY..IFR CIG RA WND. WINDY..MVFR CIG
RA.
PORTAGE..IFR CIG RA WND.
...TURB...
***SIGMET***KILO 1 VALID 251607/252000 PANC-
OCNL SEV TURB FCST BLW 080 WI AN AREA FM TKA-JOH-MDO-AKN-SQA-TKA.
THIS IS THE AREA S OF A PAHZ-PATK LN.
***AIRMET TURB/LLWS***OCNL MOD TURB BLW 120. LLWS. NC...
...ICE AND FZLVL...
***AIRMET ICE***OCNL MOD RIME/MX ICEIC 050-160. FZLVL 050. NC...
.
COPPER RIVER BASIN AC...VALID UNTIL 260200
...CLOUDS/WX...
FEW045 SCT090 BKN-OVC180 TOP FL250.
SFC WND SE G 25 KTS.
WRN MTS..ISOL BKN025 OVC045 4SM -SHRA.
OTLK VALID 260200-262000...VFR.
PASS...TAHNETA..MVFR CIG.
...TURB...
NIL SIG.
```

```
...ICE AND FZLVL...
NIL SIG. FZLVL 050.

.
CNTRL GLF CST AD...VALID UNTIL 260200
...CLOUDS/WX...
***AIRMET MT OBSC***MTS OBSCD IN CLDS/PRECIPITATION. NC...
SCT020 OVC040 LYRD ABV TO FL250 -RA.
OCNL SCT005 OVC020 VIS 3-5SM -RA BR.
SFC WND E 20G35 KTS. THRU TRRN GAPS WND E-NE 25G50 KTS.
ALONG KENAI PENINSULA..ISOL CIGS BLW 10 VIS BLW 3SM RA BR.
OTLK VALID 260200-260200..MVFR CIG RA WND.
...TURB...
***SIGMET***KILO 1 VALID 251607/252000 PANC-
OCNL SEV TURB FCST BLW 080 WI AN AREA FM TKA-JOH-MDO-AKN-SQA-TKA.
THIS IS THE AREA E OF A JOH-PAMD LN.
***AIRMET TURB/LLWS***OCNL MOD TURB BLW 120. LLWS NR TRRN. NC...
...ICE AND FZLVL...
***AIRMET ICE***OCNL MOD RIME ICEIC 050-160. FZLVL 050. NC...

.
KODIAK ISLAND AE...VALID UNTIL 260200
...CLOUDS/WX...
***AIRMET MT OBSC***MTS OBSCD IN CLDS/PRECIPITATION. NC...
SCT020 OVC040 LYRD ABV TO FL250 -RA.
OCNL SCT005 OVC020 VIS 3-5SM -RA BR.
E SIDE..ISOL CIGS BLW 10 VIS BLW 3SM RA BR.
SFC WND SE G 25 KT.
OTLK VALID 260200-262000...MVFR CIG SHRA WND. AFT 06Z..VFR.
...TURB...
NIL SIG.
...ICE AND FZ LVL...
ISOL MOD RIME ICEIC 030-120. FZLVL 030.
```

7.1.4.4 Alaska FA Issuance
The Alaskan FAs are produced four (4) times daily

Table 7-6. Area Forecast Issuance Schedule - Alaska

Alaska Area Forecast	Standard Time (UTC)	Daylight Time (UTC)
1st Issuance	0245	0145
2nd Issuance	0845	0745
3rd Issuance	1445	1345
4th Issuance	2045	1945

7.1.4.4.1 FA Amendments
FAs are under continuous review and amended at the discretion of the forecaster. An amended FA contains **AAA** after the date/time group on the WMO heading line for the first amendment, **AAB** for the second, and continuing for all subsequent amendments. **AMD** is also included after the date/time group on the FAA product line (section 7.1.4.3 line 1).

7.1.4.4.2 FA Corrections

FAs containing errors are corrected. This is identified by **COR** after the date/time group on the FAA product line (section 7.1.4.3 line 1). The first time indicated is the issuance time, with the ending valid time unchanged.

7.1.4.4.3 Routine Delayed (RTD) FAs

For FAs delayed in transmission, **RTD** is added after the date/time group on the FAA product line (section 7.1.4.3 line 1). The first time indicated is the issuance time, with the ending valid time unchanged.

7.2 Terminal Aerodrome Forecast (TAF)

A Terminal Aerodrome Forecast (TAF) is a concise statement of the expected meteorological conditions significant to aviation for a specified time period within five statute miles (SM) of the center of the airport's runway complex (terminal). The TAFs use the same weather codes found in METAR weather reports (Section 2) and can be viewed on the National Weather Service (NWS) Aviation Digital Data Service (ADDS) web site at: http://adds.aviationweather.noaa.gov/tafs/.

7.2.1 Responsibility

TAFs are issued by NWS Weather Forecast Offices (WFOs). A map of U.S. TAF locations is located on Figures 7-10, 7-11, and 7-12.

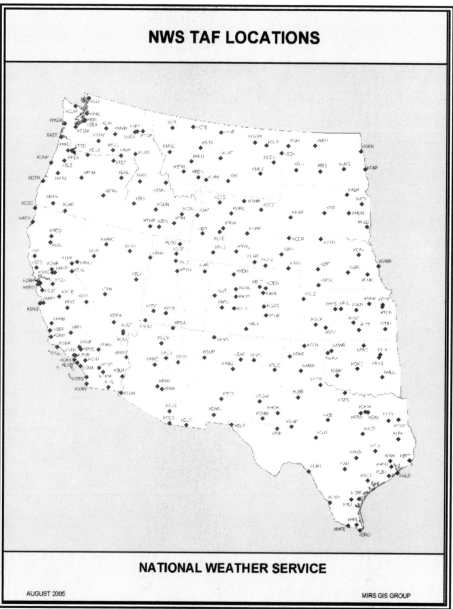

Figure 7-10. TAF Locations – Western Contiguous United States

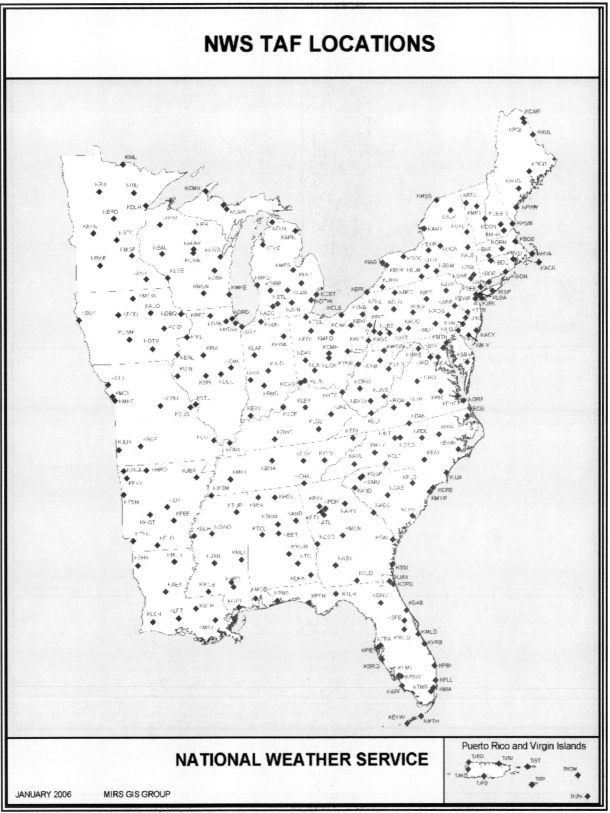

Figure 7-11. TAF Locations – Eastern Contiguous U.S., Puerto Rico and Virgin Islands

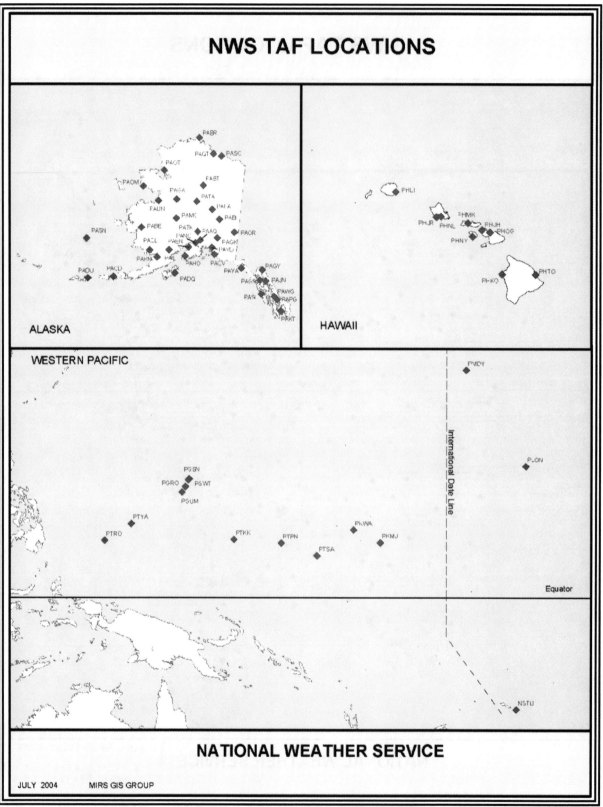

Figure 7-12. TAF Locations – Alaska, Hawaii and Western Pacific

7.2.2 Generic Format of the Forecast Text of a NWS-Prepared TAF

TAF or TAF AMD			
Type of report			

CCCC	YYGGggZ	YlY1G1G1G2G2	DddffGfmfmKT
Location identifier	Date/time of forecast origin group	Valid period	Wind group

VVVV	w'w' or NSW	NsNsNshshshs or VVhshshs or SKC	WShwshwshws/dddftKT
Visibility group	Significant weather group	Cloud and vertical obscuration groups	Non-convective low-level wind shear (LLWS) group

TTGGgg		
Forecast change indicator groups		

FMGG GGGeGe	TEMPO GGGeGe	PROB40GGGeGe
"From" group	"Temporary" group	Probability Forecast

7.2.2.1 Type of Report (TAF or TAF AMD)
The report-type header always appears as the first element in the TAF and is produced in two forms: a routine forecast, **TAF**, and an amended forecast, **TAF AMD**.

TAFs are amended whenever they become, in the forecaster's judgment, unrepresentative of existing or expected conditions, particularly regarding those elements and events significant to aircraft and airports. An amended forecast is identified by **TAF AMD** (in place of **TAF**) on the first line of the forecast text.

7.2.2.2 Location Identifier (CCCC)
After the line containing either **TAF** or **TAF AMD**, each TAF begins with its four-letter International Civil Aviation Organization (ICAO) location identifier. Figures 7-11, 7-12 and 7-13 contains the locations of NWS issued TAFs.

Examples:
KDFW – Dallas-Fort Worth
PANC – Anchorage, Alaska
PHNL – Honolulu, Hawaii

7.2.2.3 Date/Time of Forecast Origin Group (YYGGggZ)
The date/time of forecast origin group (**YYGGggZ**) follows the terminal's location identifier. It contains the day of the month in two (2) digits (**YY**) and time in four (4) digits (**GGgg** in hours

and minutes) the forecast is completed and ready for transmission, with a **Z** appended to denote UTC. This time is entered by the forecaster.

Examples
061737Z
The TAF was issued on the 6th day of the month at 1737 UTC.

121123Z
The TAF was issued on the 12th day of the month at 1123 UTC.

7.2.2.4 Valid Period (Y1Y1G1G1G2G2)

The TAF valid period (**Y1Y1G1G1G2G2**) is the next group. Scheduled 24-hour TAFs are issued four (4) times per day, at 0000, 0600, 1200, and 1800Z. The first two digits (**Y1Y1**) are the day of the month for the start of the TAF. The next two digits (**G1G1**) are the starting hour, and the last two digits (**G2G2**) are the ending hour of the valid period. A forecast period beginning at midnight UTC is annotated as **00**. If the end time of a valid period is at midnight UTC, it is annotated as **24**. For example, a 00Z TAF issued on the 9th of the month would have a valid period of **090024**.

Examples:
151212
The TAF is valid from 1200 UTC on the 15th of the month until 1200 UTC on the 16th.

230606
The TAF is valid from 0600 UTC on the 23rd of the month until 0600 UTC on the 24th of the month.

011818
The TAF is valid from 1800 UTC on the 1st of the month until 1800 UTC on the 2nd of the month.

060024
The TAF is valid from 0000 UTC on the 6th of the month until 0000 UTC on the 7th of the month.

7.2.2.5 Valid Period of Amended TAFs

An amended TAF (**TAF AMD**) covers all of the remaining valid period of the original scheduled forecast. Expired portions of the amended forecast or references to weather occurring before the issuance time are omitted from the amendment.

In an amended forecast, the date and time of the forecast origin group (**YYGGggZ**) reflects the time the amended forecast was prepared. In the forecast valid period group (**Y1Y1G1G1G2G2**), the first four digits (**Y1Y1G1G1**) reflect the UTC date and time of the beginning of the valid period of the amended TAF.

With an issuance time (**YYGGggZ**) in the first half hour of any given hour (:00 to :29), the current hour (based on UTC) is used to denote the beginning valid time. For example, an amended TAF issued at 1416Z would be valid from 1400 UTC until the standard ending time of the TAF. For the second half of any given hour (:30 to :59), the next hour (based on UTC) is used for the beginning valid time. For example, an amended TAF issued at 1639Z would be valid from 1700 UTC until the standard ending time of the TAF.

Example:

Original	Amended
TAF	**TAF AMD**
PAEN 030540Z 030606...	**PAEN 031012Z 031006...**

The scheduled forecast was sent, and 4 ½ hours later, the forecaster prepared an amendment to the forecast, at 1012Z on the 3rd day of the month.

7.2.2.6 Wind Group (dddffGfmfmKT)
The initial time period and any subsequent **FM** groups (Section 7.2.2.12.1) begin with a mean surface wind forecast (**dddffGfmfmKT**) for the period. Wind forecasts are expressed as the mean three-digit direction (**ddd** - relative to true north) rounded to the nearest ten degrees and the mean wind speed in knots (**ff**) for the time period. If wind gusts are forecast (gusts are defined as rapid fluctuations in wind speeds with a variation of 10 knots or more between peaks and lulls), they are indicated immediately after the mean wind speed by the letter **G**, followed by the peak gust speed expected. **KT** is appended to the end of the wind forecast group. Any wind speed of 100 knots or more will be encoded in three digits. Calm winds are encoded as **00000KT**.

The prevailing wind direction is forecast for any speed greater than or equal to seven (7) knots. When the prevailing surface wind direction is variable (variations in wind direction of 30 degrees or more), the forecast wind direction is encoded as **VRBffKT**. Two conditions where this can occur are very light winds and convective activity. Variable wind direction for very light winds must have a wind speed of one (1) through six (6) knots inclusive. For convective activity, the wind group may be encoded as **VRBffGfmfmKT**, where **Gfmfm** is the maximum expected wind gusts. **VRB** is not used in the non-convective LLWS group.

Squalls are forecast in the wind group as gusts (**G**), but must be identified in the significant weather group (Section 7.2.2.8) with the code **SQ**.

Examples:

23010KT
Wind from 230 degrees "true" (southwest) at 10 knots.

28020G35KT
Wind from 280 degrees "true" (west) at 20 knots gusting to 35 knots.

VRB05KT
Wind variable at 5 knots. This example depicts a forecast for light winds that are expected to variable in direction.

VRB15G30KT
Wind variable at 15 knots gusting to 30 knots. This example depicts winds that are forecast to be variable with convective activity.

00000KT
Wind calm

090105KT
Wind from 90 degrees at 105 knots

7.2.2.7 Visibility Group (VVVV)

The initial time period and any subsequent FM groups (Section 7.2.2.12.1) include a visibility forecast (**VVVV**) in statute miles appended by the contraction SM.

When the prevailing visibility is forecast to be less than or equal to six (6) SM, one or more significant weather groups (Section 7.2.2.8) are included in the TAF. However, drifting dust (**DRDU**), drifting sand (**DRSA**), drifting snow (**DRSN**), shallow fog (**MIFG**), partial fog (**PRFG**), and patchy fog (**BCFG**) may be forecast with prevailing visibility greater than or equal to seven (7) statute miles.

When a whole number and a fraction are used to forecast visibility, a space is included between them (e.g., **1 1/2SM**). Visibility greater than six (6) statute miles is encoded as **P6SM**.

If the visibility is not expected to be the same in different directions, prevailing visibility is used.

When volcanic ash (**VA**) is forecast in the significant weather group, visibility is included in the forecast, even if it is unrestricted (**P6SM**). For example, an expected reduction of visibility to 10 statute miles by volcanic ash is encoded in the forecast as **P6SM VA**.

Examples

P6SM
Visibility unrestricted

1 1/2SM
Visibility 1 and ½ statute miles.

4SM
Visibility 4 statute miles.

7.2.2.8 Significant Weather Group (w'w' or NSW)

The significant weather group (**w'w'** or **NSW**) consists of the appropriate qualifier(s) and weather phenomenon contraction(s) (Section 2) or **NSW** (No significant weather).

If the initial forecast period and subsequent **FM** groups (Section 7.2.2.12.1) are not forecast to have explicit significant weather, the significant weather group is omitted. **NSW** is **not** used in the initial forecast time period or **FM** groups.

Tornadic activity (tornadoes, waterspouts, and funnel clouds) are not forecast in terminal forecasts because the probability of occurrence at a specific site is extremely small.
One or more significant weather group(s) is (are) required when the visibility is forecast to be 6SM or less. The exceptions are: volcanic ash (**VA**), low drifting dust (**DRDU**), low drifting sand (**DRSA**), low drifting snow (**DRSN**), shallow fog (**MIFG**), partial fog (**PRFG**), and patchy fog (**BCFG**). Obstructions to vision are only forecast when the prevailing visibility is less than 7 statute miles or, in the opinion of the forecaster, is considered operationally significant.

Volcanic ash (**VA**) is always forecast when expected. When **VA** is included in the significant weather group, visibility is included in the forecast as well, even if the visibility is unrestricted (**P6SM**).

NSW is used in place of significant weather only in a **TEMPO** group (Section 7.2.2.12.2) to indicate when significant weather (including in the vicinity (**VC**), see below) included in a previous sub-divided group is expected to end.

Multiple precipitation elements are encoded in a single group (e.g., **-TSRASN**). If more than one type of precipitation is forecast, up to three appropriate precipitation contractions can be combined in a single group (with no spaces) with the predominant type of precipitation being first. In this single group, the intensity refers to the total precipitation and can be used with either one or no intensity qualifier, as appropriate. In TAFs, the intensity qualifiers (light, moderate, and heavy) (Section 2.1.3.8.1) refer to the intensity of the precipitation and not to the intensity of any thunderstorms associated with the precipitation.

Intensity is coded with precipitation types, except ice crystals and hail, including those associated with thunderstorms and those of a showery nature (**SH**). No intensity is ascribed to blowing dust (**BLDU**), blowing sand (**BLSA**), or blowing snow (**BLSN**). Only moderate or heavy intensity is ascribed to sandstorm (**SS**) and duststorm (**DS**).

7.2.2.8.1 Exception for Encoding Multiple Precipitation Types
When more than one type of precipitation is forecast in a time period, any precipitation type associated with a descriptor (e.g., **FZRA**) (Section 2.1.3.8.3) is encoded first in the precipitation group, regardless of the predominance or intensity of the other precipitation types. Descriptors are not encoded with the second or third precipitation type in the group. The intensity is associated with the first precipitation type of a multiple precipitation type group. For example, a forecast of moderate snow and light freezing rain is coded as **-FZRASN** although the intensity of the snow is greater than the freezing rain.

Examples:

Combinations of one precipitation and one non-precipitation weather phenomena:

`-DZ FG`
Light drizzle and fog (obstruction which reduces visibility to less than 5/8 SM – See Section 7.2.2.8.3)

`RA BR`
Moderate rain and mist (obstruction which reduces visibility to less than 7 SM but greater than or equal to 5/8 SM – See Section 7.2.2.8.3)

`-SHRA FG`
Light rain showers and fog (visibility less than 5/8 statute miles)

`+SN FG`
Heavy snow and fog

Combinations of more than one type of precipitation:

`-RASN FG HZ`
Light rain and snow (light rain predominant), fog and haze

`TSSNRA`
Thunderstorm with moderate snow and rain (moderate snow predominant)

FZRASNPL
Moderate freezing rain, snow, and ice pellets (freezing rain mentioned first due to the descriptor, followed by other precipitation types in order of predominance)

SHSNPL
Moderate snow showers and ice pellets

7.2.2.8.2 Thunderstorm Descriptor
The TS descriptor is treated differently than other descriptors in the following cases:

- When non-precipitating thunderstorms are forecast, TS may be encoded as the sole significant weather phenomenon; and

- When forecasting thunderstorms with freezing precipitation (**FZRA** or **FZDZ**), the **TS** descriptor is included first, followed by the intensity and weather phenomena.

Example:

TS -FZRA
When a thunderstorm is included in the significant weather group (even using vicinity - **VCTS**), the cloud group (**NsNsNshshshs**) includes a forecast cloud type of CB. See the following example for encoding **VCTS**.

Example

-FZRA VCTS BKN010CB

7.2.2.8.3 Fog Forecast
A visibility threshold must be met before a forecast for fog (FG) is included in the TAF. When forecasting a fog-restricted visibility from 5/8SM to 6SM, the phenomena is coded as **BR** (mist). When a fog-restricted visibility is forecast to result in a visibility of less than 5/8SM, the code **FG** is used. The forecaster never encodes weather obstruction as mist (**BR**) when the forecast visibility is greater than 6 statute miles (P6SM).

The following fog-related terms are used as described below:

Table 7-7. TAF Fog Terms

TERM	DESCRIPTION
Freezing Fog (**FZFG**)	Any fog (visibility less than 5/8 SM) consisting predominantly of water droplets at temperatures less than or equal to 32° F/0°C, whether or not rime ice is expected to be deposited. **FZBR** is not a valid significant weather combination and will not be used in TAFs.
Shallow Fog (**MIFG**)	The visibility at 6 feet above ground level is greater than or equal to 5/8 SM and the apparent visibility in the fog layer is less than 5/8 SM.
Patchy Fog (**BCFG**)	Fog patches covering part of the airport. The apparent visibility in the fog patch or bank is less than 5/8 SM, with the foggy patches extending to at least 6 feet above ground level.
Partial Fog (**PRFG**)	A substantial part of the airport is expected to be covered by fog while the remainder is expected to be clear of fog (e.g., a fog bank). NOTE: **MIFG**, **PRFG** and **BCFG** may be forecast with prevailing visibility of P6SM.

Examples:

1/2SM FG
Fog is reducing visibilities to less than 5/8SM, therefore FG is used to encode the fog.

3SM BR
Fog is reducing visibilities to between 5/8 and 6SM, therefore BR is used to encode the fog.

7.2.2.9 Vicinity (VC)

In the United States, vicinity (**VC**) is defined as a donut-shaped area between 5 and 10SM from the center of the airport's runway complex. The FAA requires TAFs to include certain meteorological phenomena which may directly affect flight operations to and from the airport. Therefore, NWS TAFs may include a prevailing condition forecast of fog, showers and thunderstorms in the airport's vicinity. A prevailing condition is defined as a greater than or equal to 50% probability of occurrence for more than ½ of the sub-divided forecast time period. **VC** is not included in **TEMPO** or **PROB** groups.

The significant weather phenomena in Table 7-8 are valid for use in prevailing portions of NWS TAFs in combination with **VC**:

Table 7-8: TAF Use of Vicinity (VC)

Phenomenon	Coded
Fog*	VCFG
Shower(s)**	VCSH
Thunderstorm	VCTS

* Always coded as **VCFG** regardless of visibility in the obstruction, and without qualification as to intensity or type (frozen or liquid)

** The **VC** group, if used, should be the last entry in any significant weather group (**w'w'**).

7.2.2.10 Cloud and Vertical Obscuration Groups

The initial time period and any subsequent **FM** groups include a cloud or obscuration group (**NsNsNshshshs**, **VVhshshs** or **SKC**), used as appropriate to indicate the cumulative amount (**NsNsNs**) of all cloud layers in ascending order and height (**hshshs**), to indicate vertical visibility (**VVhshshs**) into a surface-based obstructing medium, or to indicate a clear sky (**SKC**). All cloud layers and obscurations are considered opaque

7.2.2.10.1 Cloud Group

The cloud group (**NsNsNshshshs**) is used to forecast cloud amount in Table 7-8.

Table 7-9. TAF Sky Cover

SKY COVER CONTRACTION	SKY COVERAGE
SKC	0 oktas
FEW	0 to 2 oktas
SCT	3 to 4 oktas
BKN	5 to 7 oktas
OVC	8 oktas

When zero (0) oktas of sky coverage is forecast, the cloud group is replaced by **SKC**. The contraction **CLR**, which is used in the METAR code, is not used in TAFs. TAFs for sites with ASOS/AWOS contain the cloud amount and/or obscurations which the forecaster expects, not what is expected to be reported by an ASOS/AWOS.

Heights of clouds (**hshshs**) are forecast in hundreds of feet AGL.

The lowest level at which the cumulative cloud cover equals 5/8 or more of the celestial dome is understood to be the forecast ceiling (Section 2.1.3.9). For example, **VV008**, **BKN008** or **OVC008** all indicate an 800 ft ceiling.

7.2.2.10.2 Vertical Obscuration Group

The vertical obscuration group (**VVhshshs**) is used to forecast, in hundreds of feet AGL, the vertical visibility (**VV**) into a surface-based total obscuration (Section 2.1.3.9). **VVhshshs** is this ceiling at the height indicated in the forecast. TAFs do not include forecasts of partial obscurations (i.e., **FEW000**, **SCT000**, or **BKN000**).

Example:

```
1SM BR VV008
```
Ceiling is 800 feet due to vertical visibility into fog

7.2.2.10.3 Cloud Type

The only cloud type included in the TAF is **CB**. **CB** follows cloud or obscuration height (**hshshs**) without a space whenever thunderstorms are included in significant weather group (**w'w'**), even if thunderstorms are only forecast in the vicinity (**VCTS**). **CB** can be included in the cloud group (**NsNsNshshshs)** or the vertical obscuration group (**VVhshshs**) without mentioning thunderstorm in the significant weather group (**w'w'**). Therefore, situations may occur where nearly identical **NsNsNshshshs** or **VVhshshs** appear in consecutive time periods, with the only change being the addition or elimination of **CB** in the forecast cloud type.

Examples:

```
1/2SM TSRA OVC010CB
```
Thunderstorms are forecast at the airport

7.2.2.11 Non-Convective Low-Level Wind Shear (LLWS) Group

Wind Shear (**WS**) is defined as a rapid change in horizontal wind speed and/or direction, with distance and/or a change in vertical wind speed and/or direction with height. A sufficient difference in wind speed, wind direction, or both, can severely impact airplanes, especially within 2,000 feet AGL because of limited vertical airspace for recovery.

Forecasts of LLWS in the TAF refer only to non-convective LLWS from the surface up to and including 2,000 feet AGL. LLWS is always assumed to be present in convective activity. LLWS is included in TAFs on an "as-needed" basis to focus the aircrew's attention on LLWS problems which currently exist or are expected. Non-convective LLWS may be associated with the following: frontal passage, inversion, low-level jet, lee side mountain effect, sea breeze front, Santa Ana winds, etc.

When LLWS conditions are expected, the non-convective LLWS code **WS** is included in the TAF as the last group (after cloud forecast). Once in the TAF, the **WS** group remains the prevailing condition until the next **FM** change group or the end of the TAF valid period if there are no subsequent **FM** groups. Forecasts of non-convective LLWS are not included in **TEMPO** or **PROB** groups.

The format of the non-convective low-level wind shear group is:

```
WShwshwshws/dddffKT
```

WS – Indicator for non-convective LLWS
hwshwshws – Height of the top of the WS layer in hundreds of feet AGL
ddd – True direction in ten degree increments at the indicated height
-- **VRB** is not used for direction in the non-convective LLWS forecast group.
ff – Speed in knots of the forecast wind at the indicated height
KT – Unit indicator for wind

Example:

```
TAF...13012KT...WS020/27055KT
```
Wind shear from the surface to 2,000 feet. Surface winds from 130 (southeast) at 12 knots changes to 270 (west) at 55 knots at 2,000 feet.

In this example the indicator **WS** is followed by a three-digit number which is the top of the wind shear layer. LLWS is forecast to be present from the surface to this level. After the solidus /, the five digit wind group is the wind direction and speed at the top of the wind shear layer. It is not a value for the amount of shear.

A non-convective LLWS forecast is included in the initial time period or a **FM** group in a TAF whenever:

- One or more PIREPs are received of non-convective LLWS within 2,000 feet of the surface, at or in the vicinity of the TAF airport, causing an indicated air speed loss or gain of 20 knots or more, and the forecaster determines the report(s) reflect a valid non-convective LLWS event rather than mechanical turbulence, or

- When non-convective vertical **WS** of 10 knots or more per 100 feet in a layer more than 200 feet thick are expected or reliably reported within 2,000 feet of the surface at, or in the vicinity of, the airport.

7.2.2.12 Forecast Change Indicator Groups

Forecast change indicator groups are contractions which are used to sub-divide the forecast period (24-hours for scheduled TAFs; less for amended or delayed forecasts) according to significant changes in the weather.

The forecast change indicators, FM, TEMPO, and PROB, are used when a change in any or all of the elements forecast is expected:

7.2.2.12.1 From (FM) Group (FMGGgg)

The change group **FMGGgg** (voiced as "from") is used to indicate when prevailing conditions are expected to change significantly over a period of less than one hour. In these instances, the forecast is sub-divided into time periods using the contraction **FM**, followed, without a space, by four digits indicating the time (in hours and minutes Z) the change is expected to occur. While the use of a four-digit time in whole hours (e.g. 2100Z) is acceptable, if a forecaster can predict changes and/or events with higher resolution, then more precise timing of the change to the minute will be indicated. All forecast elements following **FMGGgg** relate to the period of time from the indicated time (**GGgg**) to the end of the valid period of the terminal forecast, or to the next **FM** if the terminal forecast valid period is divided into additional periods.

The **FM** group will be followed by a complete description of the weather (i.e., self-contained) and all forecast conditions given before the **FM** group are superseded by those following the group. All elements of the TAF (surface wind, visibility, significant weather, clouds, obscurations, and when expected, non-convective LLWS) will be included in each **FM** group, regardless if they are forecast to change or not. For example, if forecast cloud and visibility changes warrant a new **FM** group but the wind does not, the new **FM** group will include a wind forecast, even if it is the same as the most recently forecast wind.

The only exception to this involves the significant weather group. If no significant weather is expected in the **FM** time period group, then significant weather group is omitted. A TAF may include one or more **FM** groups, depending on the prevailing weather conditions expected. In the interest of clarity, each **FM** group starts on a new line of forecast text, indented five spaces.

Examples:

```
TAF
KDSM 022336Z 030024 20015KT P6SM BKN015
     FM0230 29020G35KT 1SM +SHRA OVC005
      TEMPO 0304 30030G45KT 3/4SM -SHSN
     FM0500 31010G20KT P6SM SCT025...
```

A change in the prevailing weather is expected at **0230** UTC and **0500** UTC.

```
TAF
KAPN 312330Z 010024 13008KT P6SM SCT030
```

```
FM0320 31010KT 3SM -SHSN BKN015
FM0500 31010KT 1/4SM +SHSN VV007...
```

Note the wind in the **FM0500** group is the same as the previous **FM** group, but is repeated since all elements are required to be included in a **FM** group.

7.2.2.12.2 TEMPO GGGeGe

The change-indicator group **TEMPO GGGeGe** is used to indicate temporary fluctuations to forecast meteorological conditions which are expected to:

- Have a high percentage (greater than 50%) probability of occurrence,
- Last for one hour or less in each instance and,
- In the aggregate, cover less than half of the period **GG** to **GeGe**

Temporary changes described by **TEMPO** groups occur during a period of time defined by a two-digit beginning and two-digit ending time, both in whole hours UTC.

Each **TEMPO** group is placed on a new line in the TAF. The **TEMPO** identifier is followed by a description of all the elements in which a temporary change is forecast. A previously forecast element which has not changed during the **TEMPO** period is understood to remain the same and will not be included in the **TEMPO** group. Only those weather elements forecast to temporarily change are required to be included in the **TEMPO** group.

TEMPO groups will not include forecasts of either significant weather in the vicinity (**VC**) or non-convective LLWS.

Examples:

```
TAF
KDDC 221130Z 221212 29010G25KT P6SM SCT025
        TEMPO 1517 30025G35KT 1 1/2SM SHRA BKN010...
```

In the example, all forecast elements in the TEMPO group are expected to be different than the prevailing conditions.

```
TAF
KSEA 091125Z 091212 19008KT P6SM SCT010 BKN020 OVC090
        TEMPO 1215 -RA SCT010 BKN015 OVC040...
```

In this example the visibility is **not** forecast in the TEMPO group. Therefore, the visibility is expected to remain the same (P6SM) as forecast in the prevailing conditions group. Also, note that in the TEMPO 1215 group, all three cloud layers are included, although the lowest layer is not forecast to change from the initial time period.

7.2.2.12.3 PROB30 GGGeGe

The probability group, **PROB30 GGGeGe,** is only used by NWS forecasters to forecast a low probability occurrence (30% chance) of a thunderstorm or precipitation event and its associated weather and obscuration elements (wind, visibility and/or sky condition) at an airport.

The **PROB30** group is the forecaster's assessment of probability of occurrence of the weather event which follows it. **PROB30** is followed by a space, then four digits (**GGGeGe**) stating the beginning and ending time (in hours) of the expected condition. **PROB30** is the only **PROB** group used in NWS TAFs.

NOTE: U.S. military and international TAFs may use the PROB40 (40% chance) group as well.

The **PROB30** group is located within the same line of the prevailing condition group, continuing on the line below if necessary.

The **PROB30** group is not used in the first nine (9) hours of the TAF's valid period, including amendments. **PROB30** groups are six (6) hours or less in length. Only one **PROB30** group is used following any subsequent **FM** groups.

PROB30 groups do not include forecasts of significant weather in the vicinity (**VC**) or non-convective LLWS.

Example:
```
FM2100 18015KT P6SM SCT050 PROB30 2301 2SM TSRA OVC020CB
```

7.2.2.13 TAF Examples

```
TAF
KPIR 111140Z 111212 13012KT P6SM BKN100 WS020/35035KT
     TEMPO 1214 5SM BR
     FM1500 16015G25KT P6SM SCT040 BKN250
     FM0000 14012KT P6SM BKN080 OVC150 PROB30 0004 3SM TSRA BKN030CB
     FM0400 14008KT P6SM SCT040 OVC080 TEMPO 0408 3SM TSRA OVC030CB
```

TAF	▶ Terminal Aerodrome Forecast
KPIR	▶ Pierre, South Dakota
111140	▶ prepared on the 11th at 1140 UTC
111212	▶ valid from the 11th at 1200 UTC until the 12th at 1200 UTC
13012KT	▶ wind 130 at 12 knots
P6SM	▶ visibility greater than 6 statute miles
BKN100	▶ ceiling 10,000 broken
WS020/35035KT	▶ wind shear at 2,000 feet, wind from 350 at 35 knots
TEMPO 1214	▶ temporary conditions between 1200 UTC and 1400 UTC
5SM	▶ visibility 5 statute miles
BR	▶ mist
FM1500	▶ from 1500 UTC
16015G25KT	▶ wind 160 at 15 knots gusting to 25 knots
P6SM	▶ visibility greater than 6 statute miles
SCT040 BKN250	▶ 4,000 scattered, ceiling 25,000 broken
FM0000	▶ from 0000Z
14012KT	▶ wind 140 at 12 knots
P6SM	▶ visibility greater than 6 statute miles
BKN080 OVC150	▶ ceiling 8,000 broken, 15,000 overcast
PROB30 0004	▶ 30% probability between 0000 UTC and 0400 UTC
3SM	▶ visibility 3 statute miles
TSRA	▶ thunderstorm with moderate rain showers
BKN030CB	▶ ceiling 3,000 broken with cumulonimbus
FM0400	▶ from 0400 UTC
14008KT	▶ wind 140 at 8 knots
P6SM	▶ visibility greater than 6 statute miles
SCT040 OVC080	▶ 4,000 scattered, ceiling 8,000 overcast
TEMPO 0408	▶ temporary conditions between 0400 UTC and 0800 UTC
3SM	▶ visibility 3 statute miles
TSRA	▶ thunderstorms with moderate rain showers
OVC030CB	▶ ceiling 3,000 overcast with cumulonimbus

```
TAF AMD
KEYW 131555Z 131612 VRB03KT P6SM VCTS SCT025CB BKN250
     TEMPO 1618 2SM TSRA BKN020CB
     FM1800 VRB03KT P6SM SCT025 BKN250 TEMPO 2024 1SM TSRA OVC010CB
     FM0000 VRB03KT P6SM VCTS SCT020CB BKN120 TEMPO 0812 BKN020CB
```

TAF AMD	▶ Amended Terminal Aerodrome Forecast
KEYW	▶ Key West, Florida
131555Z	▶ prepared on the 13th at 1555 UTC
131612	▶ valid from the 13th at 1600 UTC until the 14th at 1200 UTC
VRB03KT	▶ wind variable at 3 knots
P6SM	▶ visibility greater than 6 statute miles
VCTS	▶ thunderstorms in the vicinity
SCT025CB BKN250	▶ 2,500 scattered with cumulonimbus, ceiling 25,000 broken
TEMPO 1618	▶ temporary conditions between 1600 UTC and 1800 UTC
2SM	▶ visibility 2 statute miles
TSRA	▶ thunderstorms with moderate rain showers
BKN020CB	▶ ceiling 2,000 broken with cumulonimbus
FM1800	▶ from 1800 UTC
VRB03KT	▶ wind variable at 3 knots
P6SM	▶ visibility greater than 6 statute miles
SCT025 BKN250	▶ 2,500 scattered, ceiling 25,000 broken
TEMPO 2024	▶ temporary conditions between 2000 UTC and 0000 UTC
1SM	▶ visibility 1 statute mile
TSRA	▶ thunderstorms with moderate rain showers
OVC010CB	▶ ceiling 1,000 overcast with cumulonimbus
FM0000	▶ from 0000 UTC
VRB03KT	▶ variable wind at 3 knots
P6SM	▶ visibility greater than 6 statute miles
VCTS	▶ thunderstorms in the vicinity
SCT020CB BKN120	▶ 2,000 scattered with cumulonimbus, ceiling 12,000 broken
TEMPO 0812	▶ temporary conditions between 0800 UTC and 1200 UTC
BKN020CB	▶ ceiling 2,000 broken with cumulonimbus

```
TAF
KCRP 111730Z 111818 19007KT P6SM SCT030
     TEMPO 1820 BKN040
     FM2000 16011KT P6SM VCTS FEW030CB SCT250
     FM0200 14006KT P6SM FEW025 SCT250
     FM0800 VRB03KT 5SM BR SCT012
     FM1500 17007KT P6SM SCT025
```

TAF	▶ Terminal Aerodrome Forecast
KCRP	▶ Corpus Christi, Texas
111730Z	▶ prepared on the 11th at 1730 UTC
111818	▶ valid from the 11th at 1800 UTC until the 12th at 1800 UTC
19007KT	▶ wind 190 at 7 knots
P6SM	▶ visibility greater than 6 statute miles
SCT030	▶ 3,000 scattered
TEMPO 1820	▶ temporary conditions between 1800 UTC and 2000 UTC
BKN040	▶ ceiling 4,000 broken
FM2000	▶ from 2000 UTC
16011KT	▶ wind 160 at 11 knots
P6SM	▶ visibility greater than 6 statute miles
VCTS	▶ thunderstorms in the vicinity
FEW030CB SCT250	▶ 3,000 few with cumulonimbus, 25,000 scattered
FM0200	▶ from 0200 UTC
14006KT	▶ wind 140 at 6 knots
P6SM	▶ visibility greater than 6 statute miles
FEW025 SCT250	▶ 2,500 few, 25,000 scattered
FM0800	▶ from 0800 UTC
VRB03KT	▶ wind variable at 3 knots
5SM	▶ visibility 5 statute miles
BR	▶ mist
SCT012	▶ 1,200 scattered
FM1500	▶ from 1500 UTC
17007KT	▶ wind 170 at 7 knots
P6SM	▶ visibility greater than 6 statute miles
SCT025	▶ 2,500 scattered

7.2.3 Issuance

Scheduled TAFs prepared by NWS offices are issued four times a day, every six (6) hours, according to the following schedule:

Table 7-10. TAF Issuance Schedule

SCHEDULED ISSUANCE	VALID PERIOD	ISSUANCE WINDOW
0000 UTC	0000 to 2400 UTC	2320 to 2340 UTC
0600 UTC	0600 to 0600 UTC	0520 to 0540 UTC
1200 UTC	1200 to 1200 UTC	1120 to 1140 UTC
1800 UTC	1800 to 1800 UTC	1720 to 1740 UTC

7.2.3.1 Minimum Observational Requirements for Routine TAF Issuance and a Continuation

The NWS WFO forecaster must have certain information for the preparation and scheduled issuance of each individual TAF. Observations or other complementary and/or supplementary data sources must include, at a minimum:

- Wind (speed and direction)
- Visibility
- Weather and obstructions to vision
- Sky condition
- Temperature
- Dewpoint
- Altimeter setting

All weather elements need not be provided completely and/or at all times in the hourly/special observation itself. Alternative methods of obtaining the required weather elements can be utilized, at the discretion of the forecaster, in order to continue providing TAFs. However, in the event the forecaster believes the absence of one or more observed elements will lead to a degradation of the quality of the TAF, the TAF is limited (e.g., **NIL AMD**, indicating no amendments will be provided) or suspended (**NIL**).

Once a particular TAF has been suspended (**NIL**), a delayed or scheduled TAF for that airport is not issued until two consecutive observations not less than 30 minutes nor more than about one (1) hour apart have been received to establish a trend. The forecaster may also use alternative observations, such as satellite, in addition to a single surface observation to issue a TAF.

7.2.3.2 Sites with Scheduled Part-Time Observations

For TAFs with less than 24-hour observational coverage, or for which part-time TAFs are provided, the TAF is valid to the end of the routine scheduled forecast period even if observations cease prior to that time. The time observations are scheduled to end and/or resume is indicated by expanding the **AMD NOT SKED** statement. Expanded statements will include the observation ending time (**AFT 02Z**), the scheduled observation resumption time (**TIL 12Z**) or the period of observation unavailability (**02Z-12 Z**).

7.2.3.2.1 Examples of Scheduled Part-Time Observations TAFs

`TAF AMD`
`KACV 141410Z 141412 NIL=`
The TAF is suspended until a complete data source is available

`TAF AMD`
`KRWF 150202Z 150224 AMD NOT SKED 05Z-18Z=`
No amendments will be available between 0500 UTC an 1800 UTC due to lack of a complete observational set between those times.

`TAF AMD`
`KPSP 190230Z 190324`
`NIL AMD=`
No amendments will be made to the TAF.

7.2.3.3 Automated Observing Sites Requiring Part-Time Augmentation

TAFs for AWOS-III sites which have part-time augmentation are prepared using the procedures for part-time manual observation sites detailed in the previous section, with one exception. This exception is the remark used when the automated system is unattended. Specifically, the time an augmented automated system is scheduled to go into unattended operation and/or the time augmentation resumes is included in a remark unique to automated observing sites: **AMD LTD TO CLD VIS AND WIND** (**AFT aaZ**, or **TIL bbZ**, or **aaZ-bbZ**), where **aaZ** is the time of the last augmented observation and **bbZ** is the time the second complete observation is expected to be received. This remark, which does not preclude amendments for other forecast elements, is appended to the last scheduled TAF issued prior to the last augmented observation. It will also be appended to all subsequent amendments until augmentation resumes.

The **AMD LTD TO** (elements specified) remark is a flag for users and differs from the **AMD NOT SKED AFT Z** remark for part-time manual observation sites. **AMD LTD TO** (elements specified) means users should expect amendments only for those elements and the times specified.

Example:

```
TAF AMD
KCOE 150202Z 150224 text
AMD LTD TO CLD VIS AND WIND 05Z-18Z=
```

The amended forecast indicates that between 0500 and 1800Z amendments will only be issued for wind, visibility and clouds.

An amendment includes forecasts for all appropriate TAF elements, even those not reported when the automated site is not augmented. If unreported elements are judged crucial to the TAF and cannot be adequately determined (e.g., fog versus moderate snow), the TAF will be suspended (i.e. an amended TAF stating **NIL** may be issued). AWOS-III systems with part-time augmentation, which the forecaster suspects are providing unreliable information when not augmented, is reported for maintenance and treated the same as part-time manual observation sites. In such cases, the **AMD NOT SKED AFT Z** remark will be used.

7.2.3.4 Non-Augmented Automated Observing Sites

The TAF issued for a non-augmented ASOS site may be suspended in the event the forecaster is notified of, or strongly suspects, an outage or unrepresentative data. The term **NIL AMD** is appended to the end of an amendment to the existing TAF when appropriate. If the outage occurs within one (1) hour of the next scheduled issuance or if the forecaster believes the existing TAF is unrepresentative of conditions, an amendment or scheduled issuance containing only the statement **NIL** may be issued.

7.3 International Aviation Route Forecasts (ROFOR)

International ROFORs are prepared and issued several hours in advance of regularly scheduled flights. The only NWS office which routinely issues ROFORs is the Weather Forecast Office (WFO) in Honolulu in its capacity as a Meteorological Watch Office (MWO) for ICAO, for routes within its area of responsibility that are underserved by conventional aviation forecasts and products.

7.3.1 ROFOR Criteria
WFO Honolulu honors all ROFOR requests for flights beginning, ending, or having most of the flight path within its area of responsibility within the Pacific Region which is generally the Oakland Oceanic FIR south of 30N and west of 140W.

7.3.2 Issuance
ROFORs are issued for prescribed times, several hours in advance, for regularly scheduled flights. ROFOR requests for unscheduled flights are prepared as soon as time permits.

7.3.2.1 ROFOR Amendments
ROFORs are not amended.

7.3.2.2 ROFOR Corrections
ROFOR corrections are issued as soon as possible when erroneous data has been identified as being transmitted.

7.3.3 ROFOR Content
ROFORs contain some or all of the following forecast parameters:
> a. Winds and temperatures aloft
> b. Significant en-route weather
> c. Zone weather
> d. Weather Synopsis.

At a minimum, ROFORs include a. and b. above. They may contain data for multiple altitudes and include TAFs for destination points and/or alternates.

The core of a ROFOR is formatted as follows: **0iQLL 4hhhTT ddFFF**

Where   **i** = 1 for zone up to latitude **L**
                i = 2 for zone up to longitude **LL**

                Q = 1 east of the dateline in the northern hemisphere
                Q = 2 west of the dateline in the northern hemisphere
                Q = 6 east of the dateline in the southern hemisphere
                Q = 7 west of the dateline in the southern hemisphere

                hhh = height to which the temperature and wind refer
                TT = air temperature in whole degrees Celsius at hhh
                dd = true direction in tens of degrees from which the wind will blow at hhh
                fff = wind speed in know at hhh

01104 4300M31 10010

Decoded as: The 30,000 foot wind (10010) and temperature (M31) are for that zone along the flight path from the equator to 05N east of the dateline.

7.3.4 ROFOR Example

Tarawa to Majuro Route

```
FROC33 PHFO 291510  (ICAO Communication Header)

FOR PKMJYMYX
ROFOR VALID 2008 FOR ROUTE NGTA TO PKMJ
01205 4100P08 06010 4140P00 06015 4180M03 07020
01201 4100P08 09015 4140P00 09020 4180M04 10025
SIGWX...ISOL TCU/VIS 5SM SHRA
PKMJ 221120Z 221212 NIL=
```

7.4 Wind and Temperature Aloft Forecast (FB)

Wind and Temperature Aloft Forecasts (FB) are computer prepared forecasts of wind direction, wind speed, and temperature at specified times, altitudes, and locations. Forecasts are based on the North American Mesoscale (NAM) forecast model run. FBs are available on the Aviation Weather Center (AWC) web site at: http://aviationweather.gov/products/nws/winds/

7.4.1 Forecast Altitudes
The following table contains the altitudes for which winds are forecast. Altitudes up to 15,000 feet are referenced to Mean Sea Level (MSL). Altitudes at or above 18,000 feet are references to flight levels (FL).

Table 7-11. Wind and Temperature Aloft Forecast Levels

Actual Altitudes (MSL)
1,000 feet*
1,500 feet*
2,000 feet*
3,000 feet
6,000 feet
9,000 feet
12,000 feet
15,000 feet*

Pressure Altitudes (Hectopascals)
18,000 feet (500 Hectopascals)
24,000 feet (400 Hectopascals)
30,000 feet (300 Hectopascals)
34,000 feet (250 Hectopascals)
39,000 feet (200 Hectopascals)
45,000 feet (150 Hectopascals)#
53,000 feet (100 Hectopascals)#

* Hawaii and Western Pacific only.

Not available for selected locations in the Contiguous US.

Wind forecasts are not issued for altitudes within 1,500 feet of a location's elevation. Temperature forecasts are not issued for altitudes within 2,500 feet of a location's elevation. Forecasts for intermediate levels are determined by interpolation.

7.4.2 Format
The symbolic form of the forecasts is **DDff+TT** in which **DD** is the wind direction, **ff** the wind speed, and **TT** the temperature.

Wind direction is indicated in tens of degrees (two digits) with reference to true north and wind speed is given in knots (two digits). Light and variable wind or wind speeds of less than 5 knots are expressed by **9900**. Forecast wind speeds of 100 through 199 knots are indicated by subtracting 100 from the speed and adding 50 to the coded direction. For example, a forecast of 250 degrees, 145 knots, is encoded as **7545**. Forecast wind speeds of 200 knots or greater are indicated as a forecast speed of 199 knots. For example, **7799** is decoded as 270 degrees at 199 knots or greater.

Temperature is indicated in degrees Celsius (two digits) and is preceded by the appropriate algebraic sign for the levels from 6,000 through 24,000 feet. Above 24,000 feet, the sign is omitted since temperatures are always negative at those altitudes.

The product header includes the date and time observations were collected, the forecast valid date and time, and the time period during which the forecast is to be used.

Examples

1312+05
The wind direction is from 130 degree (i.e. - southeast), the wind speed is 12 knots and the temperature is 5 degrees Celsius.

9900+10
Wind light and variable, temperature +10 degrees.

7735-07
The wind direction is from 270 degrees (i.e. west), the wind speed is 135 knots and the temperature is minus 7 degrees Celsius.

7.4.2.1 Coding Example
Sample winds aloft text message:

```
DATA BASED ON 010000Z
VALID 010600Z    FOR USE 0500-0900Z. TEMPS NEG ABV 24000
FT   3000    6000    9000    12000    18000    24000    30000    34000    39000
MKC 9900 1709+06 2018+00 2130-06 2242-18 2361-30 247242 258848 550252
```

Sample message decoded:

(Line 1) **DATA BASED ON 010000Z**

Forecast data is based on computer forecasts generated the first day of the month at 0000 UTC.

(Line 2) **VALID 010600Z   FOR USE 0500-0900Z. TEMPS NEG ABV 24000**

The valid time of the forecast is the 1st day of the month at 0600 UTC. The forecast winds and temperature are to be used between 0500 and 0900 UTC. Temperatures are negative above 24,000 feet.

(Line 3)
```
FT   3000    6000    9000    12000    18000    24000    30000    34000    39000
```

FT indicates the altitude of the forecast.
(Line4)
```
MKC 9900 1709+06 2018+00 2130-06 2242-18 2361-30 247242 258848 550252
```

MKC indicates the location of the forecast. The rest of the data is the winds and temperature aloft forecast for the respective altitudes.

The following table shows data for MKC (Kansas City, MO).

Table 7-12. Wind and Temperature Aloft Forecast Decoding Examples

```
FT 3000  6000   9000   12000  18000  24000  30000 34000 39000
MKC 9900 1709+06 2018+00 2130-06 2242-18 2361-30 247242 258848 550252
```

Altitude (feet)	Coded	Wind	Temperature (°C)
3,000 FT	9900	Light and variable	Not forecast
6,000 FT	1709+06	170 degrees at 9 knots	+06 degrees Celsius
9,000 FT	2018+00	200 degrees at 18 knots	Zero degrees Celsius
12,000 FT	2130-06	210 degrees at 30 knots	-06 degrees Celsius
18,000 FT	2242-18	220 degrees at 42 knots	-18 degrees Celsius
24,000 FT	2361-30	230 degrees at 61 knots	-30 degrees Celsius
30,000 FT	247242	240 degrees at 72 knots	-42 degrees Celsius
34,000 FT	258848	250 degrees at 88 knots	-48 degrees Celsius
39,000 FT	750252	250 degrees at 102 knots	-52 degrees Celsius

7.4.2.2 Example for the Contiguous US and Alaska

```
DATA BASED ON 091200Z
VALID 091800Z    FOR USE 1400-2100Z. TEMPS NEG ABV 24000

FT   3000    6000    9000    12000   18000   24000  30000  34000  39000
ABI        1931+10 1929+10 2024+06 2331-10 2448-23 235239 246348 256056
ABQ                2213+03 2327-04 2253-17 2263-27 227242 236946 245749
ABR  2017 2312+14 2308+09 2615+02 2724-13 2527-26 273641 274051 274562
AGC (etc.)


FT   45000   53000
ABI 301049 281149
ABQ 235061 244859
ABR 224559 243756
AGC (etc.)
```

Note: 45,000- and 53,000-foot winds are not available for selected locations in the conterminous US.

7.4.2.3 Example for Hawaii and the Western Pacific

```
DATA BASED ON 091200Z
VALID 091800Z    FOR USE 1400-2100Z. TEMPS NEG ABV 24000

FT  1000 1500 2000 3000    6000    9000   12000   15000   18000   24000
LIH 9900 9900 1705 1806 1711+13 2216+10 2520+05 2523+01 2833-07 2937-19
HNL 9900 9900 9900 9900 1407+14 1908+11 2410+05 2612+01 2928-07 2930-18
LNY 9900 9900 9900 9900 1208+14 9900+11 9900+06 2909+01 3024-07 3027-18
OGG (etc.)


FT   30000   34000   39000   45000   53000
LIH 040734 990044 241055 281666 990072
HNL 051234 010543 250654 301066 990072
```

```
LNY 041433 010743 230754 260966 990072
OGG (etc)
```

Note: The altitudes forecast in the Hawaii and western Pacific bulletins are different than those forecast in the Contiguous US and Alaska

Note: The Hawaii and western Pacific bulletins are separated at the 24,000 foot level instead of 39,000 feet because of the additional, lower levels noted in Table 7-10.

7.4.3 Issuance

The NWS National Centers for Environmental Prediction (NCEP) produces scheduled Wind and Temperature Aloft Forecasts (**FB**) four (4) times daily for specified locations in the Continental United States (CONUS), the Hawaiian Islands, Alaska and coastal waters, and the western Pacific Ocean (Figures 7-13 through 7-16).

Amendments are not issued to the forecasts.

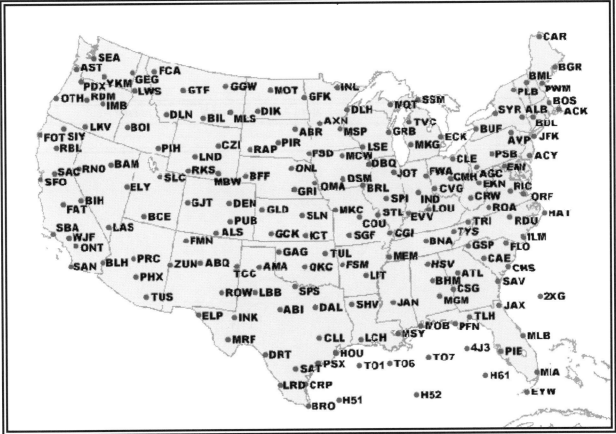

Figure 7-13. Wind and Temperature Aloft Forecast Network - Contiguous US

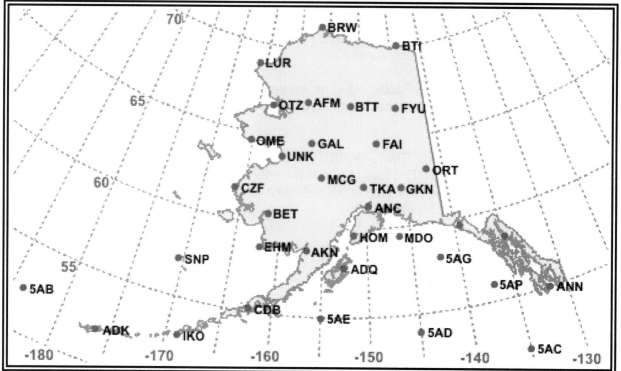

Figure 7-14. Wind and Temperature Aloft Forecast Network - Alaska

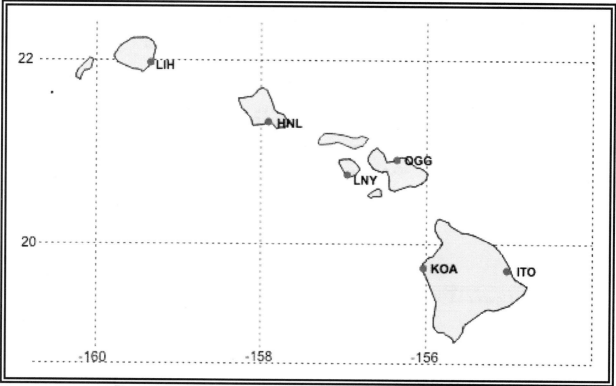

Figure 7-15. Wind and Temperature Aloft Forecast Network - Hawaii

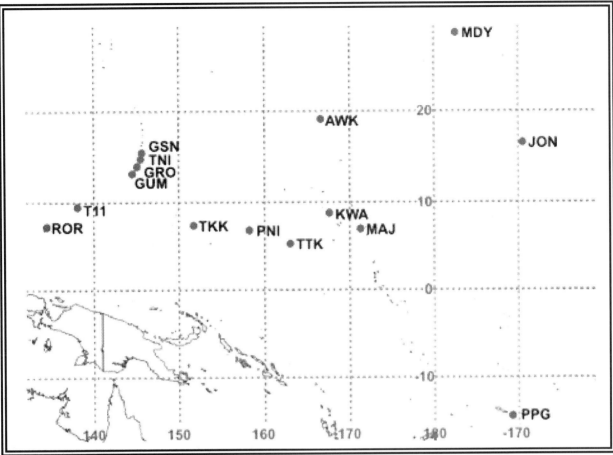

Figure 7-16. Wind and Temperature Aloft Forecast Network - Western Pacific

Table 7-13. Wind and Temperature Aloft Forecast (FB) Periods

Model Run	Product Available	6 hour Forecast		12 hour Forecast		24 hour Forecast	
		Valid	For Use	Valid	For Use	Valid	For Use
0000Z	~0200Z	0600Z	0200-0900Z	1200Z	0900-1800Z	0000Z	1800-0600Z
0600Z	~0800Z	1200Z	0800-1500Z	1800Z	1500-0000Z	0600Z	0000-1200Z
1200Z	~1400Z	1800Z	1400-2100Z	0000Z	2100-0600Z	1200Z	0600-1800Z
1800Z	~2000Z	0000Z	2000-0300Z	0600Z	0300-1200Z	1800Z	1200-0000Z

7.4.4 Delayed Forecasts

If the scheduled forecast transmission is delayed, the existing valid forecast based on the earlier 6-hourly data can be used until a new forecast is transmitted.

8 FORECAST CHARTS

8.1 Short-Range Surface Prognostic (Prog) Charts

Short-Range Surface Prognostic (Prog) Charts (Figure 8-1) provide a forecast of surface pressure systems, fronts and precipitation for a 2-day period. The forecast area covers the 48-contiguous states, the coastal waters and portions of adjacent countries. The forecasted conditions are divided into four forecast periods, 12-, 24-, 36-, and 48-hours. Each chart depicts a "snapshot" of weather elements expected at the specified valid time.

The Surface Prognostic (Prog) Charts are available at the Aviation Digital Data Services (ADDS) web site at: http://adds.aviationweather.noaa.gov/progs/.

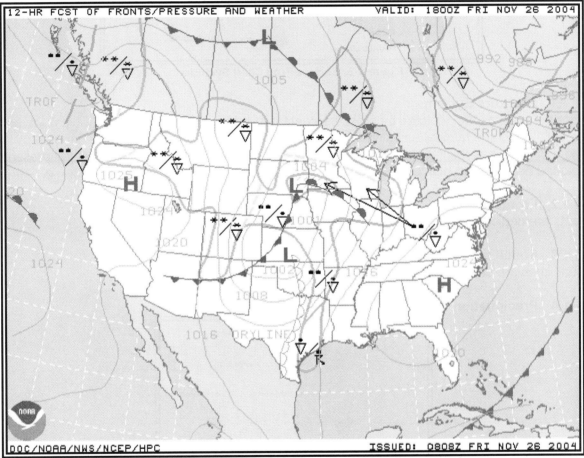

Figure 8-1. Surface Prog Chart Example

8.1.1 Content

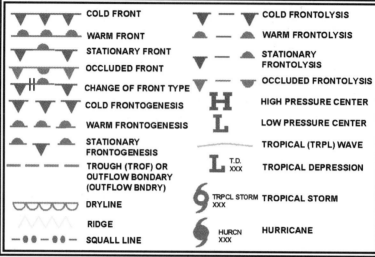

Figure 8-2. Surface Prog Chart Symbols

8.1.1.1 Pressure Systems

Pressure systems are depicted by pressure centers, troughs, isobars, drylines, tropical waves, tropical storms and hurricanes using standard symbols (Figure 8-2). Isobars are denoted by solid thin gray lines and labeled with the appropriate pressure in millibars. The central pressure is plotted near the respective pressure center.

8.1.1.2 Fronts

Fronts are depicted using the standard symbols in Figure 8-2.

8.1.1.3 Squall Lines

Squall lines are denoted using the standard symbol in Figure 8-2.

8.1.1.4 Precipitation

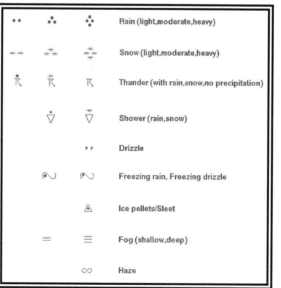

Figure 8-3. Surface Prog Chart Precipitation Symbols

Precipitation areas are enclosed by thick, solid, green lines (Figure 8-4). Standard precipitation symbols are used to identify precipitation types (Figure 8-3). These symbols are positioned within or adjacent to the associated area of precipitation. If adjacent to the area, an arrow will point to the area with which they are associated. A mix of precipitation is indicated by the use of two pertinent symbols separated by a slash (Figure 8-4). A bold, dashed, grey line is used to separate precipitation within an outlined area with contrasting characteristics (Figure 8-4). For instance, a dashed line would be used to separate an area of snow from an area of rain.

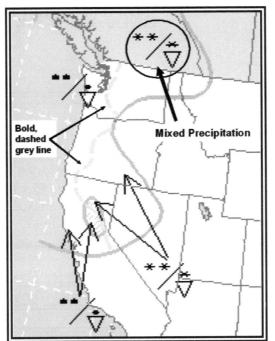

Figure 8-4. Surface Prog Chart Precipitation Example

Precipitation characteristic are further described by the use of shading (Figure 8-5). Shading or lack of shading indicates the expected coverage of the precipitation. Shaded areas indicate the precipitation is expected to have more than 50% (broken) coverage. Unshaded areas indicate 30-50% (scattered) coverage.

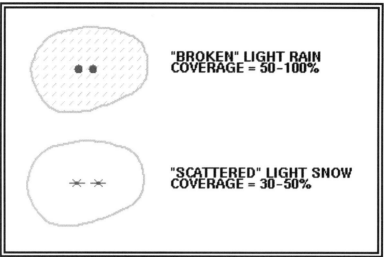

Figure 8-5. Surface Prog Chart Precipitation Coverage

8.1.2 Issuance

Short-Range Surface Prognostic (Prog) Charts are issued by the Hydrometeorological Prediction Center (HPC) in Camp Springs, MD. Table 8-1 provides the product schedule. The 12- and 24-Hour Surface Prognostic (Prog). Charts are issued four times a day and are termed "Day 1" progs. The 36- and 48- Hour Surface Prog Charts are issued twice daily and are termed "Day 2" progs. They are available on the HPC web site at: http://adds.aviationweather.noaa.gov/progs/.

Table 8-1. Short-Range Surface Prog Charts Schedule

	Issuance Time (UTC)			
	~1720	~2310	~0530	~0935
	Valid Time (UTC)			
12-Hour Surface Prog	0000	0600	1200	1800
24-Hour Surface Prog	1200	1800	0000	0600
36-Hour Surface Prog	0000	NA	1200	NA
48-Hour Surface Prog	1200	NA	0000	NA

8.1.3 Use

Short-Range Surface Prognostic (Prog) Charts can be used to obtain an overview of the progression of surface weather features during the next 48 hours. The progression of weather is the change in position, size, and intensity of weather with time. Progression analysis is accomplished by comparing charts of observed conditions to the 12-, 24-, 36-, and 48-hour progs. Short-Range Surface Prognostic (PROG) Charts make the comprehension of weather details easier and more meaningful. For example, in Figures 8-6 through 8-9, the cold front located from the eastern Great Lakes to Missouri is forecast to move southeastward and the High pressure center just north of the Minnesota/North Dakota boarder is also forecast to move southeast and weaken.

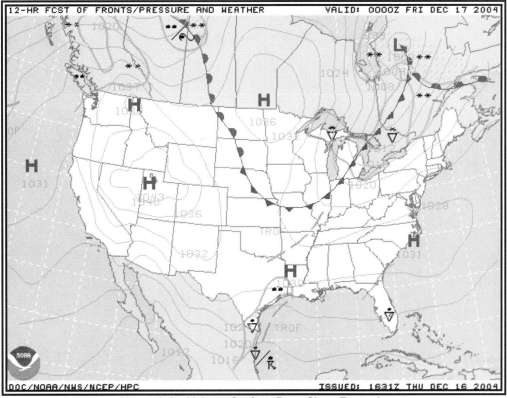

Figure 8-6. 12-hour Surface Prog Chart Example

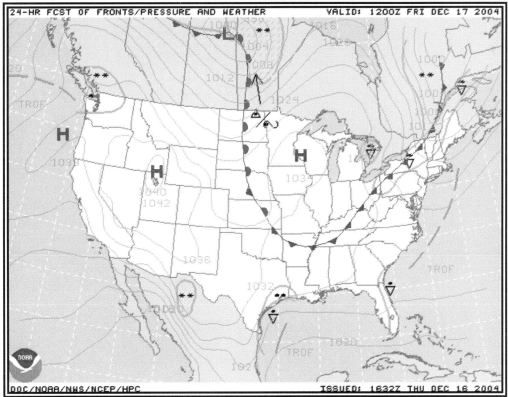

Figure 8-7. 24-hour Surface Prog Chart Example

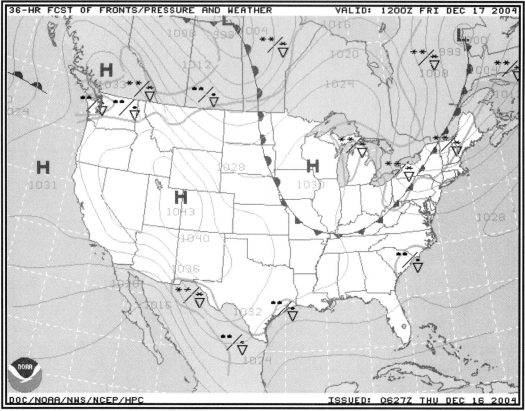

Figure 8-8. 36-hour Surface Prog Chart Example

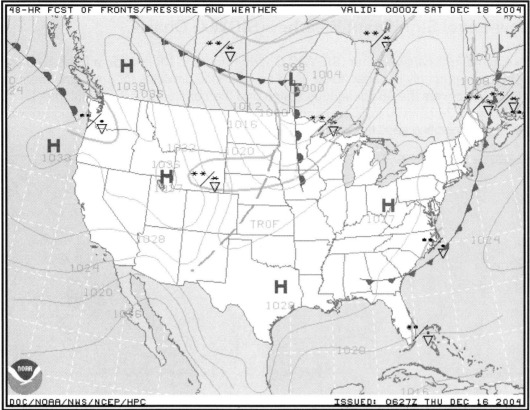

Figure 8-9. 48-hour Surface Prog Chart Example

8.2 Low-Level Significant Weather (SIGWX) Charts

The Low-Level Significant Weather (SIGWX) Charts (Figure 8-10) provide a forecast of aviation weather hazards primarily intended to be used as guidance products for pre-flight briefings. The forecast domain covers the 48 contiguous states and the coastal waters for altitudes 24,000 ft MSL (Flight Level 240 or 400 millibars) and below. Each chart depicts a "snapshot" of weather expected at the specified valid time.

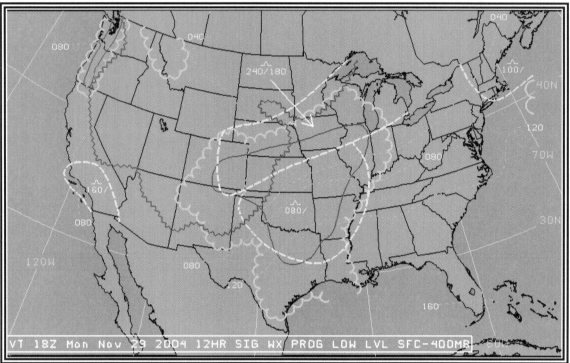

Figure 8-10. 12-Hour Low-Level SIGWX Chart Example

8.2.1 Content
Low-Level Significant Weather (SIGWX) Charts depict weather flying categories, turbulence, and freezing levels (Figure 8-11). Icing is not specifically forecast.

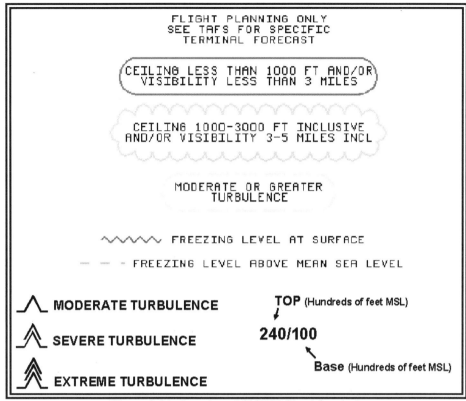

Figure 8-11. Low-Level SIGWX Chart Symbols

8.2.1.1 Flying Categories

Instrument Flight Rules (IFR) areas are outlined with a solid red line, Marginal Visual Flight Rules (MVFR) areas are outlined with a scalloped blue line, Visual Flight Rules (VFR) areas are not depicted (Figure 8-12).

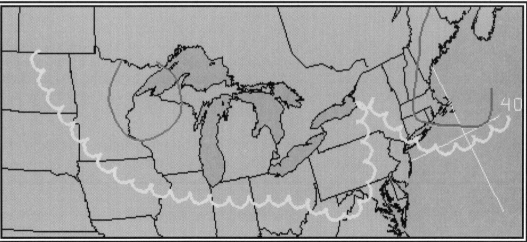

Figure 8-12. Low-Level SIGWX Chart Flying Categories Example

8.2.1.2 Turbulence

Areas of moderate or greater turbulence are enclosed by bold, dashed, yellow lines (Figure 8-13). Turbulence intensities are identified by standard symbols (Figure 8-11). The vertical extent of turbulence layers is specified by top and base heights separated by a slant. The intensity symbols and height information may be located within or adjacent to the forecasted areas of turbulence. If located adjacent to an area, an arrow will point to the associated area. Turbulence height is depicted by two numbers separated by a solidus /. For example, an area on the chart with turbulence indicated as **240/100** indicates the turbulence can be expected from the top at FL240 to the base at 10,000 feet MSL. When the base height is omitted, the turbulence is forecast to reach the surface. For example, **080/** identifies a turbulence layer from the surface to 8,000 feet MSL. Turbulence associated with thunderstorms is not depicted on the chart.

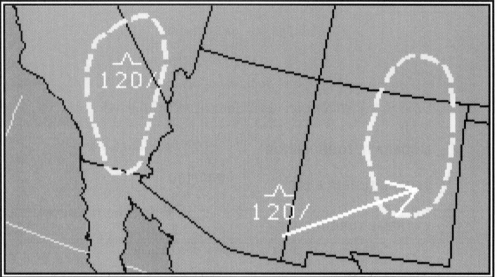

Figure 8-13. Low-Level SIGWX Chart Turbulence Forecast Example

8.2.1.3 Freezing Levels

The freezing level at the surface is depicted by a blue, saw-toothed symbol (Figure 8-11). The surface freezing level separates above-freezing from below-freezing temperatures at the Earth's surface.

Freezing levels above the surface are depicted by fine, green, dashed lines labeled in hundreds of feet MSL beginning at 4,000 feet using 4,000 foot intervals (Figure 8-11). If multiple freezing levels exist, these lines are drawn to the <u>highest</u> freezing level. For example, **80** identifies the 8,000-foot freezing level contour (Figure 8-14). The lines are discontinued where they intersect the surface.

The freezing level for locations between lines is determined by interpolation. For example, the freezing level midway between the 4,000 and 8,000 foot lines is 6,000 feet.

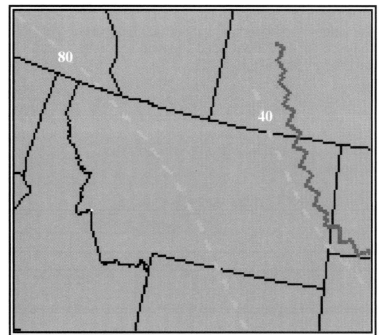

Figure 8-14. Low-Level SIGWX Chart Freezing Level Forecast Example

Multiple freezing levels occur when the temperature is zero degrees Celsius at more than one altitude aloft. Multiple freezing levels can be forecasted on the Low-Level Significant Weather Prog Charts in situations where the temperature is below-freezing (negative) at the surface with multiple freezing levels aloft.

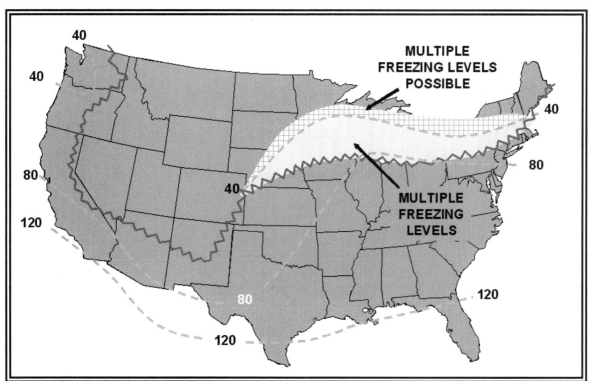

Figure 8-15. Low-Level SIGWX Chart Multiple Freezing Levels Example

On the chart, areas with multiple freezing levels are located on the below-freezing side of the surface freezing level contour and bounded by the 4,000 foot freezing level. Multiple freezing levels are **possible** beyond the 4,000 feet freezing level (i.e., below 4,000 feet MSL), but the exact cutoff cannot be determined (Figure 8-15).

8.2.2 Issuance

Low-Level Significant Weather (SIGWX) Charts are issued four times per day by the Aviation Weather Center (AWC) in Kansas City, Missouri (Table 8-2). Two charts are issued; a 12-hour and a 24-hour prog. Both are available on the AWC web site: http://aviationweather.gov/products/swl/.

Table 8-2. Low-Level SIGWX Chart Issuance Schedule

	Issuance Time			
	~1720Z	~2310Z	~0530Z	~0935Z
Chart	**Valid Time**			
12-Hour Prog	00Z	06Z	12Z	18Z
24-Hour Prog	12Z	18Z	00Z	06Z

8.2.3 Use

The Low-Level Significant Weather (SIGWX) Charts provide an overview of selected aviation weather hazards up to 24,000 feet MSL (FL240 or 400 millibars) at 12- and 24-hours into the future.

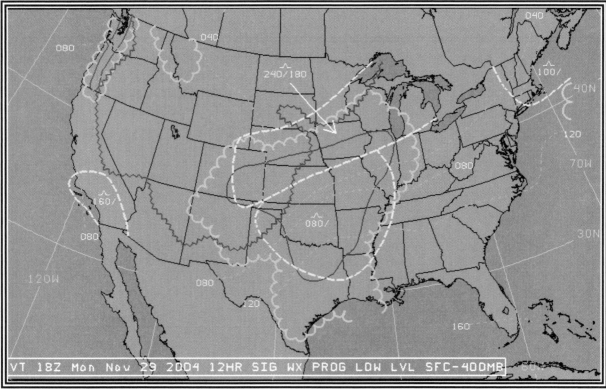

Figure 8-16. 12-Hour Low-Level SIGWX Chart Example

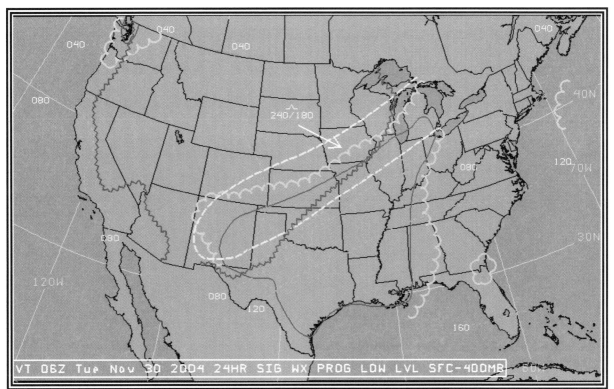

Figure 8-17. 24-hour Low-Level SIGWX Chart Example

8.3 Mid-Level Significant Weather (SIGWX) Chart

The Mid-Level Significant Weather (SIGWX) Chart (Figure 8-18) provides a forecast of significant en route weather phenomena over a range of flight levels from 10,000 ft MSL to FL450, and associated surface weather features. The chart depicts a "snapshot" of weather expected at the specified valid time.

The Mid-Level Significant Weather (SIGWX) Chart is available on the Aviation Weather Center web site at: http://aviationweather.gov/products/swm/.

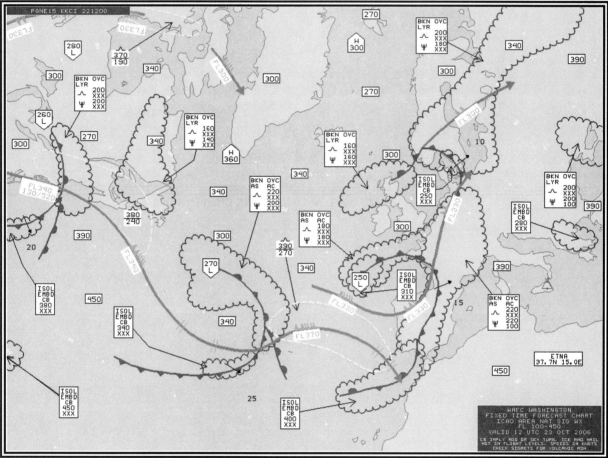

Figure 8-18. Mid-Level SIGWX Chart Example

8.3.1 Content
The Mid-Level Significant Weather (SIGWX) Chart depicts numerous weather elements that can be hazardous to aviation.

8.3.1.1 Thunderstorms
The abbreviation **CB** is only included where it refers to the expected occurrence of an area of widespread cumulonimbus clouds, cumulonimbus along a line with little or no space between individual clouds, cumulonimbus embedded in cloud layers, or cumulonimbus concealed by haze. It does not refer to isolated or scattered cumulonimbus not embedded in cloud layers or concealed by haze.

Each cumulonimbus area is identified with **CB** and characterized by coverage, bases and tops.

Table 8-3. Mid-Level SIGWX Chart Cumulonimbus Coverage

CODED	CHARACTERIZATION	MEANING
ISOL	Isolated	Less than 1/8th coverage
OCNL	Occasional	1/8th to 4/8ths coverage
FRQ	Frequent	More than 4/8ths coverage
EMBD	Embedded	CBs concealed by other cloud layers, haze, dust, etc.

Coverage, Table 8-3, is identified as isolated (**ISOL**) meaning less than 1/8th, occasional (**OCNL**) meaning 1/8th to 4/8ths, and frequent (**FRQ**) meaning more than 4/8ths coverage. Isolated and occasional **CB**s are further characterized as embedded (**EMBD**). The chart does not display isolated or scattered cumulonimbus clouds unless they are embedded in other clouds, haze, or dust.

The vertical extent of cumulonimbus layer is specified by top and base heights. Bases that extend below 10,000 feet (the lowest altitude limit of the chart) are encoded **XXX**.

Cumulonimbus clouds (**CB**s) are depicted by enclosed (red) scalloped lines (Figure 8-19). The identification and characterization of each cumulonimbus area appears within or adjacent to the outlined area. If the identification and characterization is adjacent to an outlined area, an arrow points to the appropriate cumulonimbus area.

On significant weather (SIGWX) charts, the inclusion of **CB** or the thunderstorm symbol (Figure 8-3) should be understood to include all weather phenomena normally associated with cumulonimbus or thunderstorm, namely, moderate or severe icing, moderate or severe turbulence, and hail.

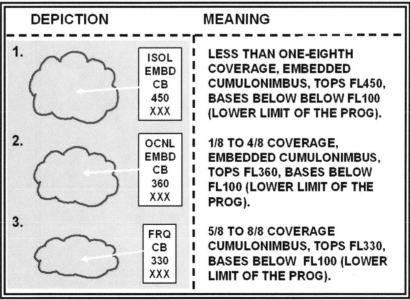

Figure 8-19. Mid-Level SIGWX Chart Thunderstorm Examples

8.3.1.2 Surface Frontal Positions and Movements

Surface fronts are depicted using the standard symbols found on the Surface Analysis Chart. (Figure 8-2). An arrow identifies the direction of frontal movement with the speed indicated in knots plotted near the arrow head (Figure 8-20).

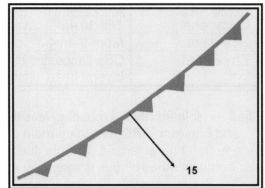

Figure 8-20. Mid-Level SIGWX Chart Surface Frontal Position and Movement Example

8.3.1.3 Jet Streams

A jet stream axis with a wind speed of more than 80 knots is identified by a bold green line (Figure 8-21). An arrowhead is used to indicate wind direction. Double-hatched, light green lines positioned along a jet stream axis identify 20 knot wind speed changes.

Symbols and altitudes are used to further characterize a jet stream axis. A standard wind symbol (light green) is placed at each pertinent position to identify wind velocity. The flight level "FL" in hundreds of feet MSL is placed adjacent to each wind symbol to identify the altitude of the jet stream axis.

Jet stream vertical depth (jet depth) forecasts are included when the maximum speed is 120 knots or more. Jet depth is defined as the vertical depths to the 80 knot wind field above and below the jet stream axis using flight levels.

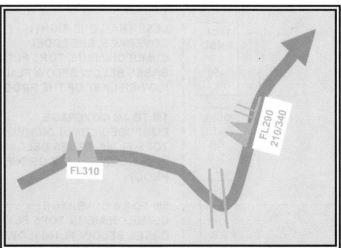

Figure 8-21. Mid-Level SIGWX Chart Jet stream Example.
Forecast maximum speeds of 100 knots at FL310 at one location and 120 knots at FL290 at another location. At the latter location, the base of the 80 knot wind field is FL210, and the top of the 80 knot wind field is FL340.

8.3.1.4 Tropopause Heights

Tropopause heights are plotted at selected locations on the chart (Figure 8-22). They are enclosed by rectangles and plotted in hundreds of feet MSL. Centers of high (**H**) and low (**L**) tropopause heights are enclosed by polygons and plotted in hundreds of feet MSL.

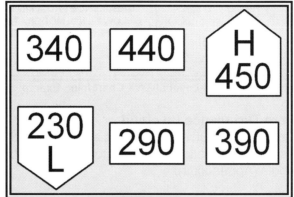

Figure 8-22. Mid-Level SIGWX Chart Tropopause Height Examples

8.3.1.5 Tropical Cyclones

Tropical cyclones are depicted by the appropriate symbol (Figure 8-23) with the storm's name positioned adjacent to the symbol. Cumulonimbus clouds meeting chart criteria are identified and characterized relative to each storm.

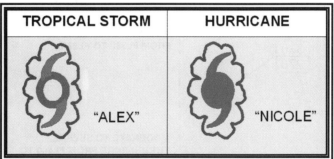

Figure 8-23. Mid-Level SIGWX Chart Tropical Cyclone Examples

8.3.1.6 Moderate or Severe Icing

Areas of moderate or severe icing are depicted by enclosed (red) scalloped lines (Figure 8-24). The identification and characterization of each area appears within or adjacent to the outlined area. If the identification and characterization is adjacent to an outlined area, an arrow points to the appropriate area.

The identification box uses the standard icing symbol (Appendix J). The vertical extent of the icing layer is specified by top and base heights. Bases which extend below the layer of the chart are identified with **XXX**.

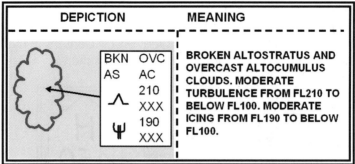

Figure 8-24. Mid-Level SIGWX Chart Icing Examples

8.3.1.7 Moderate or Severe Turbulence (in cloud or in clear air)

Forecast areas of moderate or severe turbulence associated with wind shear zones and/or mountain waves are enclosed by bold yellow dashed lines (Figure 8-25). Intensities are identified by standard symbols (Appendix J).

The vertical extent of a turbulence layer is specified by top and base heights, separated by a horizontal line. A turbulence base which extends below the layer of the chart is identified with **XXX**.

Thunderstorm turbulence is not identified.

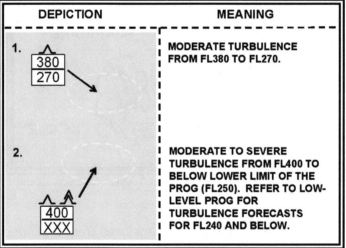

Figure 8-25. Mid-Level SIGWX Chart Turbulence Examples

Areas of moderate or severe turbulence are also depicted by enclosed (red) scalloped lines (Figure 8-24). The identification and characterization of each area appears within or adjacent to the outlined area. If the identification and characterization is adjacent to an outlined area, an arrow points to the associated area.

Standard turbulence symbols are used (Appendix J). The vertical extent of the turbulence layer is specified by top and base heights. Bases which extend below the layer of the chart are identified with **XXX**.

8.3.1.8 Cloud Coverage (non-cumulonimbus)

Clouds are enclosed within (red) scalloped lines (Figure 8-26). Cloud coverage (non-cumulonimbus) appears within or adjacent to the outlined area. If the cloud coverage is adjacent to an outlined area, an arrow points to the appropriate area.

The cloud coverage symbols are listed in Table 8-4. See Table 8-3 for cumulonimbus cloud coverage.

Table 8-4. Mid-Level SIGWX Chart Cloud Coverage (Non-cumulonimbus)

CODED	MEANING	COVERAGE
SKC	Sky Clear	$0/8^{ths}$
FEW	Few clouds	$1/8^{th}$ to $2/8^{ths}$
SCT	Scattered	$3/8^{ths}$ to $4/8^{ths}$
BKN	Broken	$5/8^{ths}$ to $7/8^{ths}$
OVC	Overcast	$8/8^{ths}$

8.3.1.9 Cloud Type

Table 8-5 shows the contractions used to identify cloud type.

Table 8-5. Mid-Level SIGWX Chart Cloud Types

CODED	MEANING
CI	Cirrus
CC	Cirrocumulus
CS	Cirrostratus
AC	Altocumulus
AS	Altostratus
NS	Nimbostratus
SC	Stratocumulus
ST	Stratus
CU	Cumulus
CB	Cumulonimbus

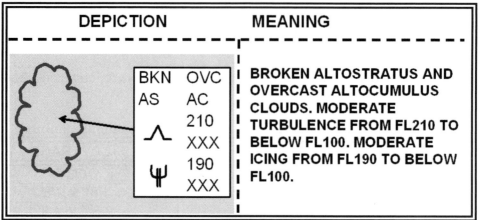

Figure 8-26. Mid-Level SIGWX Chart - Example of Moderate or Severe Icing, Moderate or Severe Turbulence (in cloud or in clear air), Clouds, and Cloud Types

8.3.1.10 Volcanic Eruptions

Volcanic eruption sites are identified by a trapezoidal symbol (Figure 8-27). The dot on the base of the trapezoid identifies the location of the volcano. The name of the volcano, as well as the latitude and longitude are noted adjacent to the symbol.

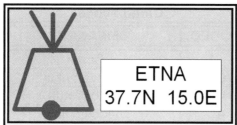

Figure 8-27. Mid-Level SIGWX Chart Volcanic Eruption Example

8.3.1.11 Release of Radioactive Materials

Radioactive materials in the atmosphere are depicted by the standard symbol shown in Figure 8-28. Information on the chart regarding the radioactive material includes the latitude/longitude of the accident site, the date and time of the accident, and a reference to check NOTAMs for further information.

Figure 8-28. Mid-Level SIGWX Chart Release of Radioactive Materials Example

8.3.2 Issuance

The Aviation Weather Center (AWC) in Kansas City has the responsibility, as part of the World Area Forecast Center (WAFC), Washington, to provide global weather forecasts of significant weather phenomena. The AWC issues a 24-hour Mid-Level Significant Weather chart, four times daily, for the North Atlantic Ocean Region (NAT) (Table 8-6). The Mid Level Significant (WIGWX) Chart is found online at: http://aviationweather.gov/products/swm/

Table 8-6. Mid-Level SIGWX Chart Issuance Schedule

North Atlantic Ocean Region (NAT)	Valid Times (UTC)			
	Issued 1015	Issued 1615	Issued 2215	Issued 0415
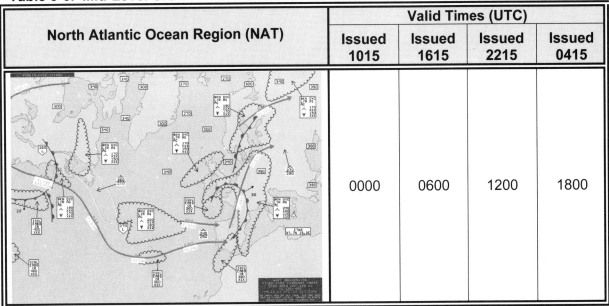	0000	0600	1200	1800

8.3.3 Use

The Mid-Level Significant Weather (SIGWX) Chart is used to determine an overview of selected flying weather conditions between 10,000 feet MSL and FL450. It can be used by airline dispatchers for flight planning and weather briefings before departure and by flight crew members during flight.

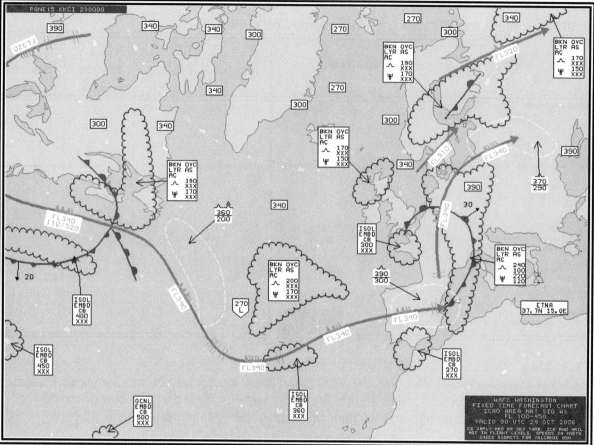

Figure 8-29. Mid-Level SIGWX Chart Example

8.4 High-Level Significant Weather (SIGWX) Charts

High-Level Significant Weather (SIGWX) Charts (Figure 8-30) provide a forecast of significant en route weather phenomena over a range of flight levels from FL250 to FL630, and associated surface weather features. Each chart depicts a "snap-shot" of weather expected at the specified valid time. They are available on the Aviation Weather Center (AWC) web site at: http://aviationweather.gov/products/swh/.

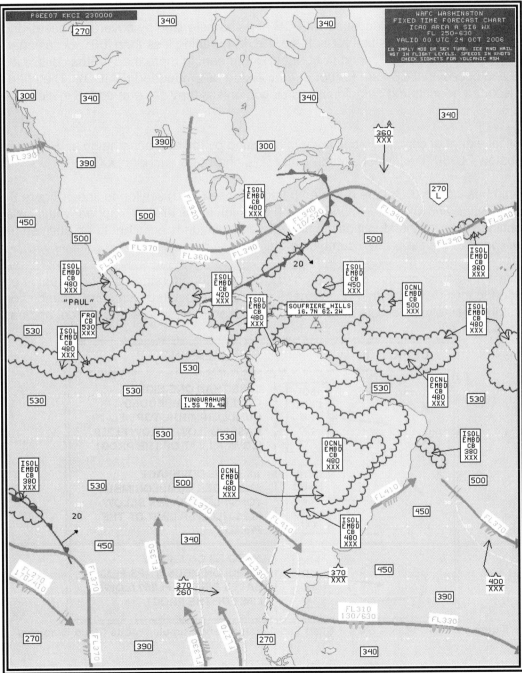

Figure 8-30. High-level SIGWX Chart Example

8.4.1 Content

8.4.1.1 Thunderstorms and Cumulonimbus Clouds

The abbreviation **CB** is only included where it refers to the expected occurrence of an area of widespread cumulonimbus clouds, cumulonimbus along a line with little or no space between individual clouds, cumulonimbus embedded in cloud layers, or cumulonimbus concealed by haze. It does not refer to isolated or scattered cumulonimbus not embedded in cloud layers or concealed by haze.

Each cumulonimbus area is identified with **CB** and characterized by coverage, bases and tops. Coverage (Table 8-3) is identified as isolated (**ISOL**) meaning less than 1/8th, occasional (**OCNL**) meaning 1/8th to 4/8^{ths}, and frequent (**FRQ**) meaning more than 4/8ths coverage. Isolated and occasional CBs are further characterized as embedded (**EMBD**). The chart will not display isolated or scattered cumulonimbus clouds unless they are embedded in clouds, haze, or dust.

The vertical extent of cumulonimbus layer is specified by top and base heights. Bases that extend below FL250 (the lowest altitude limit of the chart) are encoded **XXX**.

Cumulonimbus clouds (CBs) are depicted by an enclosed (red) scalloped lines (Figure 8-31). The identification and characterization of each cumulonimbus area will appear within or adjacent to the outlined area. If the identification and characterization is adjacent to an outlined area, an arrow will point to the associated cumulonimbus area.

On significant weather charts, the inclusion of **CB** or the thunderstorm symbol should be understood to include all weather phenomena normally associated with cumulonimbus or thunderstorm, namely, moderate or severe icing, moderate or severe turbulence, and hail.

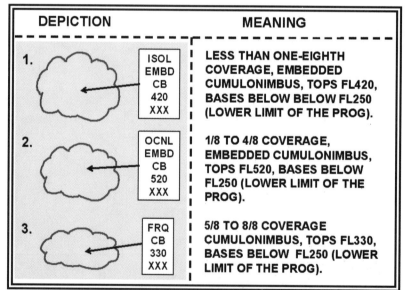

Figure 8-31. High-Level SIGWX Chart Thunderstorm and Cumulonimbus Cloud Examples

8.4.1.2 Moderate or Severe Turbulence

Forecast areas of moderate or severe turbulence (Figure 8-32) associated with wind shear zones and/or mountain waves are enclosed by bold yellow dashed lines. Intensities are identified by standard symbols (Appendix J).

The vertical extent of turbulence layers is specified by top and base heights, separated by a horizontal line. Turbulence bases which extend below the layer of the chart are identified with **XXX**.

Thunderstorm turbulence is not identified.

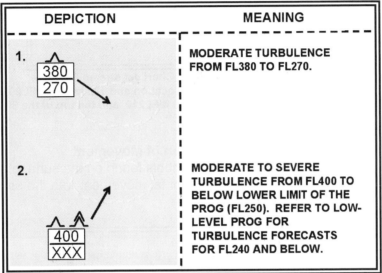

Figure 8-32. High-Level SIGWX Chart Turbulence Examples

8.4.1.3 Moderate or Severe Icing

Moderate and severe icing (outside of thunderstorms) above FL240 is rare and is not generally forecasted on High-Level Significant Weather Prog charts.

8.4.1.4 Jet Streams

A jet stream axis with a wind speed of more than 80 knots is identified by a bold green line. An arrowhead is used to indicate wind direction. Wind change bars (double-hatched, light green lines) positioned along a jet stream axis identifies 20 knot wind speed changes (Figure 8-33).

Symbols and altitudes are used to further characterize a jet stream axis. A standard wind symbol (light green) is placed at each pertinent position to identify wind velocity. The flight level **FL** in hundreds of feet MSL is placed adjacent to each wind symbol to identify the altitude of the jet stream axis.

Jet stream vertical depth (jet depth) forecasts are included when the maximum speed is 120 knots or more. Jet depth is defined as the vertical depths to the 80 knot wind field above and below the jet stream axis using flight levels. Jet depth information is placed at the maximum speed point only, normally at one point on each jet stream. When the jet stream is very long and there are several wind maxima, then each maximum should include forecasts of the vertical depth.

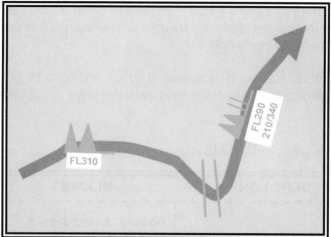

Figure 8-33. High-Level SIGWX Chart Jet stream Example
Forecast maximum speeds of 100 knots at FL310 at one location and 120 knots at FL290 at another location. At the latter location, the base of the 80 knot wind field it FL210, and the top of the 80 knot wind field is FL340.

8.4.1.5 Surface Fronts with Speed and Direction of Movement

Surface fronts are depicted using the standard symbols found on the surface analysis chart. (Figure 8-2). An arrow identifies the direction of frontal movement with the speed in knots plotted near the arrow head (Figure 8-34).

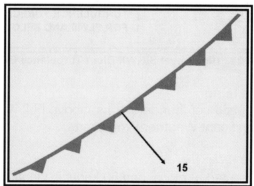

Figure 8-34. High Level SIGWX Chart Surface Front with Speed and Direction of Movement Example

8.4.1.6 Tropopause Heights

Tropopause heights are plotted at selected locations on the chart. They are enclosed by rectangles and plotted in hundreds of feet MSL (Figure 8-35). Centers of high (**H**) and low (**L**) tropopause heights are enclosed by polygons and plotted in hundreds of feet MSL.

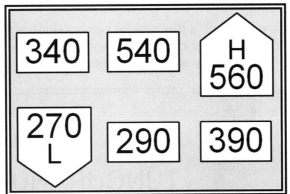

Figure 8-35. High-Level SIGWX Chart Tropopause Height Examples

8.4.1.7 Tropical Cyclones

Tropical cyclones are depicted by the appropriate symbol (Figure 8-36) with the storm's name positioned adjacent to the symbol. Cumulonimbus clouds meeting chart criteria are identified and characterized relative to each storm.

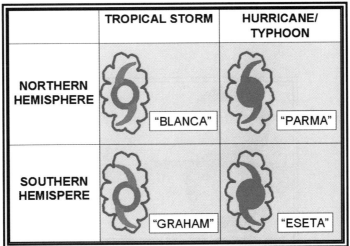

Figure 8-36. High Level SIGWX Chart Tropical Cyclone Examples

8.4.1.8 Severe Squall Lines

Severe squall lines are lines of CBs with 5/8 coverage or greater. They are identified by long dashed (white) lines with each dash separated by a **V** (Figure 8-37).Cumulonimbus clouds meeting chart criteria are identified and characterized with each squall line.

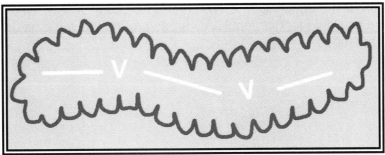

Figure 8-37. High-Level SIGWX Chart Severe Squall Line Example

8.4.1.9 Volcanic Eruption Sites

Volcanic eruption sites are identified by a trapezoidal symbol (Figure 8-38). The dot on the base of the trapezoid identifies the location of the volcano. The name of the volcano, its latitude, and its longitude are noted adjacent to the symbol.

Figure 8-38. High-Level SIGWX Chart Volcanic Eruption Site Example

8.4.1.10 Widespread Sandstorms and Dust storms

Widespread sandstorms and dust storms are labeled with the appropriate symbol (Appendix I). The vertical extent of sand or dust is specified by top and base heights, separated by a horizontal line. Sand or dust which extends below the lower limit of the chart (FL240) is identified with **XXX** (Figure 8-39).

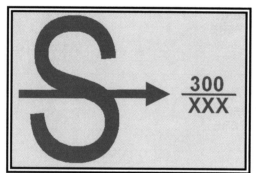

Figure 8-39. High-Level SIGWX Chart Widespread Sandstorm and Dust Storm Example

8.4.2 Issuance

In accordance with the World Meteorological Organization (WMO) and the World Area Forecast System (WAFS) of the International Civil Aviation Organization (ICAO), High-Level significant weather (SIGWX) forecasts are provided for the en-route portion of international flights. The National Weather Service (NWS) Aviation Weather Center (AWC) in Kansas City, MO provides a suite of SIGWX forecast products for the World Area Forecast Center (WAFC) in Washington, D.C. The charts are available for different ICAO areas around the world as defined in Table 8-7. The charts are not amended.

Table 8-7. High-Level SIGWX Chart Issuance Schedule – WAFC Washington

ICAO Area	Chart Type	Chart Area	Valid Times (UTC)			
			Issued 1100	Issued 1700	Issued 2300	Issued 0500
A Americas	Mercator		0000	0600	1200	1800
B1 Americas/ Africa	Mercator		0000	0600	1200	1800
F Pacific	Mercator		0000	0600	1200	1800
H N America/ Europe	Polar Stereographic		0000	0600	1200	1800
I N Pacific	Polar Stereographic		0000	0600	1200	1800
J S Pacific	Polar Stereographic		0000	0600	1200	1800
M Pacific	Mercator		0000	0600	1200	1800

The WAFC in London, England also issues High-Level Significant Weather (SIGWX) Charts for other geographical areas of the world. Both Washington and London WAFC charts are available online at: http://aviationweather.gov/iffdp/sgwx.shtml

8.4.3 Use
High-Level Significant Weather (SIGWX) Charts are provided for the en route portion of international flights. These products are used directly by airline dispatchers for flight planning and weather briefings before departure and by flight crew members during flight.

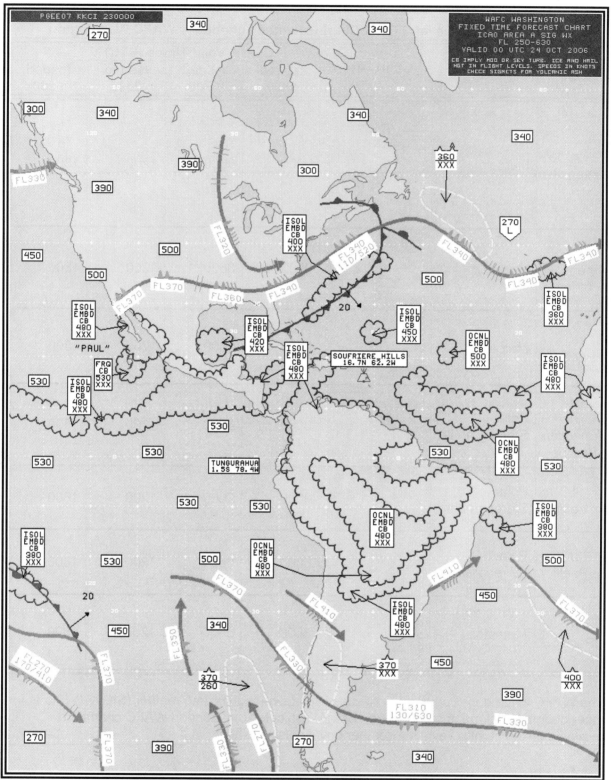

Figure 8-40. High-Level SIGWX Chart - ICAO Area A Example

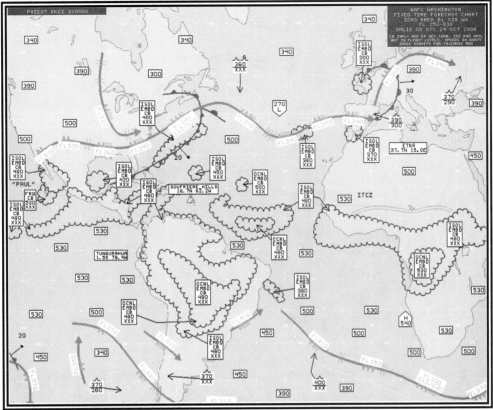

Figure 8-41. High-Level SIGWX Chart - ICAO Area B1 Example

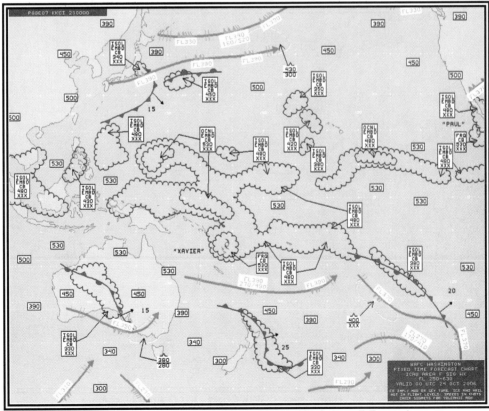

Figure 8-42. High-Level SIGWX Chart - ICAO Area F Example

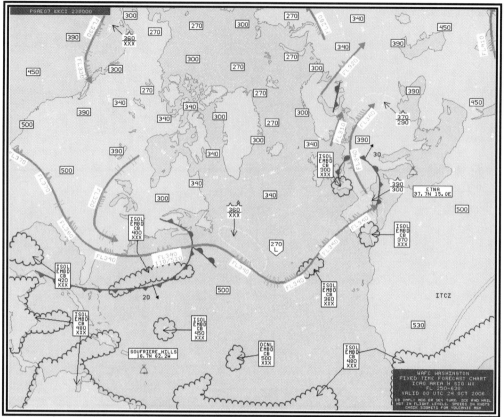

Figure 8-43. High-Level SIGWX Chart - ICAO Area H Example

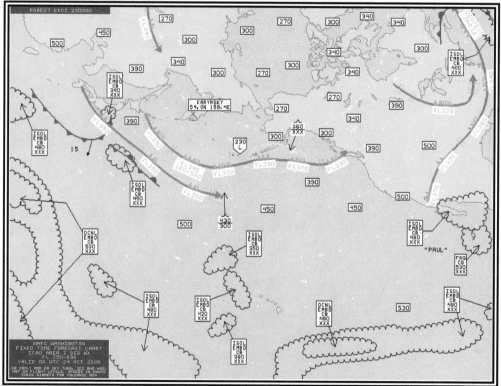

Figure 8-44. High-Level SIGWX Chart - ICAO Area I Example

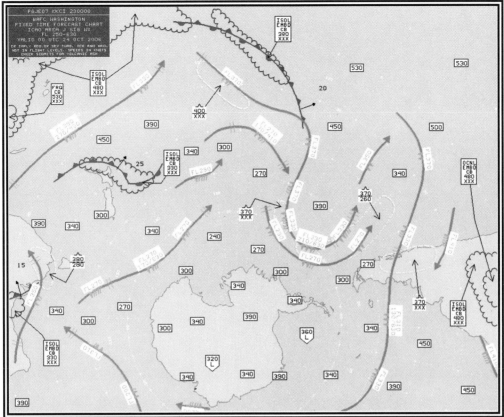

Figure 8-45. High-Level SIGWX Chart - ICAO Area J Example

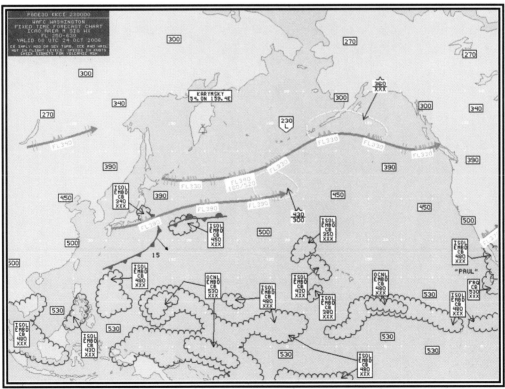

Figure 8-46. High-Level SIGWX Chart - ICAO Area M Example

9 SUPPLEMENTARY PRODUCTS

9.1 Collaborative Convective Forecast Product (CCFP)

The Collaborative Convective Forecast Product (CCFP) is a graphical representation of forecast convective occurrence verifying at 2-, 4 -, and 6-hours after issuance time (Figure 9-1). Convection, for the purposes of the CCFP forecast, is defined as a polygon of at least 3,000 square miles containing all of the following threshold criteria:

- A coverage of at least 25 percent of echoes with at least 40 dBZ <u>composite</u> reflectivity,
- A coverage of at least 25 percent of echoes with echo tops of FL250 or greater, and
- A forecaster confidence of at least 25 percent.

All three threshold criteria must be met for any area of convection 3,000 square miles or greater to be included in a CCFP forecast. This is defined as the minimum CCFP criteria. Any area of convection, which is forecasted to NOT meet all three of these criteria, is NOT included in a CCFP forecast.

The CCFP is intended to be used as a strategic planning tool for air traffic flow management. It aids in the reduction of air traffic delays, reroutes and cancellations due to significant convection. It is **not** intended to be used for tactical air traffic flow decisions, in the airport terminal environment, or for pilot weather briefing purposes. The graphical representation is subject to annual revision.

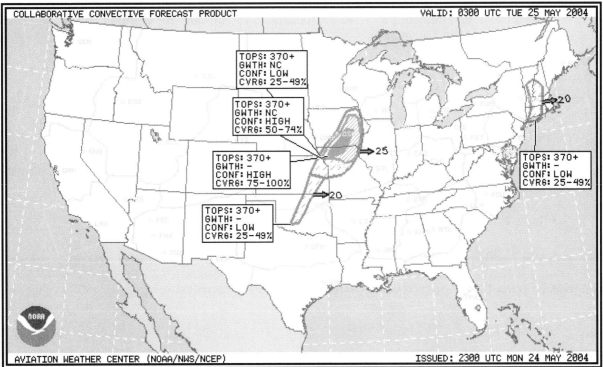

Figure 9-1. CCFP Example

9.1.1 Issuance

The CCFP is issued by the Aviation Weather Center (AWC) from March through October for the 48-contiguous states. Canadian forecasts are included on the product are available for southern Ontario and Quebec between April through September. This area is roughly from north of Wisconsin extending eastward to north of Maine.

The CCFP is issued every two hours, eleven times per day. Issuance times are from 08Z to 04Z during standard time and from 07Z to 03Z during daylight savings time. The product can be found on the AWC web page at http://aviationweather.gov/products/ccfp/.

9.1.2 Collaboration

The CCFP is produced from a collaborative effort between public and private meteorologists. The collaboration occurs between meteorologists from the Aviation Weather Center (AWC), Center Weather Service Units (CWSU), Meteorological Services of Canada (MSC), commercial airlines offices, and other private weather companies.

9.1.3 Content

Data graphically displayed on the CCFP consist of coverage of convection within a defined polygon, forecaster confidence of convective occurrence, and forecast movement of the convective areas. A data block also displays text information about coverage and confidence as well as forecast echo tops and convective growth information.

9.1.3.1 Coverage

The convective coverage within the forecast polygon is represented by the amount of fill within the polygon (Figure 9-2).

- Solid coverage, depicted by solid fill, means 75 to 100 percent of the polygon is forecast to contain convection.

- Medium coverage, defined by medium fill, indicates 50-74 percent of the polygon is forecast to contain convection.

- Sparse coverage, represented by sparse fill, means 25-49 percent of the polygon is forecast to contain convection.

A line of forecast convection, either within a forecast area or alone, is depicted by a solid purple line. For a line of convection to be forecast, it must meet the flowing criteria:

- Its length must be at least 100 miles long,

- The width of the line must be 40NM wide, and

- 75 percent of the line must be expected to contain convection.

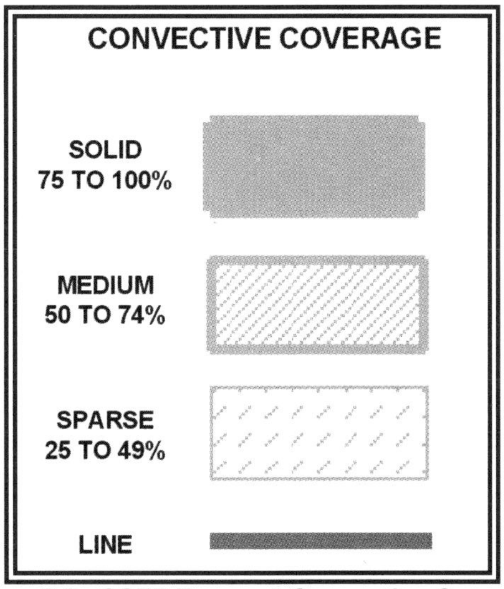

CONVECTIVE COVERAGE

SOLID
75 TO 100%

MEDIUM
50 TO 74%

SPARSE
25 TO 49%

LINE

Figure 9-2. CCFP Forecast Convective Coverage

9.1.3.2 Confidence

Confidence represents the subjective opinion of the forecasters that the polygon will meet the minimum CCFP threshold criteria. The forecaster's confidence is represented by the color used to depict the polygon (Figure 9-3).

• A blue color represents high forecaster confidence (50-100 percent) the forecast convection will meet the minimum criteria.

• A gray color indicates low forecaster confidence (25-49 percent) the forecast convection will meet the minimum criteria.

Confidence is not to be associated with probability of occurrence.

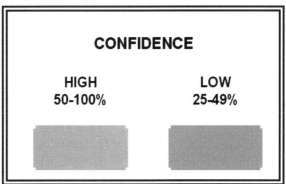

CONFIDENCE

HIGH
50-100%

LOW
25-49%

Figure 9-3. CCFP Forecast Confidence

9.1.3.3 Movement

Forecast movement for each polygon or line is indicated with a gray or blue arrow (Figure 9-4). The arrow points in the direction of forecast movement. A number at the tip of the arrow represents the speed in knots.

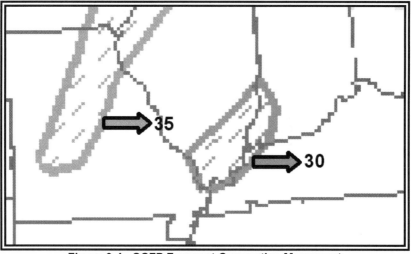

Figure 9-4. CCFP Forecast Convective Movement

9.1.3.4 Data Block

A data block is located adjacent to every polygon forecast (Figure 9-5 and Figure 9-6). A thin line connects the data block to the associated forecast area. Each data block contains information about forecast maximum echo tops (**TOPS**), convective growth rates (**GWTH**), forecaster confidence (**CONF**), and convection coverage (**CRVG**).

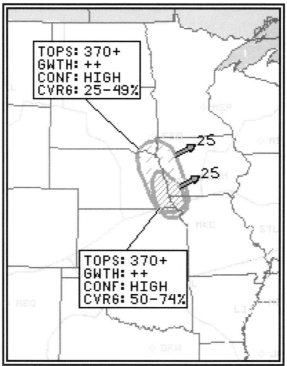

Figure 9-5. CCFP Data Block

Figure 9-6. CCFP Data Block Example

9.1.3.4.1 Tops
The word, **TOPS**, is used to depict the forecast maximum echo tops, in thousands of feet MSL, specified by three selected layers listed in Table 9-1. The heights of the forecast echo tops must cover at least 25 percent of the polygon. The exact location of the highest echo top within the polygon cannot be determined.

Table 9-1. CCFP Tops

FL250 – FL310	Flight level 250 to flight level 310
FL310 – FL370	Flight level 310 to flight level 370
370+	Above flight level 370

9.1.3.4.2 Growth
The contraction, **GWTH**, is used to depict the forecast average growth rate of the convection. The growth rate applies to both the height and areal coverage of the echoes (Table 9-2).

Table 9-2. CCFP Growth

++	Fast positive growth
+	Moderate positive growth
NC	No change in the growth
-	Negative Growth

9.1.3.4.3 Confidence
The contraction, **CONF,** depicted on the chart is the confidence the depicted polygon will meet the minimum CCFP criteria (Table 9-3).

Table 9-3. CCFP Confidence

| LOW | 25 to 49 percent |
| HIGH | 50 to 100 percent |

9.1.3.4.4 Coverage
The contraction, **CVRG**, is used to depict the forecast convective coverage within the polygon (Table 9-4). The coverage represents the percentage of the area forecast to be covered by convection.

Table 9-4. CCFP Coverage

25 – 49%	Sparse
50 – 74%	Medium
75 –100%	Solid

9.1.4 Strengths and Limitations

The primary strength of the CCFP is it relies on the vital collaborative efforts between several meteorological units in the private and public sector. The process helps produce the best possible convective forecast to assist in strategic air traffic decision-making.

The limitation of the CCFP is it does **not** include a forecast for all convection. If the convection does not meet the threshold criteria, it is not included in the CCFP. It is not intended to be used as a tactical short-term decision tool.

9.1.5 Use

The CCFP is to be used as a strategic planning tool for air traffic flow management in the 2- to 6-hour forecast period.

The product is not intended to be used as a pilot weather briefing tool.

9.2 National Convective Weather Forecast (NCWF)

The National Convective Weather Forecast (NCWF) is a near real-time, high resolution display of current and one-hour extrapolated forecasts of selected hazardous convective conditions for the conterminous United States. The NCWF is a supplement to, but does not substitute for, the report and forecast information contained within Convective SIGMETs. The NCWF is intended for use by general aviation, airline dispatchers, and Traffic Management Units.

9.2.1 Issuance
The NCWF is issued by the Aviation Weather Center (AWC) and is updated every <u>five</u> minutes. The product is available on the Aviation Digital Data Service (ADDS) web page at: http://adds.aviationweather.noaa.gov/convection/java/ and the AWC web site at: http://aviationweather.gov/products/ncwf/

9.2.2 Content
The NCWF displays current convective hazard fields, one-hour extrapolated forecast polygons, forecast speed and directions, and echo tops. Previous performance polygons can also be selected for display (Figure 9-7).

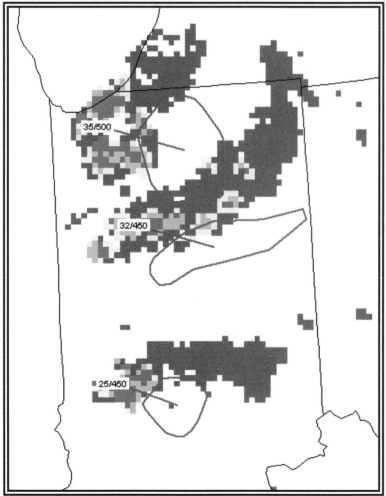

Figure 9-7. NCWF Example

9.2.2.1 Current Convective Hazard Fields

The current convective hazard field is a high-resolution display that identifies selected hazards associated with convective precipitation. The field is created from WSR-88D national reflectivity and echo-top mosaics and cloud-to-ground lightning data.

WSR-88D radar reflectivity data are filtered to identify locations having significant convective precipitation. Reflectivity data with echo tops of less than 17,000 feet MSL are eliminated from the data. This process removes ground clutter and anomalous propagation as well as significantly reduces the amount of stratiform (non-convective) precipitation from the data. Most stratiform precipitation tops are below 17,000 feet. The filter also removes shallow convection with tops below 17,000 feet. Shallow convection is often short-lived but can contain conditions hazardous to aviation and may be embedded in stratiform convection.

Frequencies of cloud-to-ground lightning are added to the filtered radar data to provide a more accurate picture of current hazardous convective conditions.

Current convective hazard fields are color coded according to the convective hazard scale for display on the NCWF (Figure 9-8).

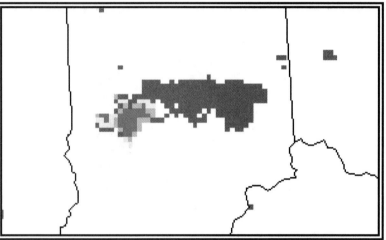

Figure 9-8. NCWF Current Convective Hazard Fields Example

9.2.2.1.1 Convective Hazard Field Scale

The convective hazard field scale uses six hazard levels (Figure 9-9) to characterize hazardous convection conditions.

The six hazard levels are determined by two factors:

- Intensities and maximum tops of WSR-88D reflectivity data, and

- Frequencies of cloud-to-ground lightning.

Higher hazard levels are associated with higher radar reflectivity intensities and higher frequencies of lightning strikes.

The six hazard levels are reduced to four-color codes for display on the NCWF. The relationships between the six hazard levels and four-color codes are summarized in Figure 9-9.

NCWF Hazard Scale		
Level	Color	Effect
6	RED	Thunderstorms may contain any or all
5	RED	of the following: severe turbulence, severe icing,
4	Orange	hail, frequent lightning, tornadoes, and low-level wind shear.
3	Yellow	The risk of hazardous weather generally increases
2	Green	with levels on the NCWF hazard scale
1	Green	

Figure 9-9. NCWF Hazard Scale

9.2.2.2 One-Hour Extrapolated Forecast Polygons

One-hour extrapolated forecast polygons are high-resolution polygons outlining areas expected to be filled by selected convective hazard fields in one hour. Extrapolated forecasts depict new locations for the convective hazard fields based on their past movements. Extrapolation forecasts do **not** forecast the development of new convective hazard conditions or the dissipation of existing conditions. Forecasts are provided **only** for convective hazard scale levels 3 or higher. The forecast polygons do not depict specific forecast hazard levels. On Figure 9-10, the light blue polygon denotes the location of the one-hour forecast convective hazard field.

9.2.2.3 Forecast Speed and Direction

Forecast speed and direction are assigned to current convective hazard fields having a one-hour extrapolated forecast. A line (or arrow on the AWC JavaScript product) is used to depict the direction of movement (Figure 9-10). The speed in knots is depicted by the first group of two numbers located near the current convective hazard field. The second group of three numbers identifies echo tops.

Forecast speed and direction is only updated every 10 minutes. The larger update time-interval (compared to five-minute updates for the NCWF) smoothes erratic forecast velocities. On Figure 9-10, the forecast direction (depicted by an arrow) is pointing to the southeast and the speed is 25 knots.

9.2.2.4 Echo Tops

Echo tops are assigned to current convective hazard fields having a one-hour extrapolated forecast. Echo tops are depicted by a group of three numbers located near the current convective hazard field and is plotted in hundreds of feet MSL (Figure 9-10). The first number of the group identifies forecast speed of movement. On Figure 9-10, the echo tops are 45,000 feet MSL.

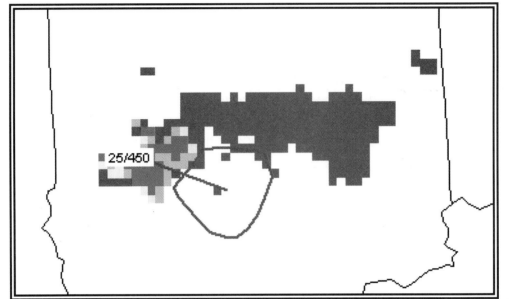

Figure 9-10. NCWF One-Hour Extrapolated Forecast Polygon, Forecast Movement Velocity, and Echo Tops Example

9.2.2.5 Previous Performance Polygons

Previous performance polygons are magenta polygons displaying the previous hour's extrapolated forecast polygons **with** the current convective hazard fields. A perfect forecast would have the polygons filled with convective hazard scale levels 3 or higher data. Levels 1 and 2 would be outside the polygons. The display of previous performance polygons allows the user to review the accuracy of the previous hour's forecast.

Figure 9-11 depicts current convective hazard fields and previous performance polygons (magenta) valid at 1500Z. The previous performance polygons are the one-hour extrapolated forecasts made at 1400Z. Although the polygons do not perfectly match the current level 3 and higher hazard fields, the forecasts are still fairly accurate.

Newly developed convective hazard levels 3 and higher do not have previous performance polygons. Extrapolated forecasts do not forecast developing hazardous convective.

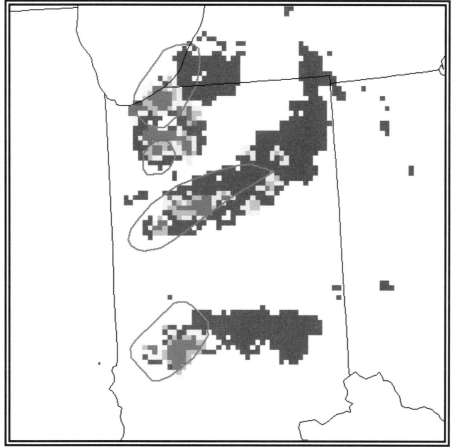

Figure 9-11. NCWF Previous Performance Polygons Example

9.2.3 The ADDS and AWC JAVA display

The ADDS web site allows for the display of all previously discussed attributes of the NCWF. This site also allows many overlay options including METARs, TAFs, VORs, ARTCC boundaries, counties, highways and rivers. Product animation is also possible on the AWC JavaScript image.

9.2.4 Strengths and Limitations

Strengths of the NCWF include:

- Convective hazard fields that agree very well with radar and lightning data,

- Updated every five minutes,

- High-resolution forecasts of convective hazards, and

- Long-lived convective precipitation is well forecast.

Limitations of the NCWF include:

- Initiation, growth, and decay of convective precipitation are not forecast,

- Short-lived or embedded convection may not be accurately displayed or forecast,

• Low-topped convection that contains little or no lightning may not be depicted,

• Erroneous motion vectors are occasionally assigned to storms, and

• Convective hazard field scales are not identified within the forecast polygons.

9.2.5 Uses of the NCWF

The purpose of the National Convective Weather Forecast (NCWF) is to produce a convective hazard field diagnostic and forecast product based on radar data, echo top mosaics, and lightning data. The target audience includes the FAA and other government agencies, pilots, airline dispatchers, aviation meteorologists, and other interested aviation users in the general public. The NCWF is a supplement to, but does not substitute for, the report and forecast information contained in Convective SIGMETs.

9.3 Current Icing Product (CIP)

The Current Icing Product (CIP) product combines sensor and numerical data to provide a hourly three-dimensional diagnosis of the icing environment. This information is displayed on a suite of twelve graphics which are available for the 48 contiguous United States, much of Canada and Mexico, and their respective coastal waters.

The CIP product suite is automatically produced with no human modifications. Information on the graphics is determined from observational data including WSR-88D radar, satellite, pilot weather reports, surface weather reports, lightning and computer model output.

FAA policy states the CIP is a supplementary weather product for enhanced situational awareness only and **must** be used with one or more primary products such as an AIRMET or SIGMET (see AIM 7-1-3).

9.3.1 Issuance
The CIP product suite is issued hourly 15 minutes after the hour by the Aviation Weather Center (AWC). The products are available through the Aviation Digital Data Service (ADDS) web site at: http://adds.aviationweather.noaa.gov/icing/icing_nav.php.

9.3.2 Content
The CIP product suite consists of 10 graphics including:

- Icing Probability,

- Icing Probability Maximum (Max),

- Icing Severity,

- Icing Severity Max,

- Icing Severity – Probability > 25%,

- Icing Severity – Probability > 25% Max,

- Icing Severity – Probability > 50%,

- Icing Severity – Probability > 50% Max,

- Icing Severity plus Supercooled Large Droplets (SLD), and

- Icing Severity plus Supercooled Large Droplets (SLD) Max.

The CIP products are generated for individual altitudes from 1,000 feet MSL to Flight Level (FL) 300 at intervals of 1,000 feet.

The CIP Max products are a composite product which displays information about icing at **all** altitudes from 1,000 feet MSL to FL300. Single altitudes are referenced to MSL from the 1,000 to 17,000 feet and Flight Levels above 17,000 feet. The ADDS web site allows for access to every other altitude (1,000 FT, 3,000 FT, 5,000 FT, etc...). However, all altitudes can be accessed by use of the Flight Path Tool on the ADDS site.

Icing PIREPs are plotted on a single altitude graphic if the PIREP is within 1,000 feet of the selected altitude and has been observed within 75 minutes of the chart's valid time. On the CIP Max graphics, PIREPs for all altitudes (i.e. 1,000 feet MSL to FL300) are displayed. However, negative reports of icing are not plotted on the CIP Max products in an effort to reduce clutter. The PIREP legend is located on the bottom of each graphic.

9.3.2.1 Icing Probability
The Icing Probability product (Figure 9-12) displays, at a single altitude, the probability of icing. Probabilities range from 0% (no icing expected) to 85% or greater (nearly certain icing.)

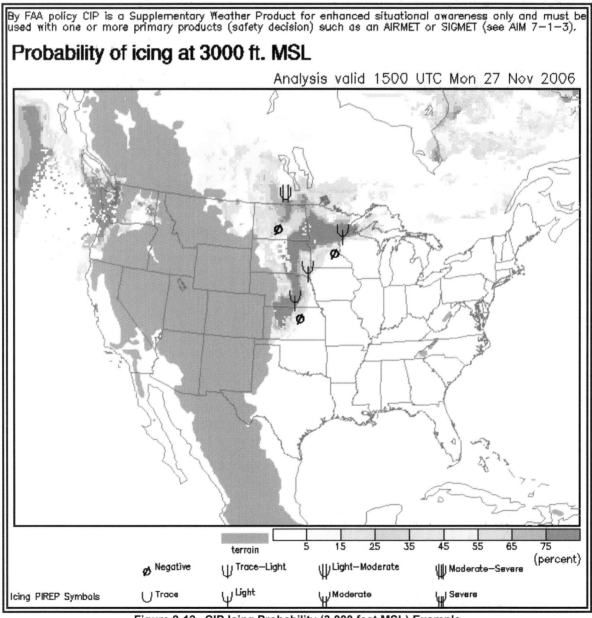

Figure 9-12. CIP Icing Probability (3,000 feet MSL) Example

"Cool" colors represent low probabilities and "warm" colors represent higher probabilities. Probabilities do not reach 100% because the data used to determine the probability of icing can not diagnose, with absolute certainty, the presence of icing conditions at any location and altitude. White regions indicate that the probability of icing is zero. Brown regions indicate where higher-elevation terrain extends above the altitude of the particular graphic.

9.3.2.2 Icing Probability -- Maximum

The Icing Probability - Maximum graphic (Figure 9-13) displays the probability of icing at **all** altitudes from 1,000 feet MSL to FL300. Probabilities range from 0% (no icing expected) to 85% or greater (nearly certain icing.)

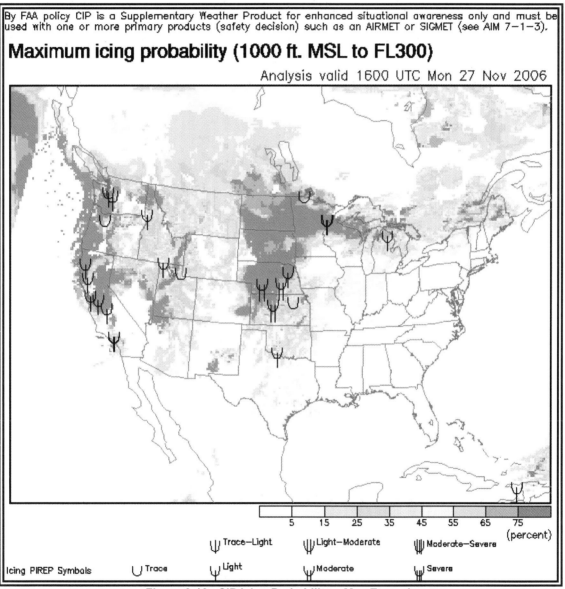

Figure 9-13. CIP Icing Probability – Max Example

"Cool" colors represent low probabilities and "warm" colors represent higher probabilities. Probabilities do not reach 100% because the data used to determine the probability of icing cannot diagnose, with absolute certainty, the presence of icing conditions at any location and altitude. White regions indicate the probability of icing is zero.

9.3.2.3 Icing Severity

The Icing Severity product (Figure 9-14) depicts, at a single altitude, the intensity of icing expected at locations where the Icing Probability product depicts possible icing. Icing intensity is displayed using standard icing intensity categories: trace, light, moderate and heavy.

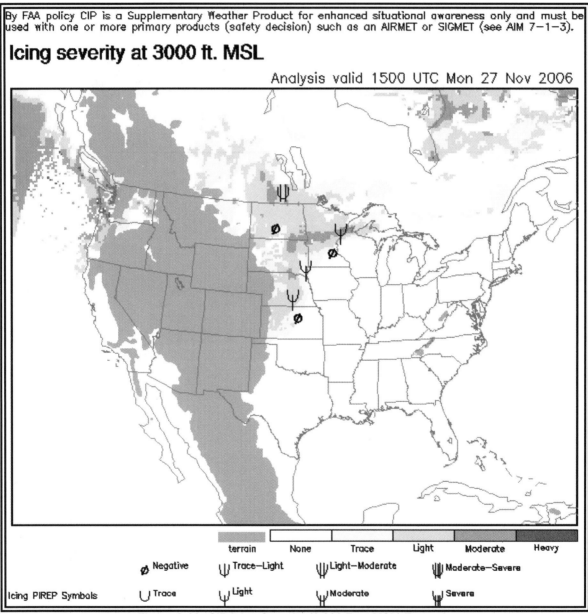

Figure 9-14. CIP Icing Severity (3,000 Feet MSL) Example

The lightest blue color represents trace icing. As the blue-color shades become darker, the icing intensity increases. The darkest blue color represents heavy icing. White regions indicate where no probability of icing exists and, therefore, no intensity is necessary. Brown regions indicate where higher-elevation terrain extends above the altitude of the particular graphic.

9.3.2.4 Icing Severity -- Maximum

The Icing Severity - Maximum product (Figure 9-15) displays the intensity of icing at **all** altitudes from 1,000 feet MSL to FL300. Icing intensity is displayed using standard icing intensity categories: trace, light, moderate and heavy.

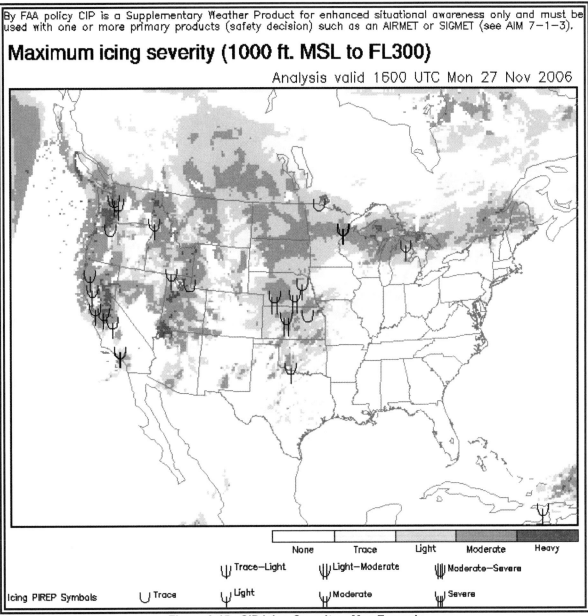

Figure 9-15. CIP Icing Severity – Max Example

The lightest blue color represents trace icing. As the blue color shades become darker, the icing intensity increases. The darkest blue color represents heavy icing. White regions indicate where no probability of icing exists and, therefore, no intensity is necessary.

9.3.2.5 Icing Severity – Probability > 25%

The Icing Severity – Probability > 25% product (Figure 9-16) depicts, at a single altitude, where a 26 to 100 percent probability exists for the indicated icing intensity. Icing intensity is displayed using standard icing intensity categories: trace, light, moderate and heavy.

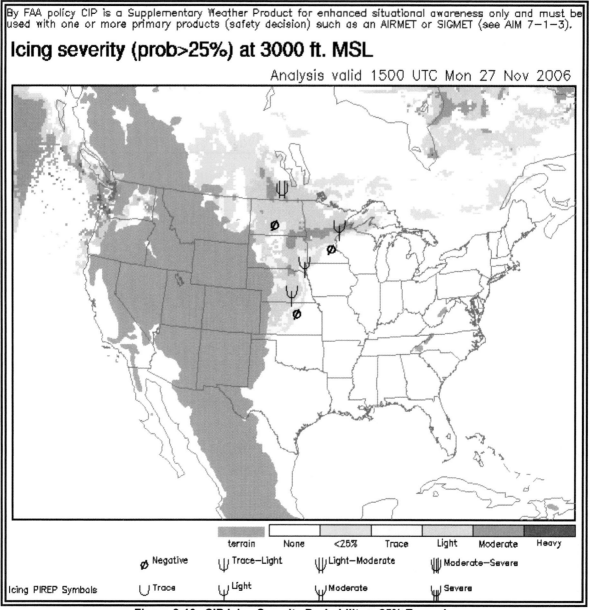

By FAA policy CIP is a Supplementary Weather Product for enhanced situational awareness only and must be used with one or more primary products (safety decision) such as an AIRMET or SIGMET (see AIM 7-1-3).

Icing severity (prob>25%) at 3000 ft. MSL

Analysis valid 1500 UTC Mon 27 Nov 2006

terrain | None | <25% | Trace | Light | Moderate | Heavy

Icing PIREP Symbols

Negative | Trace–Light | Light–Moderate | Moderate–Severe
Trace | Light | Moderate | Severe

Figure 9-16. CIP Icing Severity Probability > 25% Example

The lightest blue color represents trace icing. As the blue color shades become darker, the icing intensity increases. The darkest blue color represents heavy icing. White regions indicate where no probability of icing exists and, therefore, no intensity is necessary. Brown regions indicate higher-elevation terrain extending above the altitude of the particular graphic. A gray color is used to mask the intensity pixels where the probability of icing is 25% or less.

9.3.2.6 Icing Severity – Probability > 25% - Maximum
The Icing Severity – Probability > 25% - Maximum product (Figure 9-17) depicts, at **all** altitudes from 1,000 feet MSL to FL300, where the probability of the indicated icing intensity is 26 to 100 percent. Icing intensity is displayed using standard icing intensity categories: trace, light, moderate, heavy.

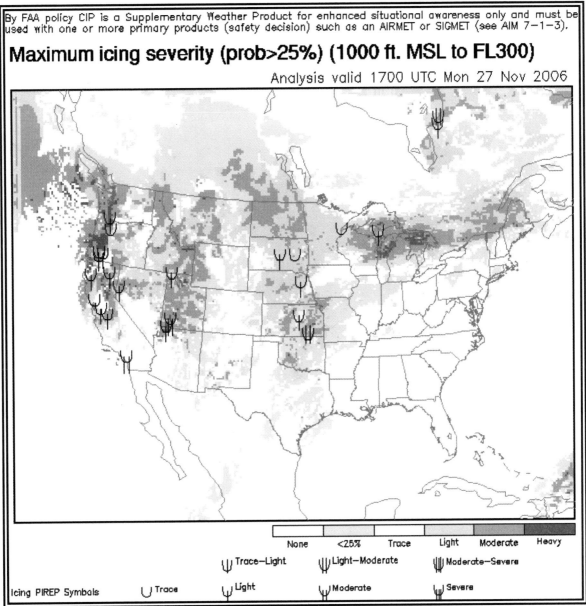

Figure 9-17. CIP Icing Severity Probability > 25% - Max Example

The lightest blue color represents trace icing. As the blue color shades become darker, the icing intensity increases. The darkest blue color represents heavy icing. White regions indicate where no probability of icing exists and, therefore, no intensity is necessary. A gray color is used to mask the intensity pixels where the probability of icing is 25% or less.

9.3.2.7 Icing Severity – Probability > 50%
The Icing Severity – Probability > 50% product (Figure 9-18) depicts, at a single altitude, where the probability of the indicated icing intensity 51 to 100 percent. Icing intensity is displayed using standard icing intensity categories: trace, light, moderate and heavy.

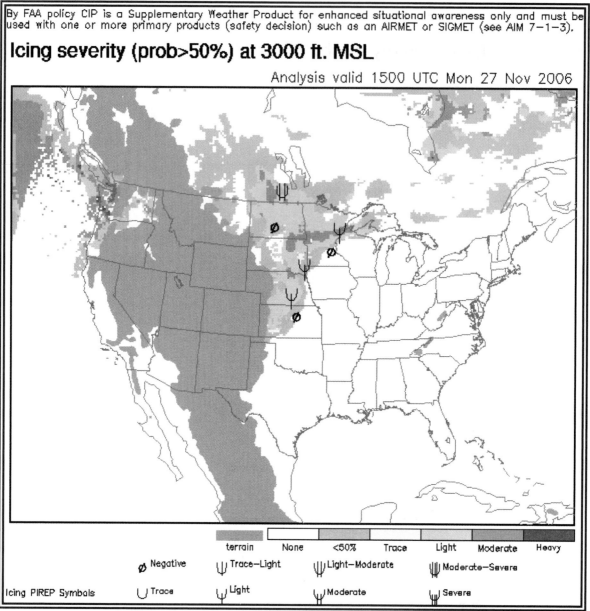

Figure 9-18. CIP Icing Severity Probability > 50% Example

The lightest blue color represents trace icing. As the blue color shades become darker, the icing intensity increases. The darkest blue color represents heavy icing. White regions indicate where no probability of icing exists and, therefore, no intensity is necessary. Brown regions indicate where higher-elevation terrain extends above the altitude of the particular graphic. A gray color is used to mask the intensity pixels where the probability of icing is 50% or less.

9.3.2.8 Icing Severity – Probability > 50% - Maximum

The Icing Severity – Probability > 50% - Maximum product (Figure 9-19) depicts, at **all** altitudes from 1,000 feet MSL to FL300, where the probability of the indicated icing intensity is 51 to 100 percent. Icing intensity is displayed using standard icing intensity categories: trace, light, moderate and heavy.

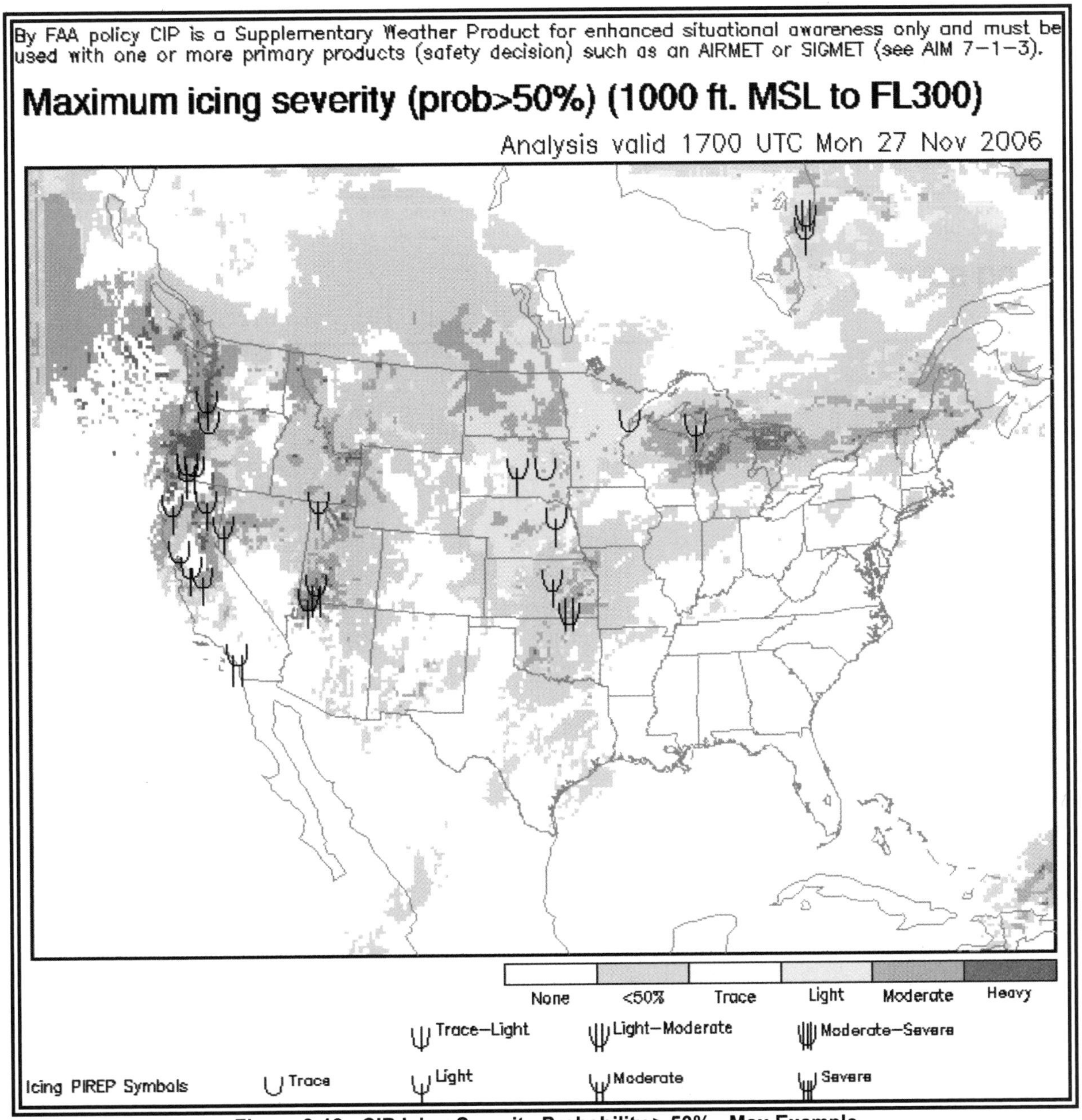

Figure 9-19. CIP Icing Severity Probability > 50% - Max Example

The lightest blue color represents trace icing. As the blue color shades become darker, the icing intensity increases. The darkest blue color represents heavy icing. White regions indicate where no probability of icing exists and, therefore, no intensity is necessary. A gray color is used to mask the intensity pixels where the probability of icing is 50% or less.

9.3.2.9   Icing Severity plus Supercooled Large Droplets (SLD)

The Icing Severity plus Supercooled Large Droplets (SLD) product (Figure 9-20) depicts, at a single altitude, the intensity of icing expected as well as locations where a threat for SLD exists.

SLD is defined as supercooled water droplets larger than 50 micrometers in diameter. These size droplets include freezing drizzle and/or freezing rain aloft. SLD, which are outside the icing certification envelopes (FAR Part 25 Appendix C), can be particularly hazardous to some aircraft.

Icing intensity is displayed using standard icing intensity categories: trace, light, moderate and heavy.

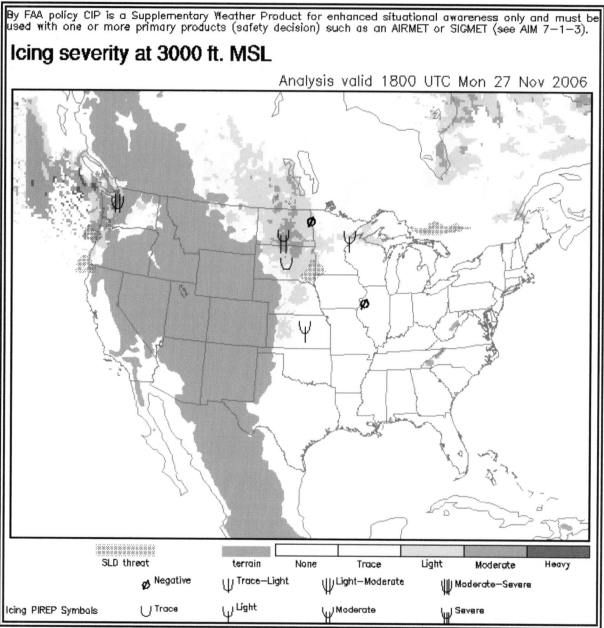

Figure 9-20. CIP Icing Severity plus Supercooled Large Droplets (SLD) Example

The lightest blue color represents trace icing. As the blue color shades become darker, the icing intensity increases. The darkest blue color represents heavy icing. White regions indicate where no probability of icing exists and, therefore, no intensity is necessary. Brown regions indicate where higher-elevation terrain extends above the altitude of the particular graphic. Locations where a threat for SLD exists are depicted with red hatching.

9.3.2.10 Icing Severity plus Supercooled Large Droplets (SLD) - Maximum

The Icing Severity plus Supercooled Large Droplets (SLD) product (Figure 9-21) depicts at all altitudes, between 1,000 feet MSL and FL300, the intensity of icing expected as well as locations where a threat for SLD exists.

SLD is defined as supercooled water droplets larger than 50 micrometers in diameter. These size droplets include freezing drizzle and/or freezing rain aloft. SLD, which are outside the icing certification envelopes (FAR Part 25 Appendix C), can be particularly hazardous to some aircraft.

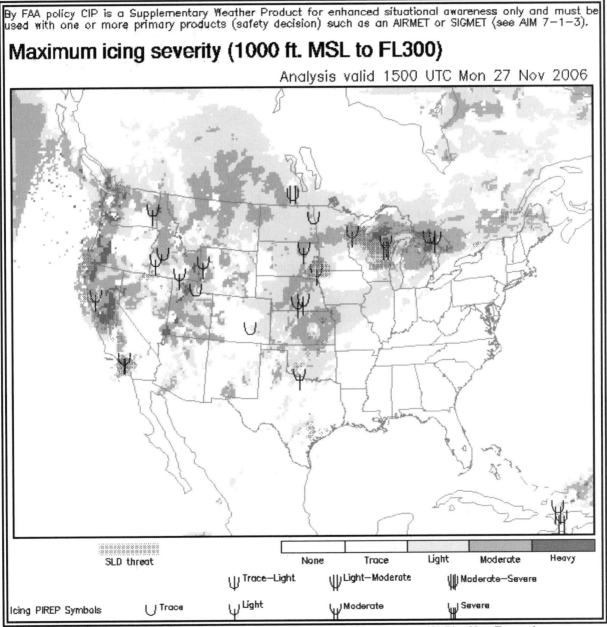

Figure 9-21. CIP Icing Severity plus Supercooled Large Droplets (SLD) – Max Example

The lightest blue color represents trace icing. As the blue color shades become darker, the icing intensity increases. The darkest blue color represents heavy icing. White regions indicate where no probability of icing exists and, therefore, no intensity is necessary. Locations where a threat for SLD exists are depicted with red hatching.

9.3.3 Strengths and Limitations

9.3.3.1 Strengths
The CIP product suite is updated hourly and provides a diagnostic tool to assist in determining the probability for icing, the intensity of icing and the threat for SLD.

9.3.3.2 Limitations
Actual icing severity may be different than what is depicted on the CIP graphics and plotted PIREPS because:

- Different aircraft types experience different severities of icing in the same atmospheric environments. Severity definitions are currently pilot-based and thus are a function of the aircraft type, flight phase (takeoff/landing, cruise, etc.), aircraft configuration, as well as the pilot's experience and perception of the icing hazard.

- Assessing the amount and drop size of supercooled liquid water (SLW) in the atmosphere is difficult.

- The Icing Severity products depict the severity of the meteorological icing environment and **not** the resultant icing that may occur on the aircraft.

9.3.4 Uses
The CIP Icing Probability product can be used to identify the current three-dimensional probability of icing.

The CIP Icing Severity product can be used to determine the intensity of icing. The CIP Icing Severity – Probability > 25% or Probability > 50% depicts the probability of a given intensity of icing occurring.

Finally the Icing Severity plus SLD product can help in determining the threat of SLD which is particularly hazardous to some aircraft.

Icing PIREPs are plotted on single altitude graphics if the PIREP is within 1,000 feet of the graphic's altitude and has been observed within 75 minutes of the chart's valid time. On CIP Max product, PIREPs for all altitudes (i.e. 1,000 feet MSL to FL300) are displayed. However, negative reports of icing are not plotted on the CIP Max product in an effort to reduce clutter. The PIREP legend is located on the bottom of each graphic.

9.4 Forecast Icing Potential (FIP)

The Forecast Icing Potential (FIP) provides a three-dimensional forecast of icing potential (or likelihood) using numerical weather prediction model output (Figure 9-22). The FIP product suite is automatically generated with no human modifications. It may be used as a higher resolution supplement to AIRMETs and SIGMETs but is **not** a substitute for them. It is authorized for operational use **only** by meteorologists and dispatchers. The forecast area covers the 48-contiguous states, much of Canada and Mexico and their respective coastal waters.

9.4.1 Issuance

The FIP is issued every hour and generates hourly forecast for 3 hours into the future. For example, forecasts issued at 1300Z would be valid for 1400Z, 1500Z and 1600Z. Six-, 9-, and 12-hour forecasts are issued every three hours beginning at 00Z. For example, a forecast suite issued at 0300Z would have valid times at 0900Z, 1200Z and 1500Z respectively. The product is issued by the Aviation Weather Center (AWC) and is available through the Aviation Digital Data Service (ADDS) web site at: http://adds.aviationweather.noaa.gov/icing/icing_nav.php.

9.4.2 Content

The FIP forecasts the likelihood of icing from super-cooled liquid water droplets. The likelihood field ranges from 0 (no icing) to 100 (icing likely). The scale depicts likelihood of icing using "cool" and "warm" colors, with warmer colors indicating a higher likelihood of icing. Regions depicted in white indicate zero icing potential according to the CIP. Brown regions indicate areas of terrain.

The scale is not calibrated as a true probability value. It does, however, have value in pointing out differences in the likelihood of encountering icing at a given location. For example, a value of 70 does not indicate there is a 70 percent chance of encountering icing. However, when comparing it to other higher or lower values will indicate if there is a greater or lesser likelihood of encountering icing. No information is provided as to the severity of icing and none should be inferred. FIP output is available for 1,000 foot vertical intervals. ADDS displays every third level except on the Flight Path Tool which provides access to all levels.

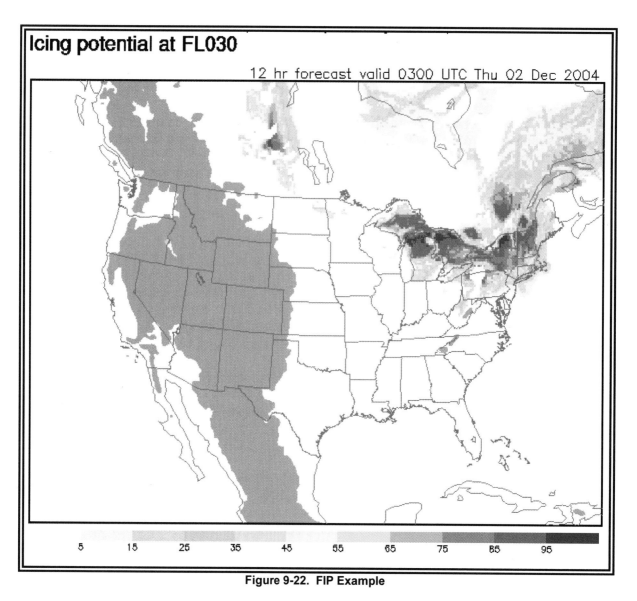

Figure 9-22. FIP Example

9.4.3 Strengths and Limitations

Strengths

- The FIP can be used to help determine the forecast for potential for icing through the entire vertical depth of the atmosphere.

- The product is updated hourly.

Limitations

- The product does not display any information about the severity of icing.

- The product only displays the forecast potential for icing, not the absolute probability.

- It is only approved for use by meteorologists and dispatchers.

• The product is generated without human modification. Therefore, the forecasts are only as accurate the computer model output used to create them.

9.4.4 Use

The FIP is primarily used to help determine the likelihood of icing at the specified forecast valid times.

9.5 Graphical Turbulence Guidance (GTG)

The Graphical Turbulence Guidance (GTG) product suite provides a three-dimensional diagnosis and forecast of Clear Air Turbulence (CAT) potential. The GTG is created using a combination of computer model output and turbulence observations. The GTG product suite, which consists of the GTG analysis and forecast and Composite analysis and forecast, is automatically generated with no human modifications. It may be used as a supplement to AIRMETs and SIGMETs but is **not** a substitute for them. It is authorized for operational use **only** by meteorologist and dispatchers. The GTG is available for the 48-contiguous states, much of Canada and Mexico and their respective coastal waters.

9.5.1 Issuance
The GTG analysis is issued hourly. The 3-, 6-, 9-, and 12-hour forecasts are issued every three hours beginning at 00Z. For example, a forecast suite issued at 0600Z would have valid times at 0900Z, 1200Z, 1500Z, and 1800Z respectively. The GTG Analysis and Forecasts, along with the Composite products, are issued by the Aviation Weather Center (AWC) and are available through the Aviation Digital Data Service (ADDS) web site at: http://adds.aviationweather.noaa.gov/turbulence/turb_nav.php.

9.5.2 Content

9.5.2.1 GTG Analysis and Forecast
The GTG Analysis and Forecast graphics depict the location and intensity of potential Clear Air Turbulence (CAT) (Figure 9-23). Standard intensity terminology is used. The GTG output is available for 1,000 foot vertical intervals between FL200 and FL450. ADDS turbulence page displays every third level starting at FL210 level, while the ADDS Flight Path Tool provides access to all levels.

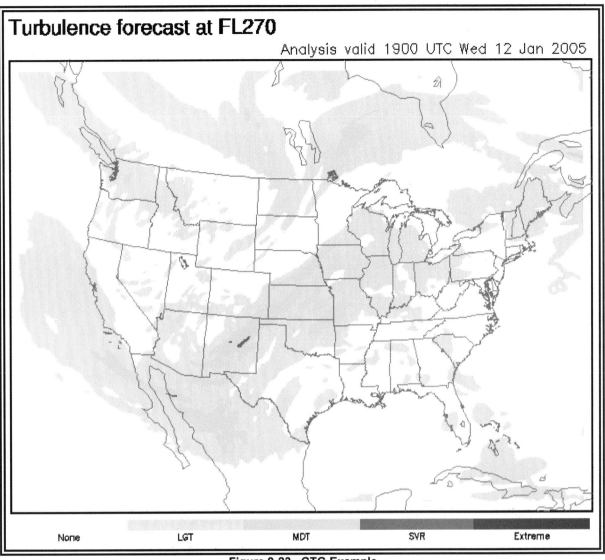

Figure 9-23. GTG Example

9.5.2.2 GTG Composite Analysis and Forecast

The GTG Composite products display the **maximum** intensity of potential turbulence between FL200 and FL450 (Figure 9-24). In other words, at any given location, the displayed value represents the maximum potential turbulence between FL200 and FL450. Single altitude graphics must be examined to determine the altitude of the potential turbulence.

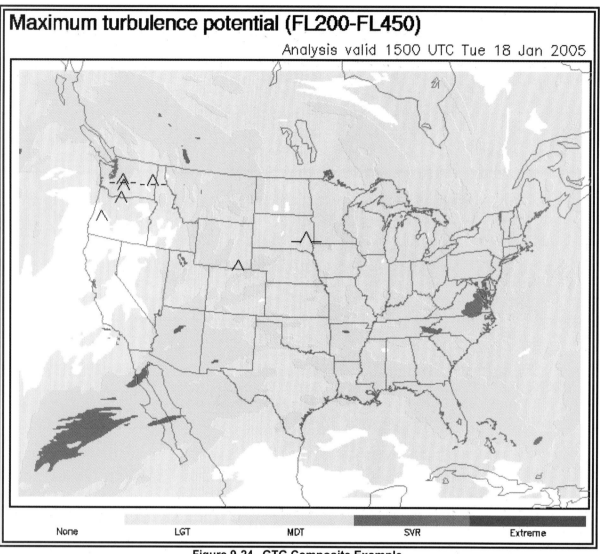

Figure 9-24. GTG Composite Example

9.5.3 Strengths and Limitations

The GTG provides an hourly, high resolution analysis and forecast of clear-air turbulence potential at and above FL200. However, the product is only for turbulence associated with upper level fronts and jet streams while other known causes of turbulence are not forecasted.

Strengths

- The product is issued hourly

- Turbulence is plotted to a high resolution.

Limitations

- The accuracy of the analysis is dependent on the number of PIREPs available. Typically at night fewer pilot reports are received, so the accuracy decreases.

• For the GTG forecast, the product is only as accurate the computer model output used to create them.

• The product only displays clear air turbulence (CAT) for upper-level fronts and the jet stream.   Other known causes of turbulence are not included in the product.

• Data and forecast are only available for FL200 and above.

• It is only approved for use by meteorologists and dispatchers.

9.5.4 Use

The Composite product can provide a quick method to determine what the greatest potential of current or forecast turbulence is at a given location. However, to determine the turbulence potential at any given altitude, the individual altitude products must be viewed.

The ADDS web site allows users to overlay turbulence PIREPS on the single altitude graphics. For the PIREP to be plotted on the single altitude product, it must be located within 1,000 feet vertically of the altitude and have been reported within 90 minutes of the chart time. For example, if a user viewed the FL240 GTG product with a valid time of 1400Z the displayed PIREPS could be located between FL230 and FL250 and reported between 1230Z and 1400Z.

9.6 Meteorological Impact Statement (MIS)

A Meteorological Impact Statement (MIS) is an unscheduled flow control and flight operations planning forecast issued by Center Weather Service Units (CWSUs) (Figures 9-25, 9-26, and 9-27). It is a forecast and briefing product for personnel at Air Route Traffic Control Centers (ARTCCs), Air Traffic Control System Command Center (ATCSCC), Terminal Radar Approach Control Facilities (TRACONS) and Airport Traffic Control Towers (ATCTs) responsible for making flow control-type decisions.

A MIS may be tailored to meet the unique requirements of the host ARTCC. These special requirements will be coordinated between the host ARTCC and the CWSU.

MISs are available on the Aviation Weather Center (AWC) web site at: http://aviationweather.gov/products/cwsu/.

9.6.1 Valid Period
A MIS is valid up to 12 hours after issuance time and details weather conditions expected to adversely impact air traffic flow in the CWSU area of responsibility. The MIS can be immediately effective for existing conditions when CWSU operations begin or for rapidly deteriorating conditions or be effective up to two hours in advance of expected conditions.

9.6.2 MIS Criteria
A MIS enables Air Traffic Control (ATC) facility personnel to include the impact of specific weather conditions in their flow control decision-making. At a minimum, a MIS should be issued when:

- Any of the following conditions occur, are forecast to occur, and, if previously forecast, are no longer expected:

    o Conditions meeting convective SIGMET criteria (Section 5.1.8)

    o Icing - moderate or greater

    o Turbulence - moderate or greater

    o Heavy precipitation

    o Freezing precipitation

    o Conditions at or approaching Low IFR

    o Surface winds/gusts >30 knots

    o Low Level Wind Shear (surface - 2,000 feet)

    o Volcanic ash, dust storms, or sandstorms; and

- In the forecaster's judgment, the conditions listed above, or any others, will adversely impact the flow of air traffic within the ARTCC area of responsibility.

9.6.3 MIS Issuance

MIS phenomena forecasts use the location reference point identifiers depicted on the In-Flight Advisory Plotting Chart (Appendix F), and include the height, extent, and movement of the conditions. MIS product issuances are numbered sequentially beginning at Midnight local time each day. The MIS is disseminated and stored as a "replaceable" product. Therefore, each issuance will contain the details of all pertinent known conditions meeting MIS issuance criteria, including ongoing conditions described in previously issued MISs.

The MIS is distributed to ARTCC personnel, including Traffic Management Unit (TMU) personnel.

9.6.4 Format

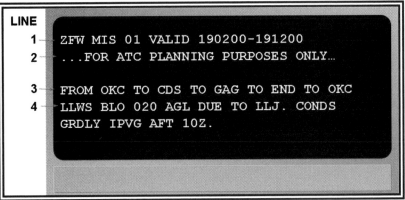

Figure 9-25. Meteorological Impact Statement (MIS) Decoding Example

Table 9-5. Meteorological Impact Statement (MIS) – Decoding

LINE	CONTENT	DESCRIPTION
1	ZFW	ARTCC Identification
	MIS	Product type
	01	Issuance number
	VALID 190200-191200	Valid period UTC date/time
2	...FOR ATC PLANNING PURPOSES ONLY...	Product use statement
3	FROM OKC TO CDS TO GAG TO END TO OKC	Phenomenon location
4	LLWS BLO 020 AGL DUE TO LLJ. CONDS GRDLY IPVG AFT 10Z	Phenomenon description

Any remarks such as "SEE CONVECTIVE SIGMET 8W"; "NO UPDATES AVBL AFT 0230Z"; and forecaster initials and/or facility identifier may be placed at the end of the MIS.

If the phenomenon described in a MIS is no longer expected, a cancellation MIS message may be issued. The FAA header does not contain an issuance number. If the phenomenon described in the MIS is expected to continue beyond the operating hours of the CWSU, then the remark "NO UPDATES AFT ttttZ" (where "ttttZ" is the UTC closing time of the CWSU) is added at the text end.

9.6.5 Examples

```
ZOA MIS 01 VALID 041415-041900
...FOR ATC PLANNING PURPOSES ONLY...
FOR SFO BAY AREA
BR/FG WITH CEILING BLW 005 AND VIS OCNL BLW 1SM.
ZOA CWSU
```

Meteorological Impact Statement issued by the Freemont, California CWSU. First MIS issuance of the day, valid from the 4th day of the month at 1415 UTC, to the 4th day of the month at 1900 UTC. For air traffic control planning purposes only. For the San Francisco Bay Area...mist and fog with ceilings below 500 feet MSL and visibility occasionally below 1 statute mile.

```
ZOA MIS 02 VALID 041650
...FOR ATC PLANNING PURPOSES ONLY...
FOR SFO BAY AREA
CNL ZOA MIS 01. CONDS HAVE IMPRD.
ZOA CWSU
```

Meteorological Impact Statement issued by the Freemont, California CWSU. The second MIS issuance of the day, valid the 4th day of the month at 1650 UTC. For air traffic control planning purposes only. For the San Francisco Bay Area. Cancel Freemont, California Meteorological Impact Statement number 1. Conditions have improved.

```
ZID MIS 03 VALID 041200-042330
...FOR ATC PLANNING PURPOSES ONLY...
FROM IND TO 17WSW APE TO LOZ TO 13NE PXV TO IND
TIL 21Z MOD TURB FL310-390 DUE TO JTST WS.
ZID W OF A LINE FM FWA TO BWG
AFT 18Z OCNL SEV TSGR TOPS TO FL450. MOV FM 24035KT. MAX SFC WINDS
60KT.
ZID E OF A LINE FM FWA TO 35SE BKW
MOD MXD ICE IN CLDS/PRECIPITATION 020-120. CONDS ENDING W OF
A 40S CLE TO 20NE BKW LINE BY 19Z.
ZID CWSU
```

Meteorological Impact Statement issued by the Indianapolis, Indiana CWSU. The third MIS issuance of the day, valid from the 4th day of the month at 1200 UTC to the 4th day of the month at 2130 UTC. For air traffic control planning purposes only. From Indianapolis, Indiana to 17 nautical miles west-southwest of Appleton, Ohio to London, Kentucky to 13 nautical miles northeast of Pocket City, Indiana to Indianapolis, Indiana. Until 21Z, moderate turbulence between flight level 310 and flight level 390 due to jet stream wind shear.

For the Indianapolis ARTCC airspace west of a line from Fort Wayne, Indiana to Bowling Green, Kentucky. After 18Z, occasional severe thunderstorms, hail, tops to flight level 450. Moving from 240 degrees at 35 knots. Maximum surface winds 60 knots.

For the Indianapolis, Indiana ARTCC airspace east of a line form Fort Wayne, Indiana to 35 nautical miles southeast of Beckley, West Virginia. Moderate mixed icing in clouds and precipitation between 2,000 feet to 12,000 feet MSL. Conditions ending west of a line from 40 nautical miles south of Cleveland, Ohio to 20 nautical miles northeast of Beckley, West Virginia by 1900Z.

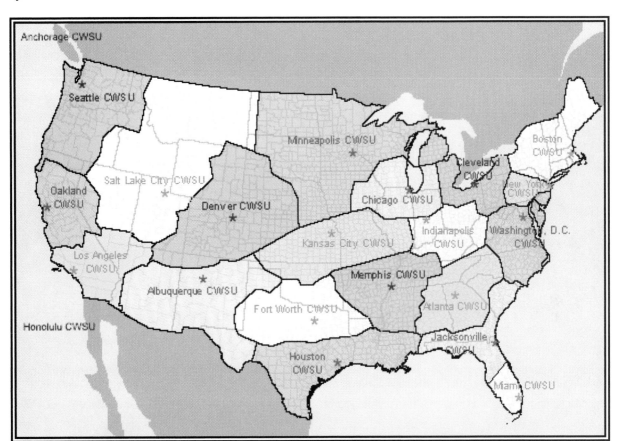

Figure 9-26. Center Weather Service Unit (CWSU) Areas of Responsibility - CONUS

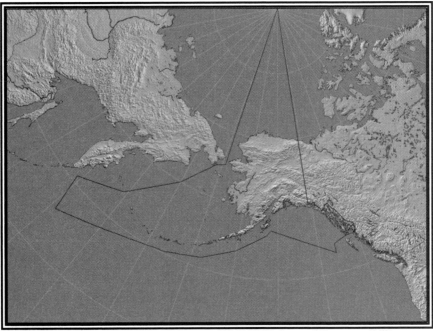

Figure 9-27. CWSU Anchorage, AK (PAZA) Area of Responsibility

10 APPENDIX A: ASOS AND AWOS LOCATIONS

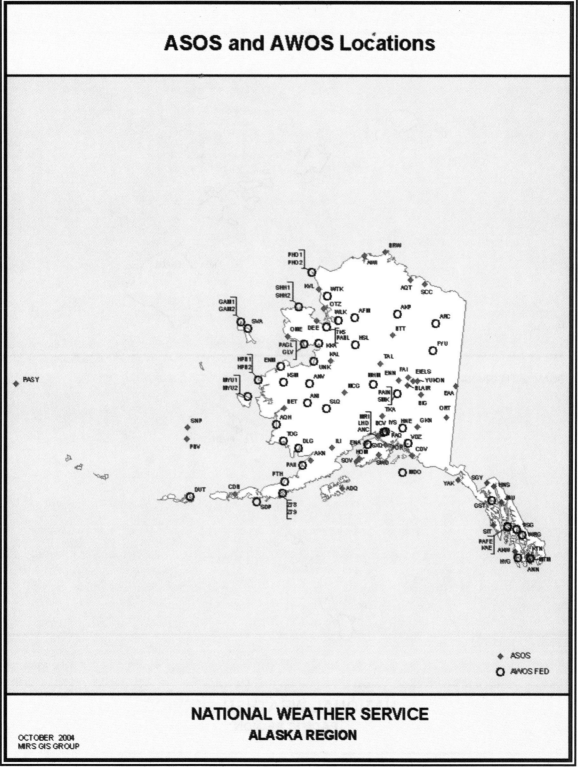

Figure A-1. ASOS and AWOS Locations – NWS Alaska Region

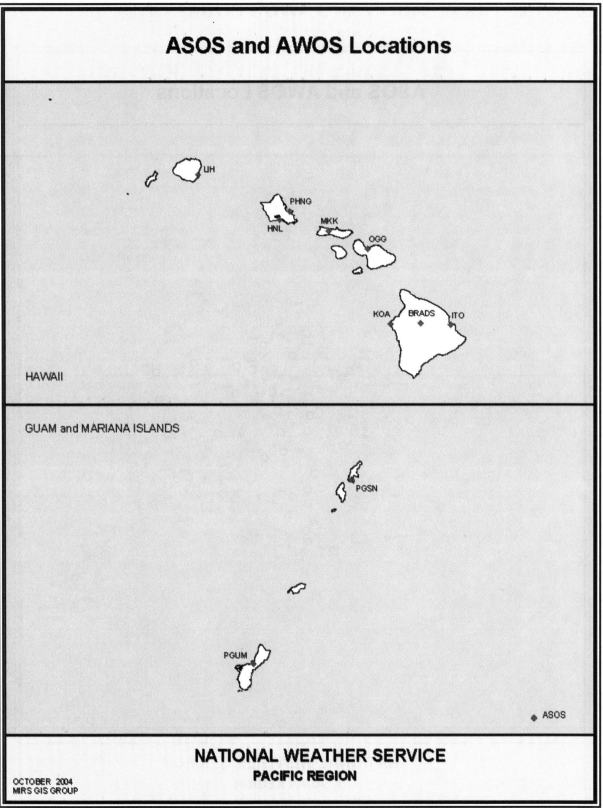

Figure A-2. ASOS and AWOS Locations – NWS Pacific Region

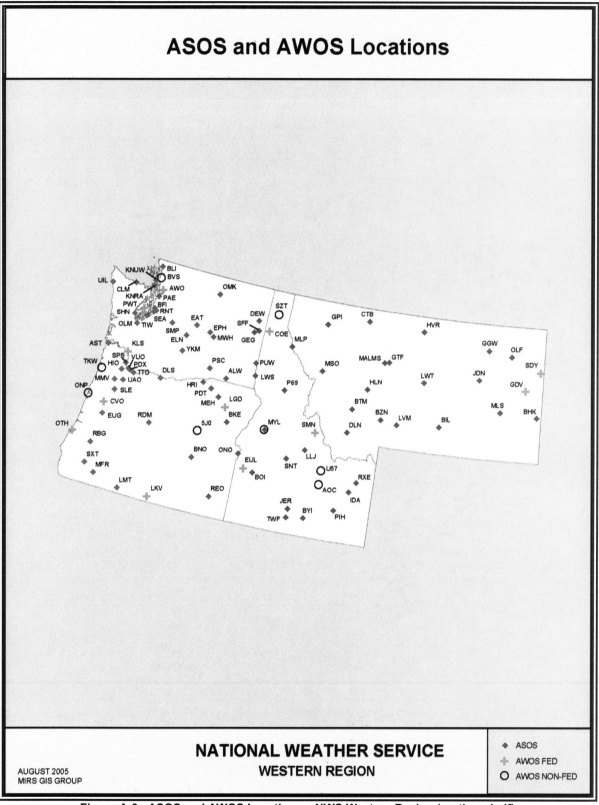

Figure A-3. ASOS and AWOS Locations – NWS Western Region (northern half)

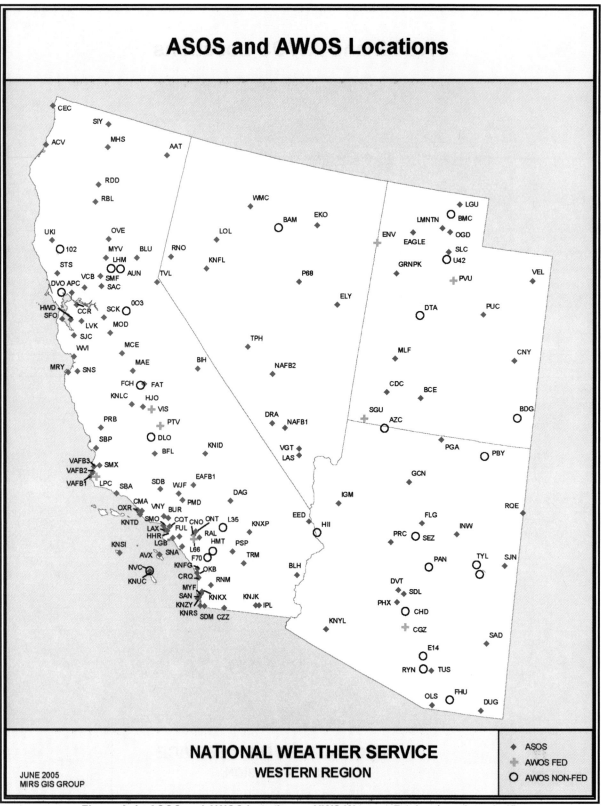

Figure A-4. ASOS and AWOS Locations – NWS Western Region (southern half)

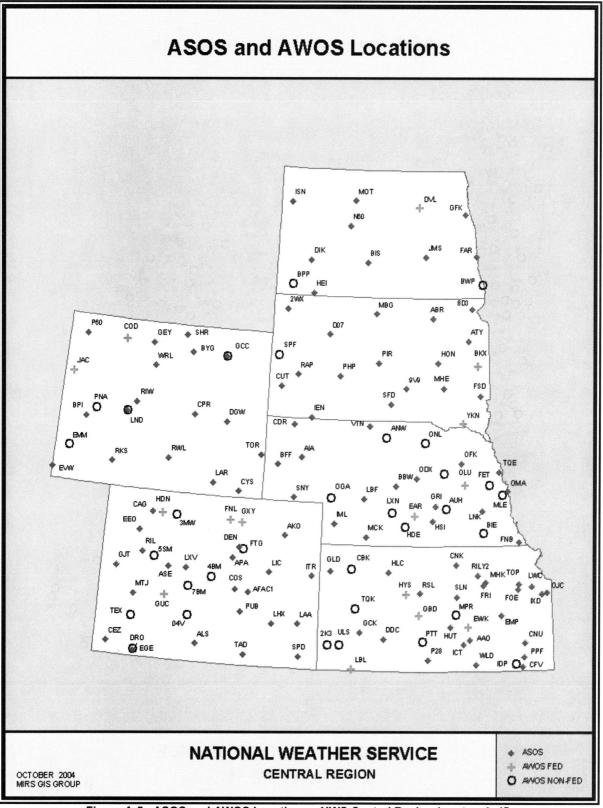

Figure A-5. ASOS and AWOS Locations – NWS Central Region (western half)

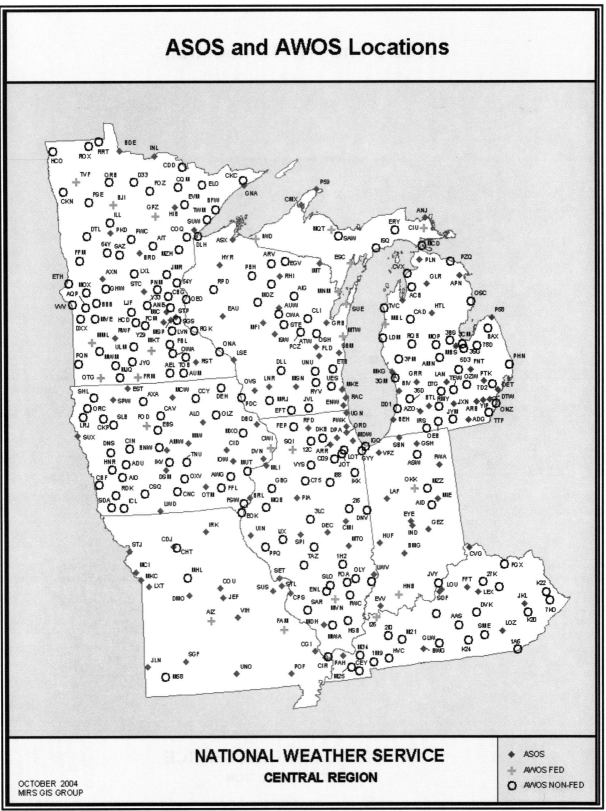

Figure A-6. ASOS and AWOS Locations – NWS Central Region (eastern half)

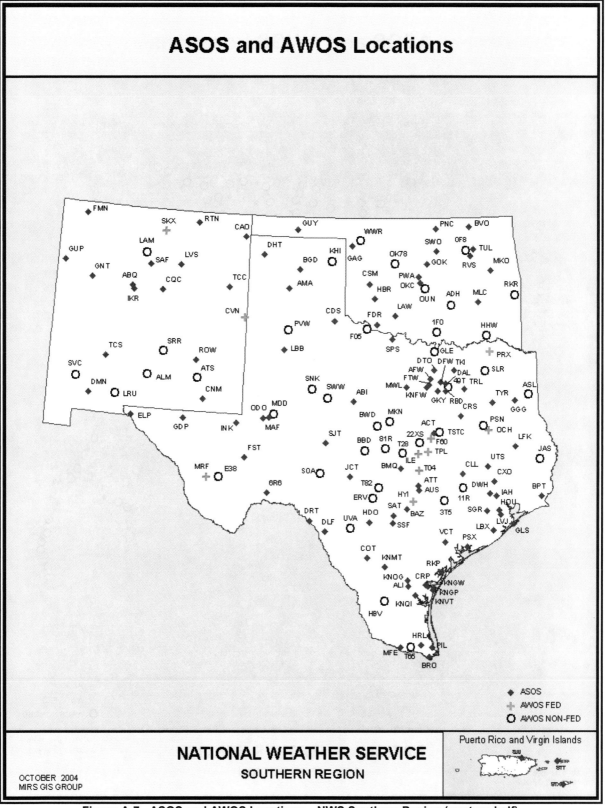

Figure A-7. ASOS and AWOS Locations – NWS Southern Region (western half)

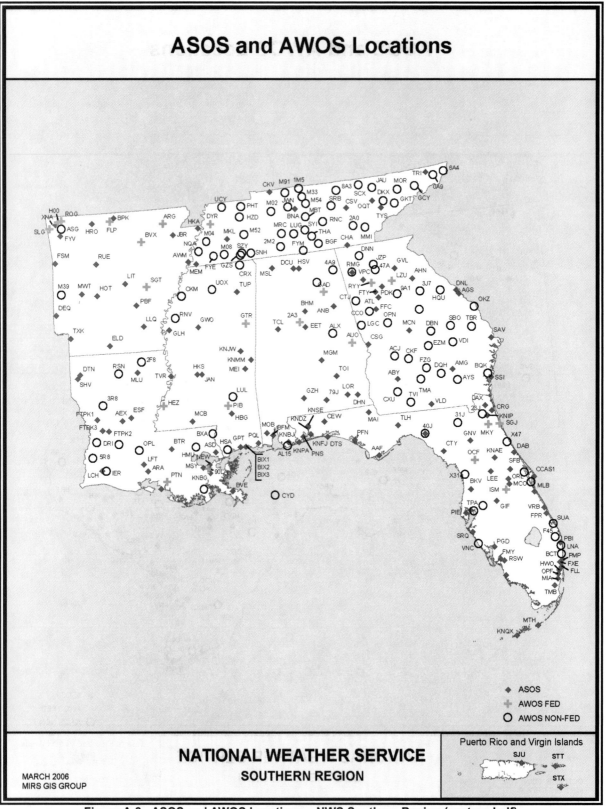

Figure A-8. ASOS and AWOS Locations – NWS Southern Region (eastern half)

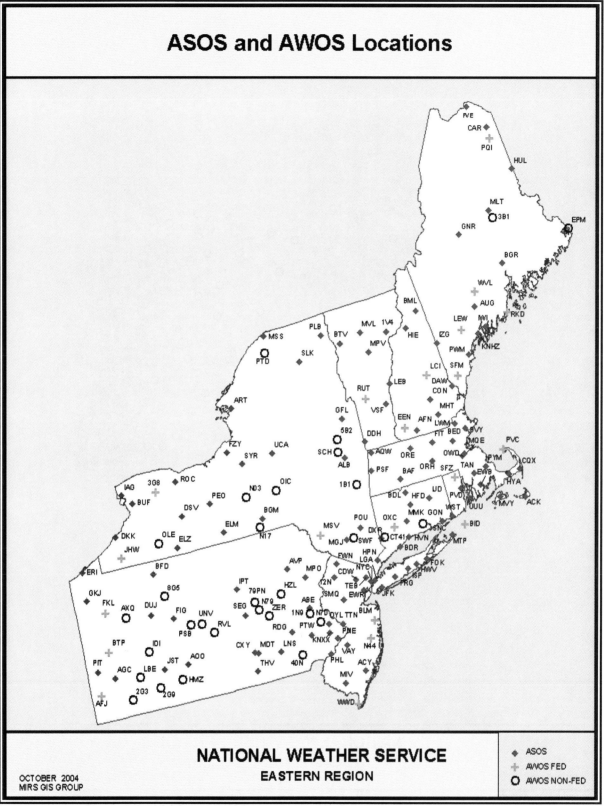

Figure A-9. ASOS and AWOS Locations – NWS Eastern Region (northern half)

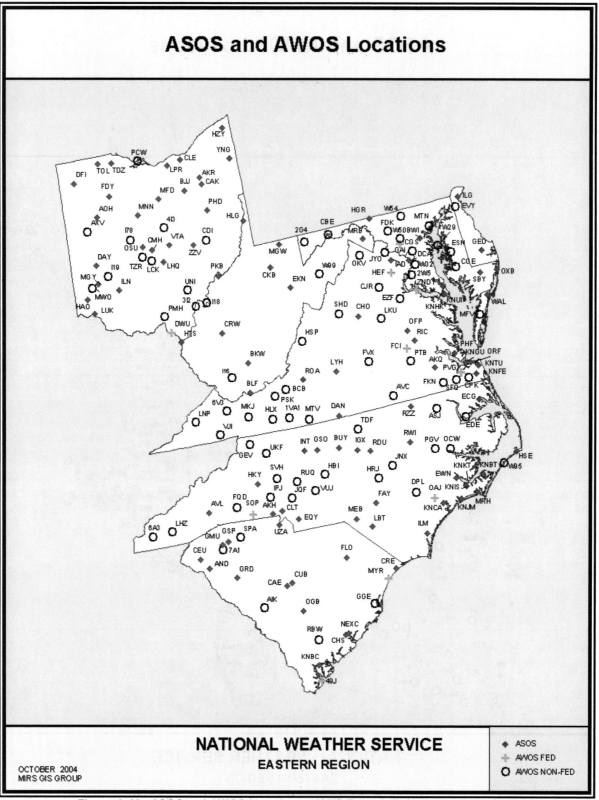

Figure A-10. ASOS and AWOS Locations – NWS Eastern Region (southern half)

11 APPENDIX B: CONTRACTIONS AND ACRONYMS

A

AAA	(or AAB, AAC…etc., in sequence) Amended meteorological message (message type designator)
AAWU	Alaskan Aviation Weather Unit
ABNDT	Abundant
ABNML	Abnormal
ABT	About
ABV	Above
AC	Altocumulus or Convective Outlook
ACARS	Aircraft communication and addressing system
ACC	Altocumulus Castellanus
ACCID	Notification of an aircraft accident
ACCUM	Accumulate
ACFT	Aircraft
ACPY	Accompany
ACRS	Across
ACSL	Altocumulus Standing Lenticular
ACT	Active or activated or activity
ACYC	Anticyclone
ADA	Advisory area
ADDN	Addition or additional
ADJ	Adjacent
ADQT	Adequate
ADQTLY	Adequately
ADRNDCK	Adirondack
ADVCT	Advect
ADVCTD	Advected
ADVCTG	Advecting
ADVCTN	Advection
ADVCTS	Advects
ADVN	Advance
ADVNG	Advancing
ADVY	Advisory
ADVYS	Advisories
AFCT	Affect
AFCTD	Affected
AFCTG	Affecting

AFT	After…(time or place)
AFTN	Afternoon
AFSS	Automated Flight Service Station
AGL	Above ground level
AGN	Again
AGRD	Agreed
AGRS	Agrees
AGRMT	Agreement
AHD	Ahead
AIREP	Air-report
AIRMET	Airman's Meteorological Information (Information concerning en-route weather phenomena which may affect the safety of low-level aircraft operations)
AK	Alaska
AL	Alabama
ALF	Aloft
ALG	Along
ALGHNY	Allegheny
ALQDS	All quadrants
ALSTG	Altimeter setting
ALT	Altitude
ALTA	Alberta
ALTHO	Although
ALTM	Altimeter
ALUTN	Aleutian
AMD	Amend or amended (used to indicate amended meteorological message; message type designator)
AMDG	Amending
AMDT	Amendment
AMP	Amplify
AMPG	Amplifying
AMPLTD	Amplitude
AMS	
AMSL	Above mean sea level
AMT	Amount
ANLYS	Analysis
ANS	Answer
AOA	At or above
AOB	At or below

AP	Airport or anomalous propagation
APCH	Approach
APCHG	Approaching
APCHS	Approaches
APLCN	Appalachian
APLCNS	Appalachians
APPR	Appear
APPRG	Appearing
APPRS	Appears
APR	April
APRNT	Apparent
APRX	Approximate or approximately
AR	Arkansas
ARFOR	Area Forecast (in aeronautical meteorological code)
ARND	Around
AS	Altostratus
ASC	Ascend to or ascending to
ASSOCD	Associated
ASSOCN	Association
AT…	At (followed by time at which weather change is forecast to occur)
ATLC	Atlantic
ATP	At…(time or place)
ATTM	At this time
ATTN	Attention
AUG	August
AVBL	Available or availability
AVG	Average
AWC	Aviation Weather Center
AWT	Awaiting
AZ	Arizona
AZM	Azimuth

B

BACLIN	Baroclinic
BAJA	Baja, California
BASE	Cloud base
BATROP	Barotropic
BC	British Columbia
BCFG	Fog patches
BCH	Beach
BCKG	Backing

BDA	Bermuda
BDRY	Boundary
BECMG	Becoming
BFR	Before
BGN	Begin
BGNG	Beginning
BGN	Begins
BHND	Behind
BINOVC	Breaks in overcast
BKN	Broken
BL…	Blowing (followed by DU = dust, SA = sand or SN = snow)
BLD	Build
BLDG	Building
BLDUP	Buildup
BLKHLS	Black Hills
BLKT	Blanket
BLKTG	Blanketing
BLKTS	Blankets
BLO	Below clouds
BLW	Below…
BLZD	Blizzard
BND	Bound
BNTH	Beneath
BR	Mist
BRF	Brief
BRK	Break
BRKG	Breaking
BRKHIC	Breaks in higher clouds
BRKS	Breaks
BRKSHR	Berkshire
BRM	Barometer
BLDU	Blowing Dust
BLSA	Blowing Sand
BLSN	Blowing Snow
BTL	Between layers
BTN	Between
BYD	Beyond

C

C	Degrees Celsius (Centigrade) or centre (runway identification)
CA	California
CAA	Cold air advection
CAPE	Convective Available

	Potential Energy	CNTY	County
CARIB	Caribbean	CNTYS	Counties
CASCDS	Cascades	CNVG	Converge
CAT	Category or Clear air turbulence	CNVGG	Converging
		CNVGNC	Convergence
CAVO	Visibility, cloud and present weather better than prescribed values or conditions	CNVTN	Convection
		CNVTV	Convective
		CNVTVLY	Convectively
		CONFDC	Confidence
		CO	Colorado
CB	Cumulonimbus	COMPR	Compare
CC	Cirrocumulus	COMPRG	Comparing
CCA	(or CCB, CCC…etc., in sequence) Corrected meteorological message	COMPRD	Compared
		COMPRS	Compares
		COND	Condition
		CONS	Continuous
CCLDS	Clear of clouds	CONT	Continue(s) or continued
CC	Counterclockwise	CONTLY	Continually
CCSL	Cirrocumulus Standing Lenticular	CONTG	Continuing
		CONTRAILS	Condensation trails
CDFNT	Cold front	CONTDVD	Continental Divide
CDN	Coordination	CONUS	Continental U.S.
CFP	Cold front passage	COORD	Coordinates
CHC	Chance	COR	Correct or correction or corrected (used to indicate corrected meteorological message; message type designator)
CHCS	Chances		
CHG	Modification (message type designator)		
CHGD	Changed		
CHGG	Changing		
CHGS	Changes	COT	At the coast
CHSPK	Chesapeake	COV	Cover or covered or covering
CI	Cirrus		
CIG	Ceiling	CPBL	Capable
CIGS	Ceilings	CPC	Climate Prediction Center
CIT	Near or over large towns	CRLC	Circulate
CLA	Clear type of ice formation	CRLN	Circulation
CLD	Cloud	CRNR	Corner
CLDNS	Cloudiness	CRNRS	Corners
CLDS	Clouds	CRS	Course
CLKWS	Clockwise	CS	Cirrostratus
CLR	Clear(s) or cleared to…or clearance	CSDR	Consider
		CSDRBL	Considerable
CLRG	Clearing	CST	Coast
CLRS	Clears	CSTL	Coastal
CLSD	Close or closed or closing	CT	Connecticut
CM	Centimeter	CTSKLS	Catskills
CMPLX	Complex	CU	Cumulus
CNL	Cancel or cancelled	CUFRA	Cumulus Fractus
CNDN	Canadian	COV	Cover or covered or covering
CNTR	Center		
CNTRD	Centered	CWSU	Center Weather Service Unit
CNTRL	Central		

CYC	Cyclonic	DMG	Damage
CYCLGN	Cyclogenesis	DMGD	Damaged
		DMGG	Damaging
		DMNT	Dominant
D		DMSH	Diminish
		DMSHD	Diminished
D	Downward (tendency in RVR during previous 10 minutes)	DMSHG	Diminishing
		DMSHS	Diminishes
		DNG	Danger or dangerous
DABRK	Daybreak	DNS	Dense
DALGT	Daylight	DNSLP	Downslope
DBL	Double	DNSTRM	Downstream
DC	District of Columbia	DNWND	Downwind
DCR	Decrease	DOM	Domestic
DCRD	Decreased	DP	Dew point temperature
DCRG	Decreasing	DPND	Deepened
DCRGLY	Decreasingly	DPNG	Deepening
DCRS	Decreases	DPNS	Deepens
DE	Delaware or from (used to precede the call sign of the calling station) (to be used in AFS as a procedure signal)	DPR	Deeper
		DPT	Depth
		DR...	Low drifting (followed by DU =dust, SA = sand or SN = snow)
DEC	December	DRFT	Drift
DEG	Degrees	DRFTD	Drifted
DELMARVA	Delaware-Maryland-Virginia	DRFTG	Drifting
DFCLT	Difficult	DRFTS	Drift
DFCLTY	Difficulty	DRG	During
DFNT	Definite	DS	Duststorm
DFNTLY	Definitely	DSCNT	Descent
DFRS	Differs	DSIPT	Dissipate
DGNL	Diagonal	DSIPTD	Dissipated
DGNLLY	Diagonally	DSIPTG	Dissipating
DIF	Diffuse	DSIPTN	Dissipation
DIGG	Digging	DSIPTS	Dissipates
DIR	Direction	DSND	Descend
DISC	Discontinue	DSNDG	Descending
DISCD	Discontinued	DSNDS	Descends
DISCG	Discontinuing	DSNT	Distant
DISRE	Disregard	DSTBLZ	Destabilize
DISRED	Disregarded	DSTBLZD	Destabilized
DISREG	Disregarding	DSTBLZG	Destabilizing
DIST	Distance	DSTBLZS	Destabilizes
DKTS	Dakotas	DSTBLZN	Destabilization
DLA	Delay or delayed or delay (message type designator)	DTG	Date-time group
		DTRT	Deteriorate or deteriorating
DLT	Delete	DU	Dust
DLTD	Deleted	DUC	Dense upper cloud
DLTG	Deleting	DUR	Duration
DLY	Daily	DURC	During climb

DURD	During descent		ESERN	East-southeastern
DVLP	Develop		ESEWD	East-southeastward
DVLPD	Developed		ESNTL	Essential
DVLPG	Developing		ESTAB	Establish
DVLPMT	Development		EST	Estimate or estimated or estimate (message type designator)
DVLPS	Develops			
DVRG	Diverge			
DVRGG	Diverging		ETA	Estimated time of arrival or estimating arrival
DVRGNC	Divergence			
DVRGS	Diverges		ETC	Et cetera
DVV	Downward vertical velocity		ETD	Estimated time of departure or estimating departure
DWNDFTS	Downdrafts			
DZ	Drizzle		ETIM	Elapsed time
			EV	Every
			EVE	Evening
E			EWD	Eastward
			EXCLV	Exclusive
E	East or eastern longitude		EXCLVLY	Exclusively
EB	Eastbound		EXC	Except
EFCT	Effect		EXP	Expect or expected or expecting
ELEV	Elevation			
ELNGT	Elongate		EXTD	Extend or extending
ELNGTD	Elongated		EXTRAP	Extrapolate
ELSW	Elsewhere		EXTRAPD	Extrapolated
EMBD	Embedded in a layer (to indicate cumulonimbus embedded in layers of clouds)		EXTRM	Extreme
			EXTRMLY	Extremely
			EXTSV	Extensive
EMC	Environmental Modeling Center			
			F	
EMERG	Emergency			
ENCTR	Encounter		F	Degrees Fahrenheit or fixed
ENDG	Ending		FA	Area Forecast (U.S. domestic)
ENE	East-northeast			
ENELY	East-northeasterly		FAA	Federal Aviation Administration
ENERN	East-northeastern			
ENEWD	East-northeastward		FAM	Familiar
ENHNC	Enhance		FAX	Facsimile transmission
ENHNCD	Enhanced		FBL	Light (used to indicate the intensity of weather phenomena, interference or static reports, e.g. FBL RA = light rains)
ENHNCG	Enhancing			
ENHNCS	Enhances			
ENHNCMNT	Enhancement			
ENR	En route			
ENTR	Entire		FC	Funnel cloud (tornado or waterspout)
EQPT	Equipment			
ERN	Eastern		FCST	Forecast
ERY	Early		FEB	February
ERYR	Earlier		FEW	Few
ESE	East-southeast		FG	Fog
ESELY	East-southeasterly		FIG	Figure

FILG	Filling	G	Gust or green
FIR	Flight information region	GA	Georgia
FIRAV	First available	GAMET	Area forecast for low level flights
FL	Florida or Flight Level		
FLD	Field	GEN	General
FLRY	Flurry	GENLY	Generally
FLRYS	Flurries	GEO	Geographic or true
FLT	Flight	GEOREF	Geographical reference
FLUC	Fluctuating or fluctuation or fluctuated	GFS	Global Forecast System (model)
FLW	Follow(s) or following	GLFALSK	Gulf of Alaska
FLY	Fly or flying	GLFCAL	Gulf of California
FM	From	GLFMEX	Gulf of Mexico
FM…	From (followed by time weather change is forecast to begin)	GLFSTLAWR	Gulf of St. Lawrence
		GND	Ground
FMT	Format	GR	Hail
FNCTN	Function	GRAD	Gradient
FNTGNS	Frontogenesis	GRDL	Gradual
FNTLYS	Frontolysis	GRDLY	Gradually
FORNN	Forenoon	GRIB	Processed meteorological data in the form of grid point values expressed in binary form (aeronautical meteorological code)
FPM	Feet per minute		
FQTLY	Frequently		
FRI	Friday		
FRM	Form		
FRMG	Forming	GRT	Great
FRMN	Formation	GRTLY	Greatly
FRNG	Firing	GRTLKS	Great Lakes
FRONT	Front (relating to weather)	GS	Small hail and/or snow pellets
FROPA	Frontal passage		
FROSFC	Frontal surface	GSTS	Gusts
FRQ	Frequent	GSTY	Gusty
FRST	Frost	GTS	Global Telecommunication System
FRWF	Forecast wind factor		
FSS	Flight Service Station		
FST	First		
FT	Feet (dimensional unit)		
FTHR	Further	**H**	
FU	Smoke		
FVRBL	Favorable	HAZ	Hazard
FWD	Forward	HDFRZ	Hard freeze
FYI	For your information	HDSVLY	Hudson Valley
FZ	Freezing	HDWND	Head wind
FZ LVL	Freezing level	HGT	Height
FZDZ	Freezing drizzle	HI	Hawaii or high
FZFG	Freezing fog	HIFOR	High level forecast
FZRA	Freezing rain	HJ	Sunrise to sunset
		HLDG	Holding
		HLF	Half
G		HLTP	Hilltop
		HN	Sunset to sunrise

HND	Hundred	IMT	Immediate or immediately
HOL	Holiday	IMPT	Important
HPA	Hectopascal	INC	In cloud
HPC	Hydrometeorological Prediction Center	INCL	Include
		INCLD	Included
HR	Hours	INCLG	Including
HRZN	Horizon	INCLS	Includes
HTG	Heating	INCR	Increase
HURCN	Hurricane	INCRD	Increased
HUREP	Hurricane report	INCRG	Increasing
HVY	Heavy or heavy (used to indicate the intensity of weather phenomena, e.g. heavy rain = HVY RA)	INCRGLY	Increasingly
		INCRS	Increases
		INDC	Indicate
		INDCD	Indicated
		INDCG	Indicating
HVYR	Heavier	INDCS	Indicates
HVYST	Heaviest	INDEF	Indefinite
HWVR	However	INFO	Information
HWY	Highway	INOP	Inoperative
HYR	Higher	INPR	In progress
HZ	Haze	INSTR	Instrument
		INSTBY	Instability
		INTCNTL	Intercontinental
I		INTL	International
		INTMD	Intermediate
IA	Iowa	INTMT	Intermittent
IAO	In and out of clouds	INTMTLY	Intermittently
IC	Icing (PIREPs only) or ice crystals (very small ice crystals in suspension, also known as diamond dust)	INTR	Interior
		INTRP	Interrupt or interruption or interrupted
		INTRMTRGN	Intermountain region
ICAO	International Civil Aviation Organization	INT	Intersection
		INTS	Intense
ICE	Icing	INTSFCN	Intensification
ICGIC	Icing in clouds	INTSF	Intensify or intensifying
ICGICIP	Icing in clouds and in precipitation	INTST	Intensity
		INTVL	Interval
ICGIP	Icing in precipitation	INVRN	Inversion
ID	Idaho or identifier or identity	IOVC	In overcast
IDENT	Identification	INVOF	In vicinity of
IFR	Instrument flight rules	IPV	Improve
IGA	International general aviation	IPVG	Improving
		ISA	International standard atmosphere
IL	Illinois		
ILS	Instrument landing system	ISOL	Isolated
IMC	Instrument meteorological conditions		
IMD	Immediate or immediately	**J**	
IMPL	Impulse		
IMPLS	Impulses	JAN	January
IMPR	Improve or improving		

JCTN	Junction
JTST	Jet stream
JUL	July
JUN	June

K

KFRST	Killing frost
KG	Kilograms
KLYR	Smoke layer aloft
KM	Kilometers
KMH	Kilometers per hour
KOCTY	Smoke over city
KPA	Kilopascals
KS	Kansas
KT	Knots
KY	Kentucky

L

L	Left (runway identification)
LA	Louisiana
LABRDR	Labrador
LAN	Inland
LAT	Latitude
LAWRS	Limited aviation weather reporting station
LCTMP	Little change in temperature
LFTG	Lifting
LGT	Light or lighting
LGWV	Long wave
LI	Lifted Index
LIS	Lifted indices
LK	Lake
LKS	Lakes
LKLY	Likely
LLJ	Low level jet
LLWAS	Low-level wind shear alert system
LLWS	Low-level wind shear
LN	Line
LOC	Local or locally or location or located
LONG	Longitude
LONGL	Longitudinal
LRG	Long range
LRGLY	Largely
LRGR	Larger

LRGST	Largest
LST	Local standard time
LTD	Limited
LTG	Lightning
LTGCC	Lightning cloud-to-cloud
LTGCG	Lightning cloud-to-ground
LTGCCCG	Lightning cloud-to-cloud, cloud-to-ground
LTGCW	Lightning cloud-to-water
LTGIC	Lightning in cloud
LTL	Little
LTLCG	Little change
LTR	Later
LTST	Latest
LV	Light and variable (relating to wind)
LVE	Leave or Leaving
LVL	Level
LWR	Lower
LWRD	Lowered
LWRG	Lowering
LRY	Layer or layered

M

M	Metres (preceded by figures) or Mach number (followed by figures)
MA	Massachusetts
MAG	Magnetic
MAINT	Maintenance
MAN	Manitoba
MAR	March
MAX	Maximum
MAY	May
MB	Millibar
MBST	Microburst
MCD	Mesoscale discussion
MD	Maryland
MDFY	Modify
MDFYD	Modified
MDFYG	Modifying
MDL	Model
MDLS	Models
MDTLY	Moderately
ME	Maine
MED	Medium
MEGG	Merging
MESO	Mesoscale

MET	Meteorological or meteorology	MTW	Mountain waves
METAR	Aviation routine weather report (in aeronautical meteorological code)	MULT	Multiple
		MULTILVL	Multilevel
		MWO	Meteorological watch office
		MX	Mixed type of ice formation (white and clear)
METRO	Metropolitan		
MEX	Mexico		
MHKVLY	Mohawk Valley		
MI	Michigan	**N**	
MID	Mid-point (related to RVR)		
MIDN	Midnight	N	North or northern latitude or no distinct tendency (in RVR during previous 10 minutes)
MIFG	Shallow fog		
MIL	Military		
MIN	Minutes		
MISG	Missing	NAB	Not above
MLTLVL	Melting level	NAM	North American Mesoscale (model)
MN	Minnesota		
MNLD	Mainland	NAT	North Atlantic
MNM	Minimum	NAV	Navigation
MNLY	Mainly	NB	New Brunswick or northbound
MNT	Monitor or monitoring or monitored		
		NBFR	Not before
MNTN	Maintain	NBRHD	Neighborhood
MO	Missouri	NC	North Carolina or no change
MOD	Moderate (used to indicate the intensity of weather phenomena, interference or static reports, e.g. moderate rain = MOD RA)		
		NCEP	National Center of Environmental Prediction
		NCO	NCEP Central Operations
		NCWX	No change in weather
MOGR	Moderate or greater	ND	North Dakota
MON	Monday or above mountains	NE	Nebraska or northeast
		NEB	Northeast bound
MOPS	Minimum operational performance standards	NEC	Necessary
		NEG	No or negative or permission not granted or that is not correct
MOV	Move or moving or movement		
MPH	Miles per hour	NEGLY	Negatively
MPS	Meters per second	NELY	Northeasterly
MRG	Medium range	NERN	Northeastern
MRGL	Marginal	NEWD	Northeastward
MRGLLY	Marginally	NEW ENG	New England
MRNG	Morning	NFLD	Newfoundland
MRTM	Maritime	NGT	Night
MS	Mississippi or minus	NH	New Hampshire
MSG	Message	NHC	National Hurricane Center
MSL	Mean sea level	NIL	None or I have nothing to send to you
MST	Most		
MSTLY	Mostly	NJ	New Jersey
MSTR	Moisture	NL	No layers
MT	Montana or mountain	NLT	Not later than

NLY	Northerly		observation
NM	New Mexico or nautical miles	OBSC	Obscure or obscured or obscuring
NMBRS	Numbers	OCFNT	Occluded front
NML	Normal	OCLD	Occlude
NMRS	Numerous	OCLDS	Occludes
NNE	North-northeast	OCLDD	Occluded
NNELY	North-northeasterly	OCLDG	Occluding
NNERN	North-northeastern	OCLN	Occlusion
NNEWD	North-northeastward	OCNL	Occasional or occasionally
NNW	North-northwest	OCR	Occur
NNWLY	North-northwesterly	OCRD	Occurred
NNWRN	North-northwestern	OCRG	Occurring
NNWWD	North-northwestward	OCRS	Occurs
NNNN	End of message	OCT	October
NOAA	National Oceanic and Atmospheric Administration	OFC	Office
		OFP	Occluded frontal passage
NOPAC	Northern Pacific	OFSHR	Offshore
NOSIG	No significant change (used in trend-type landing forecasts)	OH	Ohio
		OHD	Overhead
		OK	Oklahoma or we agree or it is correct
NOV	November		
NPRS	Non-persistent	OMTNS	Over mountains
NR	Number	ONSHR	On shore
NRLY	Nearly	OPA	Opaque, white type of ice formation
NRN	Northern		
NRW	Narrow	OPC	Ocean Prediction Center
NS	Nova Scotia or nimbostratus	OPN	Open or opening or opened
		OPR	Operator or operate or operative or operating or operational
NSC	Nil significant cloud		
NSW	Nil significant weather		
NTFY	Notify	OR	Oregon
NTFYD	Notified	ORGPHC	Orographic
NTL	National	ORIG	Original
NV	Nevada	OSV	Ocean station vessel
NVA	Negative vorticity advection	OTLK	Outlook (used in SIGMET messages for volcanic ash and tropical cyclones)
NW	Northwest		
NWB	Northwest bound		
NWD	Northward	OTP	On top
NWLY	Northwesterly	OTR	Other
NWRN	Northwestern	OTRW	Otherwise
NWS	National Weather Service	OUBD	Outbound
NY	New York	OUTFLO	Outflow
NXT	Next	OVC	Overcast
		OVNGT	Overnight
		OVR	Over
O		OVRN	Overrun
		OVRNG	Overrunning
OAT	Outside air temperature	OVTK	Overtake
OBS	Observe or observed or	OVTKG	Overtaking

OVTKS	Overtakes

P

PA	Pennsylvania
PAC	Pacific
PATWAS	Pilot's automatic telephone weather answering service
PBL	Planetary boundary layer
PCPN	Precipitation
PD	Period
PDMT	Predominant
PEN	Peninsula
PERM	Permanent
PGTSND	Puget Sound
PHYS	Physical
PIBAL	Pilot balloon observation
PIREP	Pilot weather report
PL	Ice pellets
PLNS	Plains
PLS	Please
PLTO	Plateau
PLVL	Present level
PM	Postmeridian
PNHDL	Panhandle
PO	Dust/sand whirls (dust devils)
POS	Positive
POSLY	Positively
POSS	Possible
PPI	Plan position indicator
PPINA	Plan position indicator not available (U.S. Weather Radar Report)
PPINE	Plan position indicator no echoes (U.S. Weather Radar Report)
PPSN	Present position
PRBL	Probable
PRBLY	Probably
PRBLTY	Probability
PRECD	Precede
PRECDD	Preceded
PRECDG	Preceding
PRECDS	Precedes
PRES	Pressure
PRESFR	Pressure falling rapidly
PRESRR	Pressure rising rapidly
PRFG	Aerodrome partially

	covered by fog
PRI	Primary
PRIN	Principal
PRIND	Present indications are
PRJMP	Pressure jump
PROB	Probability
PROC	Procedure
PROD	Produce
PRODG	Producing
PROG	Forecast
PROGD	Forecasted
PROGS	Forecasts
PRSNT	Present
PRSNTLY	Presently
PRST	Persist
PRSTS	Persists
PRSTNC	Persistence
PRSTNT	Persistent
PRVD	Provide
PRVDD	Provided
PRVDG	Providing
PRVDS	Provides
PS	Plus
PSG	Passing
PSN	Position
PSND	Positioned
PSR	Primary surveillance radar
PTCHY	Patchy
PTLY	Partly
PTNL	Potential
PTNLY	Potentially
PTNS	Portions
PUGET	Puget Sound
PVA	Positive vorticity advection
PVL	Prevail
PVLD	Prevailed
PVLG	Prevailing
PVLS	Prevails
PVLT	Prevalent
PWB	Pilot weather briefing
PWR	Power

Q

QFE	Atmospheric pressure at aerodrome elevation
QN	Question

QNH	Altimeter sub-scale setting to obtain elevation when on the ground		RH	Relative humidity
			RI	Rhode Island
QSTNRY	Quasistationary		RITE	Right (direction of turn)
QTR	Quarter		RIOGD	Rio Grande
QUAD	Quadrant		RLBL	Reliable
QUE	Quebec		RLTV	Relative
			RLTVLY	Relatively
			RMK	Remark
R			RMN	Remain
			RMND	Remained
R	Right (runway identification) or rain (U.S. Weather Radar Reports)		RMNDR	Remainder
			RMNG	Remaining
			RMNS	Remains
			RNFL	Rainfall
RA	Rain		ROFOR	Route forecast (in aeronautical meteorological code)
RADAT	Radiosonde additional data			
RAFC	Regional area forecast centre			
			ROT	Rotate
RAG	Ragged		ROTD	Rotated
RAOB	Radiosonde observation		ROTG	Rotating
RCH	Reach or reaching		ROTS	Rotates
RCKY	Rocky		RPD	Rapid
RCKYS	Rockies		RPDLY	Rapidly
RCMD	Recommend		RPLC	Replace or replaced
RCMDD	Recommended		RPT	Repeat or I repeat (to be used in AFS as a procedure signal)
RCMDG	Recommending			
RCMDS	Recommends			
RCRD	Record		RPTG	Repeating
RCRDS	Records		RPTS	Repeats
RDC	Reduce		RQMNTS	Requirements
RDGG	Ridging		RQR	Require
RDL	Radial		RQRD	Required
RDVLP	Redevelop		RQRG	Requiring
RDVLPG	Redeveloping		RQRS	Requires
RDVLPMT	Redevelopment		RRA	(or RRB, RRC…etc., in sequence) Delayed meteorological message (message type designator)
RE…	Recent (used to qualify weather phenomena, e.g. RERA = recent rain)			
REC	Receive or receiver		RSG	Rising
RECON	Reconnaissance		RSN	Reason
REF	Reference to…or refer to…		RSNG	Reasoning
REP	Report or reporting or reporting point		RSNS	Reasons
			RSTR	Restrict
RES	Reserve		RSTRD	Restricted
REQ	Request or requested		RSTRG	Restricting
RESP	Response		RSTRS	Restricts
RESTR	Restrict		RTD	Delayed (used to indicate delayed meteorological message; message type designator)
RGLR	Regular			
RGN	Region			
RGNS	Regions			

RTE	Route	SEV	Severe (used to qualify icing and turbulence reports)
RTN	Return or returned or returning		
		SEWD	Southeastward
RTS	Return to service	SFC	Surface
RUC	Rapid Update Cycle (model)	SG	Snow grains
		SGFNT	Significant
RUF	Rough	SGFNTLY	Significantly
RUFLY	Roughly	SH...	Showers (followed by RA = rain, SN = snow, PL = ice pellets, GR = hail, GS = small hail and/or snow pellets or combinations thereof, e.g. SHRASN = showers of rain and snow)
RVR	Runway visual range		
RVS	Revise		
RVSD	Revised		
RVSG	Revising		
RVSS	Revises		
RWY	Runway		

S

		SHFT	Shift
		SHFTD	Shifted
		SHFTG	Shifting
S	South or southern latitude	SHFTS	Shifts
SA	Sand	SHLD	Shield
SAP	As soon as possible	SHLW	Shallow
SARPS	Standards and Recommended Practices (ICAO)	SHRT	Short
		SHRTLY	Shortly
		SHRTWV	Shortwave
SASK	Saskatchewan	SHUD	Should
SAT	Saturday	SIERNEV	Sierra Nevada
SATFY	Satisfactory	SIG	Signature
SB	Southbound	SIGMET	Significant Meteorological Information (Information concerning en-route weather phenomena which may affect the safety of aircraft operations)
SBSD	Subside		
SBSDD	Subsided		
SBSDNC	Subsidence		
SBSDS	Subsides		
SC	South Carolina or stratocumulus		
SCND	Second	SIGWX	Significant weather
SCSL	Stratocumulus Standing Lenticular	SIMUL	Simultaneous or simultaneously
		SKC	Sky clear
SCT	Scattered	SKED	Schedule or scheduled
SD	South Dakota	SLD	Solid
SE	Southeast	SLGT	Slight
SEB	Southeast bound	SLGTLY	Slightly
SEC	Seconds	SLP	Slope
SECT	Sector	SLPG	Sloping
SELY	Southeasterly	SLW	Slow
SEP	September	SLY	Southerly
SEPN	Separation	SM	Statute mile
SEQ	Sequence	SML	Small
SER	Service or servicing or served	SMLR	Smaller
		SMRY	Summary
SERN	Southeastern	SMTH	Smooth

SMTHR	Smoother	STBLTY	Stability
SMTHST	Smoothest	STD	Standard
SMTM	Sometime	STDY	Steady
SMWHT	Somewhat	STFR	Stratus Fractus
SN	Snow	STF	Stratiform
SNBNK	Snow bank	STG	Strong
SNFLK	Snowflake	STGLY	Strongly
SNGL	Single	STGR	Stronger
SNOINCR	Snow increase	STGST	Strongest
SNOINCRG	Snow increasing	STM	Storm
SOP	Standard operating procedure	STMS	Storms
		STN	Station
SPC	Storm Prediction Center	STNR	Stationary
SPCLY	Especially	STS	Status
SPD	Speed	SUB	Substitute
SPECI	Aviation selected special weather Report (in aeronautical meteorological code)	SUBTRPCL	Subtropical
		SUF	Sufficient
		SUFLY	Sufficiently
		SUG	Suggest
		SUGG	Suggesting
SPECIAL	Special meteorological report (in abbreviated plain language)	SUGS	Suggests
		SUN	Sunday
		SUPG	Supplying
SPKL	Sprinkle	SUPR	Superior
SPLNS	Southern Plains	SUPSD	Supersede
SPRD	Spread	SUPSDG	Superseding
SPRDG	Spreading	SUPSDS	Supersedes
SPRDS	Spreads	SVC	Service message
SPRL	Spiral	SVRL	Several
SQ	Squall	SW	Southwest
SQL	Squall line	SWB	Southwest bound
SR	Sunrise	SWD	Southward
SRG	Short range	SWWD	Southwestward
SRN	Southern	SWLY	Southwesterly
SRND	Surround	SWRN	Southwestern
SRNDD	Surrounded	SX	Stability index
SRNDG	Surrounding	SXN	Section
SRNDS	Surrounds	SYNOP	Synoptic
SRY	Secondary	SYNS	Synopsis
SS	Sunset or sandstorm	SYS	System
SSE	South-southeast		
SSELY	South-southeasterly		
SSERN	South-southeastern		
SSEWD	South-southeastward	**T**	
SSW	South-southwest		
SSWLY	South-southwesterly	T	Temperature
SSWRN	South-southwestern	TAF	Terminal aerodrome forecast
SSWWD	South-southwestward		
ST	Stratus	TAIL	Tail wind
STAGN	Stagnation	TB	Turbulence (PIREPs only)
STBL	Stable	TC	Tropical Cyclone

TCNTL	Transcontinental	TRMTS	Terminates
TCU	Towering cumulus	TRNSP	Transport
TDA	Today	TRNSPG	Transporting
TDO	Tornado	TROF	Trough
TEMPO	Temporary or temporarily	TROFS	Troughs
TEND	Trend forecast	TROP	Tropopause
THK	Thick	TRPCD	Tropical continental air mass
THKNG	Thickening		
THKNS	Thickness	TRPCL	Tropical
THKR	Thicker	TRRN	Terrain
THKST	Thickest	TRSN	Transition
THN	Thin	TS	Thunderstorm (in aerodrome reports and forecasts, TS used alone means thunder heard but no precipitation at the aerodrome)
THNG	Thinning		
THNR	Thinner		
THNST	Thinnest		
THR	Threshold		
THRFTR	Thereafter		
THRU	Through	TS…	Thunderstorm (followed by RA = rain, SN = snow, PL = ice pellets, GR = hail, GS = small hail and/or snow pellets or combinations thereof, e.g. TSRASN = thunderstorm with rain and snow)
THRUT	Throughout		
THSD	Thousand		
THTN	Threaten		
THTND	Threatened		
THTNG	Threatening		
THTNS	Threatens		
THU	Thursday		
TIL	Until	TSFR	Transfer
TL…	Till (followed by time by which weather change is forecast to end)	TSFRD	Transferred
		TSFRG	Transferring
		TSFRS	Transfers
TMW	Tomorrow	TSNT	Transient
TN	Tennessee	TUE	Tuesday
TNDCY	Tendency	TURB	Turbulence
TNDCYS	Tendencies	TURBT	Turbulent
TNGT	Tonight	TWD	Toward
TNTV	Tentative	TWDS	Towards
TNTVLY	Tentatively	TWI	Twilight
TO	To…(place)	TWR	Aerodrome control tower or aerodrome control
TOC	Top of climb		
TOP	Cloud top	TWRG	Towering
TOPS	Tops	TX	Texas
TOVC	Top of overcast	TYP	Type of aircraft
TPC	Tropical Prediction Center	TYPH	Typhoon
TPG	Topping		
TR	Track		
TRBL	Trouble	**U**	
TRIB	Tributary		
TRKG	Tracking		
TRML	Terminal	U	Upward (tendency in RVR during previous 10 minutes)
TRMT	Terminate		
TRMTD	Terminated	UA	Pilot weather report (U.S.)
TRMTG	Terminating	UDDF	Up- and downdrafts

UFN	Until further notice	VFYD	Verified
UNA	Unable	VFYG	Verifying
UNAVBL	Unavailable	VFYS	Verifies
UNEC	Unnecessary	VIS	Visibility
UNKN	Unknown	VLCTY	Velocity
UNL	Unlimited	VLCTYS	Velocities
UNREL	Unreliable	VLNT	Violent
UNRSTD	Unrestricted	VLNTLY	Violently
UNSATFY	Unsatisfactory	VLY	Valley
UNSBL	Unseasonable	VMC	Visual meteorological conditions
UNSTBL	Unstable		
UNSTDY	Unsteady	VOL	Volume
UNSTL	Unsettle	VOLMT	Meteorological information for aircraft in flight
UNSTLD	Unsettled		
UNUSBL	Unusable	VORT	Vorticity
UPDFTS	Updrafts	VR	Veer
UPR	Upper	VRB	Variable
UPSLP	Upslope	VRG	Veering
UPSTRM	Upstream	VRISL	Vancouver Island, BC
URG	Urgent	VRS	Veers
USBL	Usable	VRT	Vertical motion
UT	Utah	VRY	Very
UTC	Coordinated Universal Time	VT	Vermont
UVV	Upward vertical velocity	VV	Vertical velocity
UWNDS	Upper winds		

V

VA	Virginia or volcanic ash
VAAC	Volcanic Ash Advisory Center
VAAS	Volcanic Ash Advisory Statement
VAL	In valleys
VARN	Variation
VC	Vicinity of the aerodrome (followed by FG = fog, FC = funnel cloud, SH = showers, PO = dust/sand whirls, BLDU = blowing dust, BLSA = blowing sand or BLSN = blowing snow, e.g. VC FG = vicinity fog)
VCOT	VFR conditions on top
VCTR	Vector
VCY	Vicinity
VER	Vertical
VFR	Visual flight rules
VFY	Verify

W

W	West or western longitude
WA	Washington
WAA	Warm air advection
WAFC	World area forecast centre
WAFS	Word area forecast system
WB	Westbound
WDI	Wind direction indicator
WDLY	Widely
WDSPR	Widespread
WED	Wednesday
WEF	With effect from or effective from
WFO	Weather Forecast Office
WFP	Warm front passage
WI	Wisconsin or within
WIBIS	Will be issued
WID	Width
WIE	With immediate effect or effective immediately
WINT	Winter
WINTEM	Forecast upper wind and temperature for aviation

WK	Weak		WY	Wyoming
WKDAY	Weekday			
WKEND	Weekend			
WKN	Weaken or weakening		**X**	
WL	Will			
WLY	Westerly		X	Cross
WND	Wind		XCP	Except
WNDS	Winds		XNG	Crossing
WNW	West-northwest		XPC	Expect
WNWLY	West-northwesterly		XPCD	Expected
WNWRN	West-northwestern		XPCG	Expecting
WNWWD	West-northwestward		XPCS	Expects
WO	Without		XPLOS	Explosive
WPLTO	Western Plateau		XS	Atmospherics
WRM	Warm		XTND	Extend
WRMG	Warming		XTNDD	Extended
WRN	Western		XTNDG	Extending
WRMR	Warmer		XTRM	Extreme
WRMST	Warmest		XTRMLY	Extremely
WRMFNT	Warm front			
WRMFNTL	Warm frontal			
WRNG	Warning		**Y**	
WRS	Worse			
WS	Wind shear		YDA	Yesterday
WSPD	Wind speed		YKN	Yukon
WSHFT	Wind shift		YLSTN	Yellowstone
WSTCH	Wasatch Range			
WSW	West-southwest			
WSWLY	West-southwesterly		**Z**	
WSWRN	West-southwestern			
WSWWD	West-southwestward		Z	Coordinated Universal Time (in meteorological messages)
WTR	Water			
WTSPT	Waterspout			
WUD	Would			
WV	West Virginia		ZN	Zone
WVS	Waves		ZNS	Zones
WW	Watch notification message			
WWD	Westward			
WWW	World wide web			
WX	Weather			

12 APPENDIX C: STANDARD CONVERSION CHART

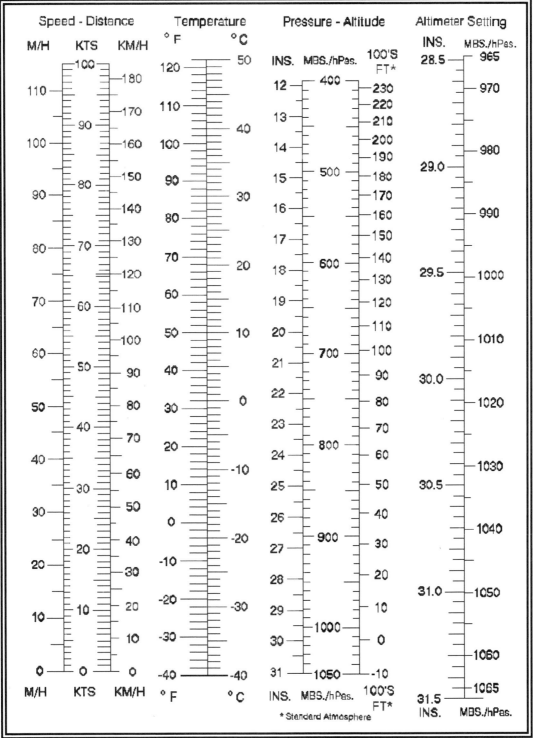

Figure C-1. Standard Conversion Chart

13 APPENDIX D:DENSITY ALTITUDE CALCULATION

To determine density altitude:

1. Set the aircraft's altimeter to 29.92 inches of Mercury. The altimeter will indicate pressure altitude.

2. Read the outside air temperature.

3. Mark the intersection of pressure altitude (horizontal) and temperature (vertical) lines on the chart.

4. Read the density altitude from the diagonal lines.

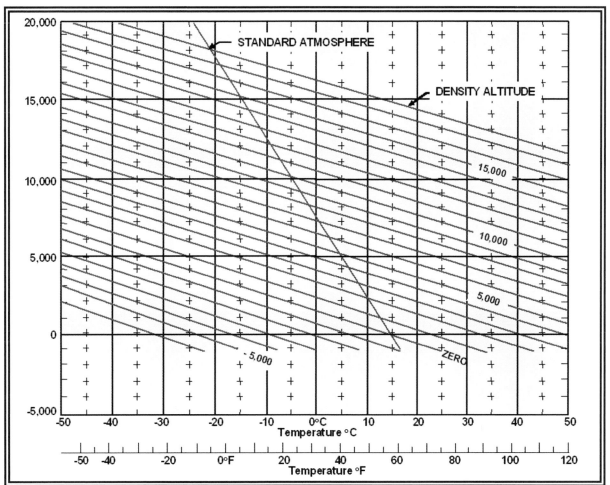

Figure D-1. Density Altitude Computation Chart

14 APPENDIX E: INTERNET LINKS

Table E-1. Selected National Weather Service (NWS) Links

SITE	WEB ADDRESS
National Weather Service (NWS)	http://weather.gov/
Aviation Digital Data Service (ADDS)	http://adds.aviationweather.noaa.gov/
Aviation Weather Center (AWC)	http://aviationweather.gov
Hydrometeorological Prediction Center (HPC)	http://www.hpc.ncep.noaa.gov/
Storm Prediction Center (SPC)	http://www.spc.noaa.gov/
Tropical Prediction Center (TPC)	http://www.nhc.noaa.gov/
Alaska Aviation Weather Unit (AAWU)	http://aawu.arh.noaa.gov/
Center Weather Service Units (CWSU)	http://aviationweather.gov/products/cwsu/
Weather Forecast Offices (WFO)	http://www.srh.noaa.gov/
Weather Forecast Office (WFO) Honolulu, HI – Aviation Products	http://www.prh.noaa.gov/hnl/pages/aviation.php
Telecommunication Operations Center - NWS Fax Charts	http://weather.noaa.gov/fax/nwsfax.html
NWS Office at the FAA Academy	http://www.srh.noaa.gov/faa/

Table E-2. Selected Federal Aviation Administration (FAA) Links

SITE	WEB ADDRESS
Federal Aviation Administration (FAA)	http://www.faa.gov/
Air Traffic Control System Command Center (ATCSCC)	http://www.fly.faa.gov/flyfaa/usmap.jsp
Automated Flight Service Station (AFSS)	http://www.afss.com/

Table E-3. Selected Links to Aviation Weather Products

PRODUCT	WEB ADDRESS
Average Surface to 500 MB Relative Humidity Chart	http://weather.noaa.gov/pub/fax/QRUA00.TIF
Collaborative Convective Weather Forecast (CFP)	http://aviationweather.gov/products/ccfp/
Constant Pressure Charts	http://weather.noaa.gov/fax/barotrop.shtml
Convective Outlooks	http://www.spc.noaa.gov/products/outlook/
Current Icing Product (CIP)	http://adds.aviationweather.noaa.gov/icing/icing_nav.php
Center Weather Advisory (CWA)	http://aviationweather.gov/products/cwsu/
Area Forecast (FA)	http://aviationweather.gov/products/fa/
Significant Meteorological Advisory (SIGMET) – US (CONUS)	http://adds.aviationweather.noaa.gov/airmets/
Significant Meteorological Advisory (SIGMET) – International	http://aviationweather.gov/products/sigmets/intl/
Airmen's Meteorological Advisory (AIRMET)	http://adds.aviationweather.noaa.gov/airmets/
Forecast Icing Potential (FIP)	http://adds.aviationweather.noaa.gov/icing/icing_nav.php
Freezing Level Graphics	http://adds.aviationweather.noaa.gov/icing/frzg_nav.php
High Level SIGWX Charts	http://aviationweather.gov/products/swh/
Lifted Index (LI) Analysis Chart	http://weather.noaa.gov/pub/fax/QXUA00.TIF
Low Level SIGWX Charts	http://aviationweather.gov/products/swl/
Mid Level SIGWX Chart	http://aviationweather.gov/products/swm/
Meteorological Impact Statement (MIS)	http://aviationweather.gov/products/cwsu/
National Convective Weather Forecast (NCWF)	http://adds.aviationweather.noaa.gov/convection/java/ http://adds.aviationweather.noaa.gov/convection/java/?appletsize=large http://aviationweather.gov/products/ncwf/

Pilot Weather Report	http://adds.aviationweather.noaa.gov/pireps/
PRODUCT	**WEB ADDRESS**
Radar Summary Chart	http://weather.noaa.gov/pub/fax/QAUA00.TIF
Aviation Routine Weather Report (METAR) / Selected Special Weather Report (SPECI)	http://adds.aviationweather.noaa.gov/metars/
Surface Analysis Charts	http://www.hpc.ncep.noaa.gov/html/sfc2.shtml http://www.hpc.ncep.noaa.gov/html/avnsfc.shtml http://www.opc.ncep.noaa.gov/ http://www.nhc.noaa.gov/marine_forecasts.shtml http://www.opc.ncep.noaa.gov/UA.shtml
Short Range Surface Prog Charts	http://adds.aviationweather.noaa.gov/progs/
Strike Probabilities of Tropical Cyclone Conditions (SPF)	http://www.nhc.noaa.gov/
Terminal Aerodrome Forecast (TAF)	http://adds.aviationweather.gov/tafs/
Aviation Tropical Cyclone Advisory (TCA)	http://www.nhc.noaa.gov/
Tropical Cyclone Public Advisory (TCP)	http://www.nhc.noaa.gov/
Volcanic Ash Advisory Statement (VAAS)	http://aviationweather.gov/iffdp/volt.shtml
Volcanic Ash Forecast Transport and Dispersion (VAFTAD) Chart	http://aviationweather.gov/iffdp/volc.shtml
Watch Notification Messages	http://www.spc.noaa.gov/products/watch/
Weather Depiction Chart	http://weather.noaa.gov/pub/fax/QGUA00.TIF
Wind and Temperature Aloft Forecast Graphics	http://adds.aviationweather.noaa.gov/winds/
Wind and Temperature Aloft Forecasts (FB) Text	http://aviationweather.gov/products/nws/winds/

15 APPENDIX F: AWC ADVISORY PLOTTING CHART - CONTIGUOUS U.S.

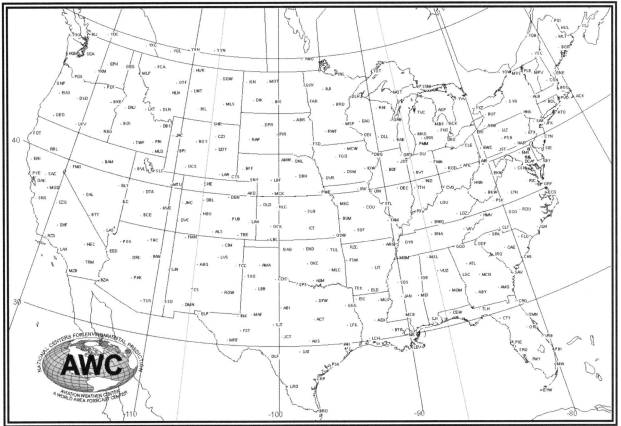

Figure F-1. AWC Advisory Plotting Chart – Contiguous U.S.

STID	NAME	ST	CO	LAT/LON (100's)		LAT/LON(deg min)			
ABI	ABILENE	TX	US	3248	-9986	32	29	-99	52
ABQ	ALBUQUERQUE	NM	US	3504	-10682	35	02	-106	49
ABR	ABERDEEN	SD	US	4542	-9837	45	25	-98	22
ABY	ALBANY	GA	US	3165	-8430	31	39	-84	18
ACK	NANTUCKET	MA	US	4128	-7003	41	17	-70	02
ACT	WACO	TX	US	3166	-9727	31	40	-97	16
ADM	ARDMORE	OK	US	3421	-9717	34	13	-97	10
AEX	ALEXANDRIA	LA	US	3126	-9250	31	16	-92	30
AIR	BELLAIRE	OH	US	4002	-8082	40	01	-80	49
AKO	AKRON	CO	US	4016	-10318	40	10	-103	11
ALB	ALBANY	NY	US	4275	-7380	42	45	-73	48
ALS	ALAMOSA	CO	US	3735	-10582	37	21	-105	49
AMA	AMARILLO	TX	US	3529	-10164	35	17	-101	38
AMG	ALMA	GA	US	3154	-8251	31	32	-82	31
ANW	AINSWORTH	NE	US	4257	-9999	42	34	-99	59
APE	APPLETON	OH	US	4015	-8259	40	09	-82	35
ARG	WALNUT_RIDGE	AR	US	3611	-9095	36	07	-90	57
ASP	OSCODA	MI	US	4445	-8339	44	27	-83	23
ATL	ATLANTA	GA	US	3363	-8444	33	38	-84	26
AUS	AUSTIN	TX	US	3030	-9770	30	18	-97	42
BAE	MILWAUKEE	WI	US	4312	-8828	43	07	-88	17
BAM	BATTLE_MOUNTAIN	NV	US	4057	-11692	40	34	-116	55
BCE	BRYCE_CANYON	UT	US	3769	-11230	37	41	-112	18
BDF	BRADFORD	IL	US	4116	-8959	41	10	-89	35
BDL	WINSOR_LOCKS	CT	US	4194	-7269	41	56	-72	41
BFF	SCOTTSBLUFF	NE	US	4189	-10348	41	53	-103	29
BGR	BANGOR	ME	US	4484	-6887	44	50	-68	52
BIL	BILLINGS	MT	US	4581	-10863	45	49	-108	38
BIS	BISMARK	ND	US	4677	-10067	46	46	-100	40
BJI	BEMIDJI	MN	US	4758	-9502	47	35	-95	01
BKE	BAKER	OR	US	4484	-11781	44	50	-117	49
BKW	BECKLEY	WV	US	3778	-8112	37	47	-81	07
BLI	BELLINGHAM	WA	US	4895	-12258	48	57	-122	35
BNA	NASHVILLE	TN	US	3614	-8668	36	08	-86	41
BOI	BOISE	ID	US	4355	-11619	43	33	-116	11
BOS	BOSTON	MA	US	4236	-7099	42	22	-70	59
BOY	BOYSEN_RESV.	WY	US	4346	-10830	43	28	-108	18
BPI	BIG_PINEY	WY	US	4258	-11011	42	35	-110	07
BRD	BRAINERD	MN	US	4635	-9403	46	21	-94	02
BRO	BROWNSVILLE	TX	US	2592	-9738	25	55	-97	23
BTR	BATON_ROUGE	LA	US	3048	-9130	30	29	-91	18
BTY	BEATTY	NV	US	3680	-11675	36	48	-116	45
BUF	BUFFALO	NY	US	4293	-7865	42	56	-78	39
BUM	BUTLER	MO	US	3827	-9449	38	16	-94	29
BVL	BONNEVILLE	UT	US	4073	-11376	40	44	-113	46
BVT	LAFAYETTE	IN	US	4056	-8707	40	34	-87	04
BWG	BOWLING_GREEN	KY	US	3693	-8644	36	56	-86	26
BZA	YUMA	AZ	US	3277	-11460	32	46	-114	36
CAE	COLUMBIA	SC	US	3386	-8105	33	52	-81	03
CDS	CHILDRESS	TX	US	3437	-10028	34	22	-100	17
CEW	CRESTVIEW	FL	US	3083	-8668	30	50	-86	41
CHE	HAYDEN	CO	US	4052	-10731	40	31	-107	19
CHS	CHARLESTON	SC	US	3289	-8004	32	53	-80	02

STID	NAME	ST	CO	LAT/LON (100's)	LAT/LON (deg min)
CIM	CIMARRON	NM	US	3649 -10487	36 29 -104 52
CLE	CLEVELAND	OH	US	4142 -8185	41 25 -81 51
CLT	CHARLOTTE	NC	US	3522 -8093	35 13 -80 56
CON	CONCORD	NH	US	4322 -7158	43 13 -71 35
COU	COLUMBIA	MO	US	3882 -9222	38 49 -92 13
CRG	JACKSONVILLE	FL	US	3034 -8151	30 20 -81 31
CRP	CORPUS_CHRISTI	TX	US	2790 -9745	27 54 -97 27
CSN	CASSANOVA	VA	US	3864 -7787	38 38 -77 52
CTY	CROSS_CITY	FL	US	2960 -8305	29 36 -83 03
CVG	COVINGTON	KY	US	3902 -8470	39 01 -84 42
CYN	COYLE	NJ	US	3982 -7443	39 49 -74 26
CYS	CHEYENNE	WY	US	4121 -10477	41 13 -104 46
CZI	CRAZY_WOMAN	WY	US	4400 -10644	44 00 -106 26
CZQ	FRESNO	CA	US	3688 -11982	36 53 -119 49
DBL	EAGLE	CO	US	3944 -10690	39 26 -106 54
DBQ	DUBUQUE	IA	US	4240 -9071	42 24 -90 43
DBS	DUBOIS	ID	US	4409 -11221	44 05 -112 13
DCA	WASHINGTON	DC	US	3886 -7704	38 52 -77 02
DDY	CASPER	WY	US	4309 -10628	43 05 -106 17
DEC	DECATUR	IL	US	3974 -8886	39 44 -88 52
DEN	DENVER	CO	US	3981 -10466	39 49 -104 40
DFW	DALLAS-FT_WORTH	TX	US	3287 -9703	32 52 -97 02
DIK	DICKINSIN	ND	US	4686 -10277	46 52 -102 46
DLF	LAUGHLIN_AFB	TX	US	2936 -10077	29 22 -100 46
DLH	DULUTH	MN	US	4680 -9220	46 48 -92 12
DLL	DELLS	WI	US	4355 -8976	43 33 -89 46
DLN	DILLON	MT	US	4525 -11255	45 15 -112 33
DMN	DEMING	NM	US	3228 -10760	32 17 -107 36
DNJ	MC_CALL	ID	US	4477 -11621	44 46 -116 13
DPR	DUPREE	SD	US	4508 -10172	45 05 -101 43
DRK	PRESCOTT	AZ	US	3470 -11248	34 42 -112 29
DSD	REDMOND	OR	US	4425 -12130	44 15 -121 18
DSM	DES_MOINES	IA	US	4144 -9365	41 26 -93 39
DTA	DELTA	UT	US	3930 -11251	39 18 -112 31
DVC	DOVE_CREEK	CO	US	3781 -10893	37 49 -108 56
DXO	DETROIT	MI	US	4221 -8337	42 13 -83 22
DYR	DYERSBURG	TN	US	3602 -8932	36 01 -89 19
EAU	EAU_CLAIRE	WI	US	4490 -9148	44 54 -91 29
ECG	ELIZABETH_CITY	NC	US	3625 -7618	36 15 -76 11
ECK	PECK	MI	US	4326 -8272	43 16 -82 43
EED	NEEDLES	CA	US	3477 -11447	34 46 -114 28
EHF	BAKERSFIELD	CA	US	3548 -11910	35 29 -119 06
EIC	SHREVEPORT	LA	US	3277 -9381	32 46 -93 49
EKN	ELKINS	WV	US	3892 -8010	38 55 -80 06
ELD	EL_DORADO	AR	US	3326 -9274	33 16 -92 44
ELP	EL_PASO	TX	US	3182 -10628	31 49 -106 17
ELY	ELY	NV	US	3930 -11485	39 18 -114 51
EMI	WESTMINSTER	MD	US	3950 -7698	39 30 -76 59
END	VANCE_AFB	OK	US	3635 -9792	36 21 -97 55
ENE	KENNEBUNK	ME	US	4343 -7061	43 26 -70 37
ENI	UKIAH	CA	US	3905 -12327	39 03 -123 16
EPH	EPHRATA	WA	US	4738 -11942	47 23 -119 25
ERI	ERIE	PA	US	4202 -8030	42 01 -80 18

STID	NAME	ST	CO	LAT/LON (100's)		LAT/LON(deg min)	
ETX	EAST_TEXAS	PA	US	4058	-7568	40 35	-75 41
EUG	EUGENE	OR	US	4412	-12322	44 07	-123 13
EWC	ELLWOOD_CITY	PA	US	4083	-8021	40 50	-80 13
EYW	KEY_WEST	FL	US	2459	-8180	24 35	-81 48
FAM	FARMINGTON	MO	US	3767	-9023	37 40	-90 14
FAR	FARGO	ND	US	4675	-9685	46 45	-96 51
FCA	KALISPELL	MT	US	4821	-11418	48 13	-114 11
FLO	FLORENCE	SC	US	3423	-7966	34 14	-79 40
FMG	RENO	NV	US	3953	-11966	39 32	-119 40
FMN	FARMINGTON	NM	US	3675	-10810	36 45	-108 06
FMY	FT_MEYERS	FL	US	2658	-8187	26 35	-81 52
FNT	FLINT	MI	US	4297	-8374	42 58	-83 44
FOD	FT_DODGE	IA	US	4261	-9429	42 37	-94 17
FOT	FORTUNA	CA	US	4067	-12423	40 40	-124 14
FSD	SIOUX_FALLS	SD	US	4365	-9678	43 39	-96 47
FSM	FT_SMITH	AR	US	3538	-9427	35 23	-94 16
FST	FT_STOCKTON	TX	US	3095	-10298	30 57	-102 59
FWA	FT_WAYNE	IN	US	4098	-8519	40 59	-85 11
GAG	GAGE	OK	US	3634	-9988	36 20	-99 53
GCK	GARDEN_CITY	KS	US	3792	-10073	37 55	-100 44
GEG	SPOKANE	WA	US	4756	-11763	47 34	-117 38
GFK	GRAND_FORKS	ND	US	4795	-9719	47 57	-97 11
GGG	LONGVIEW	TX	US	3242	-9475	32 25	-94 45
GGW	GLASGOW	MT	US	4822	-10663	48 13	-106 38
GIJ	NILES	MI	US	4177	-8632	41 46	-86 19
GLD	GOODLAND	KS	US	3939	-10169	39 23	-101 41
GQO	CHATTANOOGA	TN	US	3496	-8515	34 58	-85 09
GRB	GREEN_BAY	WI	US	4456	-8819	44 34	-88 11
GRR	GRAND_RAPIDS	MI	US	4279	-8550	42 47	-85 30
GSO	GREENSBORO	NC	US	3605	-7998	36 03	-79 59
GTF	GREAT_FALLS	MT	US	4745	-11141	47 27	-111 25
HAR	HARRISBURG	PA	US	4023	-7702	40 14	-77 01
HBU	GUNNISON	CO	US	3845	-10704	38 27	-107 02
HEC	HECTOR	CA	US	3480	-11646	34 48	-116 28
HLC	HILL_CITY	KS	US	3926	-10023	39 16	-100 14
HLN	HELENA	MT	US	4661	-11195	46 37	-111 57
HMV	HOLSTON_MOUNTAIN	TN	US	3644	-8213	36 26	-82 08
HNK	HANCOCK	NY	US	4206	-7532	42 04	-75 19
HNN	HENDERSON	WV	US	3875	-8203	38 45	-82 02
HQM	HOQUIAM	WA	US	4695	-12415	46 57	-124 09
HTO	EAST_HAMPTON	NY	US	4092	-7232	40 55	-72 19
HUL	HOULTON	ME	US	4604	-6783	46 02	-67 50
HVE	HANKSVILLE	UT	US	3842	-11070	38 25	-110 42
HVR	HAVRE	MT	US	4854	-10977	48 32	-109 46
IAH	HOUSTON_INTERNATIONAL	TX	US	2996	-9535	29 58	-95 21
ICT	WICHITA	KS	US	3775	-9758	37 45	-97 35
IGB	BIGBEE	MS	US	3348	-8852	33 29	-88 31
ILC	WILSON_CREEK	NV	US	3825	-11439	38 15	-114 23
ILM	WILMINGTON	NC	US	3435	-7787	34 21	-77 52
IND	INDIANAPOLIS	IN	US	3981	-8637	39 49	-86 22
INK	WINK	TX	US	3187	-10324	31 52	-103 14
INL	INTERNATIONAL_FALLS	MN	US	4857	-9340	48 34	-93 24
INW	WINSLOW	AZ	US	3506	-11080	35 04	-110 48

STID	NAME	ST CO	LAT/LON (100's)	LAT/LON(deg min)
IOW	IOWA_CITY	IA US	4152 -9161	41 31 -91 37
IRK	KIRKSVILLE	MO US	4014 -9259	40 08 -92 35
IRQ	COLLIERS	SC US	3371 -8216	33 43 -82 10
ISN	WILLISTON	ND US	4818 -10363	48 11 -103 38
JAC	JACKSON	WY US	4362 -11073	43 37 -110 44
JAN	JACKSON	MS US	3251 -9017	32 31 -90 10
JCT	JUNCTION	TX US	3060 -9982	30 36 -99 49
JFK	NEW_YORK/JF_KENNEDY	NY US	4063 -7377	40 38 -73 46
JHW	JAMESTOWN	NY US	4219 -7912	42 11 -79 07
JNC	GRAND_JUNCTION	CO US	3906 -10879	39 04 -108 47
JOT	JOLIET	IL US	4155 -8832	41 33 -88 19
JST	JOHNSTOWN	PA US	4032 -7883	40 19 -78 50
LAA	LAMAR	CO US	3820 -10269	38 12 -102 41
LAR	LARAMIE	WY US	4133 -10572	41 20 -105 43
LAS	LAS_VEGAS	NV US	3608 -11516	36 05 -115 10
LAX	LOS_ANGELES_INTL	CA US	3393 -11843	33 56 -118 26
LBB	LUBBOCK_INTERNATIONAL	TX US	3370 -10192	33 42 -101 55
LBF	NORTH_PLATTE	NE US	4113 -10072	41 08 -100 43
LBL	LIBERAL	KS US	3704 -10097	37 02 -100 58
LCH	LAKE_CHARLES	LA US	3014 -9311	30 08 -93 07
LEV	GRAND_ISLE	LA US	2918 -9010	29 11 -90 06
LFK	LUFKIN	TX US	3116 -9472	31 10 -94 43
LGC	LA_GRANGE	GA US	3305 -8521	33 03 -85 13
LIT	LITTLE_ROCK	AR US	3468 -9218	34 41 -92 11
LKT	SALMON	ID US	4502 -11408	45 01 -114 05
LKV	LAKEVIEW	OR US	4249 -12051	42 29 -120 31
LOU	LOUISVILLE	KY US	3810 -8558	38 06 -85 35
LOZ	LONDON	KY US	3703 -8412	37 02 -84 07
LRD	LAREDO	TX US	2748 -9942	27 29 -99 25
LVS	LAS_VEGAS	NM US	3566 -10514	35 40 -105 08
LWT	LEWISTOWN	MT US	4705 -10961	47 03 -109 37
LYH	LYNCHBURG	VA US	3725 -7923	37 15 -79 14
MAF	MIDLAND	TX US	3202 -10218	32 01 -102 11
MBS	SAGINAW	MI US	4353 -8408	43 32 -84 05
MCB	MC_COMB	MS US	3130 -9026	31 18 -90 16
MCK	MC_COOK	NE US	4020 -10059	40 12 -100 35
MCN	MACON	GA US	3269 -8365	32 41 -83 39
MCW	MASON_CITY	IA US	4309 -9333	43 05 -93 20
MEI	MERIDIAN	MS US	3238 -8880	32 23 -88 48
MEM	MEMPHIS	TN US	3506 -8998	35 04 -89 59
MGM	MONTGOMERY	AL US	3222 -8632	32 13 -86 19
MIA	MIAMI	FL US	2580 -8030	25 48 -80 18
MKC	KANSAS_CITY	MO US	3928 -9459	39 17 -94 35
MKG	MUSKEGON	MI US	4317 -8604	43 10 -86 02
MLC	MC_CALESTER	OK US	3485 -9578	34 51 -95 47
MLD	MALAD_CITY	ID US	4220 -11245	42 12 -112 27
MLP	MULLAN_PASS	ID US	4746 -11565	47 28 -115 39
MLS	MILES_CITY	MT US	4638 -10595	46 23 -105 57
MLT	MILLINOCKET	ME US	4558 -6852	45 35 -68 31
MLU	MONROE	LA US	3252 -9203	32 31 -92 02
MOD	MODESTO	CA US	3763 -12096	37 38 -120 58
MOT	MINOT	ND US	4826 -10129	48 16 -101 17
MPV	MONTPELIER	VT US	4422 -7257	44 13 -72 34

STID	NAME	ST CO	LAT/LON (100's)		LAT/LON(deg min)	
MQT	MARQUETTE	MI US	4653	-8759	46 32	-87 35
MRF	MARFA	TX US	3030	-10395	30 18	-103 57
MSL	MUSCLE_SHOALS	AL US	3470	-8748	34 42	-87 29
MSP	MINNEAPOLIS	MN US	4488	-9323	44 53	-93 14
MSS	MASSENA	NY US	4491	-7472	44 55	-74 43
MSY	NEW_ORLEANS	LA US	3000	-9027	30 00	-90 16
MTU	MYTON	UT US	4015	-11013	40 09	-110 08
MZB	MISSION_BAY	CA US	3278	-11723	32 47	-117 14
OAK	OAKLAND	CA US	3773	-12222	37 44	-122 13
OAL	COALDALE	NV US	3800	-11777	38 00	-117 46
OBH	WOLBACH	NE US	4138	-9835	41 23	-98 21
OCS	ROCKSPRINGS	WY US	4159	-10902	41 35	-109 01
ODF	TOCCOA	GA US	3470	-8330	34 42	-83 18
ODI	NODINE	MN US	4391	-9147	43 55	-91 28
OED	MEDFORD	OR US	4248	-12291	42 29	-122 55
OKC	OKLAHOMA_CITY	OK US	3536	-9761	35 22	-97 37
OMN	ORMOND_BEACH	FL US	2930	-8111	29 18	-81 07
ONL	ONEILL	NE US	4247	-9869	42 28	-98 41
ONP	NEWPORT	OR US	4458	-12406	44 35	-124 04
ORD	O'HARE_INTERNATIONAL	IL US	4198	-8790	41 59	-87 54
ORF	NORFOLK	VA US	3689	-7620	36 53	-76 12
ORL	ORLANDO	FL US	2854	-8134	28 32	-81 20
OSW	OSWEGO	KS US	3715	-9520	37 09	-95 12
OVR	OMAHA	NE US	4117	-9574	41 10	-95 44
PBI	WEST_PALM_BEACH	FL US	2668	-8009	26 41	-80 05
PDT	PENDLETON	OR US	4570	-11894	45 42	-118 56
PDX	PORTLAND	OR US	4558	-12260	45 35	-122 36
PGS	PEACH_SPRINGS	AZ US	3562	-11354	35 37	-113 32
PHX	PHOENIX	AZ US	3343	-11202	33 26	-112 01
PIE	SAINT_PETERSBURG	FL US	2791	-8268	27 55	-82 41
PIH	POCATELLO	ID US	4287	-11265	42 52	-112 39
PIR	PIERRE	SD US	4440	-10017	44 24	-100 10
PLB	PLATTSBURGH	NY US	4469	-7352	44 41	-73 31
PMM	PULLMAN	MI US	4247	-8611	42 28	-86 07
PQI	PRESQUE_ISLE	ME US	4677	-6809	46 46	-68 05
PSB	PHILLIPSBURG	PA US	4092	-7799	40 55	-77 59
PSK	DUBLIN	VA US	3709	-8071	37 05	-80 43
PSX	PALACIOS	TX US	2876	-9631	28 46	-96 19
PUB	PUEBLO	CO US	3829	-10443	38 17	-104 26
PVD	PROVIDENCE	RI US	4172	-7143	41 43	-71 26
PWE	PAWNEE_CITY	NE US	4020	-9621	40 12	-96 13
PXV	POCKET_CITY	IN US	3793	-8776	37 56	-87 46
PYE	POINT_REYES	CA US	3808	-12287	38 05	-122 52
RAP	RAPID_CITY	SD US	4398	-10301	43 59	-103 01
RBL	RED_BLUFF	CA US	4010	-12224	40 06	-122 14
RDU	RALEIGH-DURHAM	NC US	3587	-7878	35 52	-78 47
REO	ROME	OR US	4259	-11787	42 35	-117 52
RHI	RHINELANDER	WI US	4563	-8945	45 38	-89 27
RIC	RICHMOND	VA US	3750	-7732	37 30	-77 19
ROD	ROSEWOOD	OH US	4029	-8404	40 17	-84 02
ROW	ROSWELL	NM US	3334	-10462	33 20	-104 37
RWF	REDWWOD_FALLS	MN US	4447	-9513	44 28	-95 08
RZC	RAZORBACK	AR US	3625	-9412	36 15	-94 07

STID	NAME	ST	CO	LAT/LON (100's)		LAT/LON(deg min)	
RZS	SANTA_BARBARA	CA	US	3451	-11977	34 31	-119 46
SAC	SACRAMENTO	CA	US	3844	-12155	38 26	-121 33
SAT	SAN_ANTONIO	TX	US	2964	-9846	29 38	-98 28
SAV	SAVANNAH	GA	US	3216	-8111	32 10	-81 07
SAX	SPARTA	NJ	US	4107	-7454	41 04	-74 32
SBY	SALISBURY	MD	US	3835	-7552	38 21	-75 31
SEA	SEATTLE	WA	US	4744	-12231	47 26	-122 19
SGF	SPRINGFIELD	MO	US	3736	-9333	37 22	-93 20
SHR	SHERIDAN	WY	US	4484	-10706	44 50	-107 04
SIE	SEA_ISLE	NJ	US	3910	-7480	39 06	-74 48
SJI	SEMMNES	AL	US	3073	-8836	30 44	-88 22
SJN	ST_JOHNS	AZ	US	3442	-10914	34 25	-109 08
SJT	SAN_ANGELO	TX	US	3138	-10046	31 23	-100 28
SLC	SALT_LAKE_CITY	UT	US	4085	-11198	40 51	-111 59
SLN	SALINA	KS	US	3893	-9762	38 56	-97 37
SLT	SLATE_RUN	PA	US	4151	-7797	41 31	-77 58
SNS	SALINAS	CA	US	3666	-12160	36 40	-121 36
SNY	SIDNEY	NE	US	4110	-10298	41 06	-102 59
SPA	SPARTANBURG	SC	US	3503	-8193	35 02	-81 56
SPS	WICHITA_FALLS	TX	US	3399	-9859	33 59	-98 35
SQS	SIDON	MS	US	3346	-9028	33 28	-90 17
SRQ	SARASOTA	FL	US	2740	-8255	27 24	-82 33
SSM	SAULT_STE_MARIE	MI	US	4641	-8431	46 25	-84 19
SSO	SAN_SIMON	AZ	US	3227	-10926	32 16	-109 16
STL	ST_LOUIS	MO	US	3886	-9048	38 52	-90 29
SYR	SYRACUSE	NY	US	4316	-7620	43 10	-76 12
TBC	TUBA_CITY	AZ	US	3612	-11127	36 07	-111 16
TBE	TOBE	CO	US	3727	-10360	37 16	-103 36
TCC	TUCUMCARI	NM	US	3518	-10360	35 11	-103 36
TCS	TRUTH_OR_CONSEQUENCES	NM	US	3328	-10728	33 17	-107 17
TLH	TALLAHASSEE	FL	US	3056	-8437	30 34	-84 22
TOU	NEAH_BAY	WA	US	4830	-12463	48 18	-124 38
TRM	THERMAL	CA	US	3363	-11616	33 38	-116 10
TTH	TERRE_HAUTE	IN	US	3949	-8725	39 29	-87 15
TUL	TULSA	OK	US	3620	-9579	36 12	-95 47
TUS	TUCSON	AZ	US	3210	-11092	32 06	-110 55
TVC	TRAVERSE_CITY	MI	US	4467	-8555	44 40	-85 33
TWF	TWIN_FALLS	ID	US	4248	-11449	42 29	-114 29
TXK	TEXARKANA	AR	US	3351	-9407	33 31	-94 04
TXO	TEXICO	TX	US	3450	-10284	34 30	-102 50
UIN	QUINCY	IL	US	3985	-9128	39 51	-91 17
VRB	VERO_BEACH	FL	US	2768	-8049	27 41	-80 29
VUZ	VULCAN	AL	US	3367	-8690	33 40	-86 54
VXV	KNOXVILLE	TN	US	3590	-8389	35 54	-83 53
YDC	PRINCETON	BC	CN	4947	-12052	49 28	-120 31
YKM	YAKIMA	WA	US	4657	-12045	46 34	-120 27
YOW	OTTAWA	ON	CN	4532	-7567	45 19	-75 40
YQB	QUEBEC	QB	CN	4680	-7138	46 48	-71 23
YQL	LETHBRIDGE	AB	CN	4963	-11280	49 38	-112 48
YQT	THUNDER_BAY	ON	CN	4837	-8932	48 22	-89 19
YQV	YORKTON	SA	CN	5127	-10247	51 16	-102 28
YSC	SHERBROOKE	QB	CN	4543	-7168	45 26	-71 41
YSJ	ST_JOHN	NB	CN	4532	-6588	45 19	-65 53

```
YVV   WIARTON           ON CN  4475   -8110    44 45     -81 06
YWG   WINNIPEG          MB CN  4990   -9723    49 54     -97 14
YXC   CRANBROOK         BC CN  4960  -11578    49 36    -115 47
YXH   MEDICINE_HAT      AB CN  5002  -11072    50 01    -110 43
YYN   SWIFT_CURRENT     SA CN  5028  -10768    50 17    -107 41
YYZ   TORONTO           ON CN  4367   -7963    43 40     -79 38
```

16 APPENDIX G: WSR-88D WEATHER RADAR NETWORK

Figure G-1. WSR-88D Weather Radar Network Sites

Table G-1. WSR-88D Weather Radar Network

ICAO	NEXRAD SITENAME	CITY	COUNTY	STATE	AGENCY	ELEVATION
KABR	Aberdeen	Aberdeen	Brown	SD	NWS	396.85 m (1302.49 ft)
KABX	Albuquerque	Albuquerque	Bernalillo	NM	NWS	1789.18 m (5869.42 ft)
KAKQ	Norfolk	Wakefield	Sussex	VA	NWS	34.14 m (111.55 ft)
KAMA	Amarillo	Amarillo	Potter	TX	NWS	1093.32 m (3585.96 ft)
KAMX	Miami	Miami	Dade	FL	NWS	4.27 m (13.12 ft)
KAPX	Northcentral Lower Michigan	Gaylord	Alpena	MI	NWS	446.23 m (1463.25 ft)
KARX	La Crosse	La Crosse	La Crosse	WI	NWS	388.92 m (1276.25 ft)
KATX	Seattle	Everett	Island	WA	NWS	150.57 m (495.41 ft)
KBBX	Beale AFB	Oroville	Butte	CA	AFWA	52.73 m (173.88 ft)
KBGM	Binghamton	Binghamton	Broome	NY	NWS	489.51 m (1607.61 ft)
KBHX	Eureka (Bunker Hill)	Eureka	Humboldt	CA	NWS	732.13 m (2401.57 ft)
KBIS	Bismarck	Bismarck	Burleigh	ND	NWS	505.36 m (1656.82 ft)
KBLX	Billings	Billings	Yellowstone	MT	NWS	1096.67 m (3599.08 ft)
KBMX	Birmingham	Alabaster	Shelby	AL	NWS	196.6 m (646.33 ft)
KBOX	Boston	Taunton	Bristol	MA	NWS	35.97 m (118.11 ft)
KBRO	Brownsville	Brownsville	Cameron	TX	NWS	7.01m (22.97 ft)
KBUF	Buffalo	Buffalo	Erie	NY	NWS	211.23 m (692.26 ft)
KBYX	Key West	Boca Chica Key	Monroe	FL	NWS	2.44 m (6.56 ft)
KCAE	Columbia	West Columbia	Lexington	SC	NWS	70.41 m (229.66 ft)
KCBW	Caribou	Houlton	Aroostook	ME	NWS	227.38 m (744.75 ft)
KCBX	Boise	Boise	Ada	ID	NWS	932.99 m (3061.02 ft)

ICAO	NEXRAD SITENAME	CITY	COUNTY	STATE	AGENCY	ELEVATION
KCCX	State College	State College	Centre	PA	NWS	733.04 m (2404.86 ft)
KCLE	Cleveland	Cleveland	Cuyahoga	OH	NWS	232.56 m (764.44 ft)
KCLX	Charleston, SC	Grays	Beaufort	SC	NWS	29.57 m (98.43 ft)
KCRP	Corpus Christi	Corpus Christi	Nueces	TX	NWS	13.72 m (45.93 ft)
KCXX	Burlington	Colchester	Chittenden	VT	NWS	96.62 m (318.24 ft)
KCYS	Cheyenne	Cheyenne	Laramie	WY	NWS	1867.81 m (6128.61 ft)
KDAX	Sacramento	Davis	Yolo	CA	NWS	9.14 m (29.53 ft)
KDDC	Dodge City	Dodge City	Ford	KS	NWS	789.43 m (2588.58 ft)
KDFX	Laughlin AFB	Bracketville	Kinney	TX	AFWA	344.73 m (1131.89 ft)
KDGX	Jackson/ Brandon, MS	Brandon	Rankin	MS	NWS	150.92 m (495.41 ft)
KDIX	Philadelphia	Fort Dix	Burlington	NJ	NWS	45.42 m (147.64 ft)
KDLH	Duluth	Duluth	St Louis	MN	NWS	435.25 m (1427.17 ft)
KDMX	Des Moines	Johnston	Polk	IA	NWS	299.01 m (980.97 ft)
KDOX	Dover AFB	Ellendale State Forest	Sussex	DE	AFWA	15.24 m (49.21 ft)
KDTX	Detroit	White Lake	Oakland	MI	NWS	326.75 m (1072.83 ft)
KDVN	Quad Cities	Davenport	Scott	IA	NWS	229.82 m (754.59 ft)
KDYX	Dyess AFB	Moran	Shackelford	TX	AFWA	462.38 m (1515.75 ft)
KEAX	Pleasant Hill	Pleasant Hill	Cass	MO	NWS	303.28 m (994.09 ft)
KEMX	Tucson	Tucson	Pima	AZ	NWS	1586.48 m (5203.41 ft)
KENX	Albany	East Berne	Albany	NY	NWS	556.56 m (1827.43 ft)
KEOX	Ft Rucker	Echo	Dale	AL	AFWA	132.28 m (433.07 ft)
KEPZ	El Paso	Santa Teresa	Dona Ana	NM	NWS	1250.9 m (4104.33 ft)

ICAO	NEXRAD SITENAME	CITY	COUNTY	STATE	AGENCY	ELEVATION
KESX	Las Vegas	Las Vegas	Clark	NV	NWS	1483.46 m (4865.49 ft)
KEVX	Eglin AFB	Red Bay	Walton	FL	AFWA	42.67 m (141.08 ft)
KEWX	Austin/San Antonio	New Braunfels	Comal	TX	NWS	192.94 m (633.2 ft)
KEYX	Edwards AFB	Boron	San Bernadino	CA	AFWA	840.33 m (2755.91 ft)
KFCX	Roanoke	Roanoke	Floyd	VA	NWS	874.17 m (2867.45 ft)
KFDR	Altus AFB	Frederick	Tillman	OK	AFWA	386.18 m (1266.4 ft)
KFDX	Cannon AFB	Field	Curry	NM	AFWA	1417.32 m (4648.95 ft)
KFFC	Atlanta	Peachtree City	Fayette	GA	NWS	261.52 m (859.58 ft)
KFSD	Sioux Falls	Sioux Falls	Minnehaha	SD	NWS	435.86 m (1430.45 ft)
KFSX	Flagstaff	Flagstaff	Coconino	AZ	NWS	2260.7 m (7417.98 ft)
KFTG	Denver	Front Range	Arapahoe	CO	NWS	1675.49 m (5495.41 ft)
KFWS	Dallas/Ft Worth	Fort Worth	Tarrant	TX	NWS	208.18 m (682.41 ft)
KGGW	Glasgow	Glasgow	Valley	MT	NWS	693.72 m (2276.9 ft)
KGJX	Grand Junction	Grand Junction	Mesa	CO	NWS	3045.26 m (9990.16 ft)
KGLD	Goodland	Goodland	Sherman	KS	NWS	1112.82 m (3651.57 ft)
KGRB	Green Bay	Green Bay	Brown	WI	NWS	207.87 m (682.41 ft)
KGRK	Ft Hood	Granger	Bell	TX	AFWA	163.98 m (538.06 ft)
KGRR	Grand Rapids	Grand Rapids	Kent	MI	NWS	237.13 m (777.56 ft)
KGSP	Greer	Greer	Spartanburg	SC	NWS	286.51 m (941.6 ft)
KGWX	Columbus AFB	Greenwood Springs	Monroe	MS	AFWA	145.08 m (475.72 ft)
KGYX	Portland, Me	Gray	Cumberland	ME	NWS	124.66 m (410.1 ft)
KHDX	Holloman AFB	Ruidoso	Dona Ana	NM	AFWA	1286.87 m (4222.44 ft)
KHGX	Houston	Dickinson	Galveston	TX	NWS	5.49 m (16.4 ft)

ICAO	NEXRAD SITENAME	CITY	COUNTY	STATE	AGENCY	ELEVATION
KHNX	San Joaquin Valley	Hanford	Kings	CA	NWS	74.07 m (242.78 ft)
KHPX	Ft Campbell	Trenton	Todd	KY	AFWA	175.56 m (577.43 ft)
KHTX	Northeast Alabama	Hytop	Jackson	AL	NWS	537.06 m (1761.81 ft)
KICT	Wichita	Wichita	Sedgwick	KS	NWS	406.91 m (1335.3 ft)
KICX	Cedar City	Cedar City	Iron	UT	NWS	3230.88 m (10600.39 ft)
KILN	Cincinnati	Wilmington	Clinton	OH	NWS	321.87 m (1056.43 ft)
KILX	Lincoln	Lincoln	Logan	IL	NWS	177.39 m (580.71 ft)
KIND	Indianapolis	Indianapolis	Marion	IN	NWS	240.79 m (790.68 ft)
KINX	Tulsa	Inola	Rogers	OK	NWS	203.61 m (669.29 ft)
KIWA	Phoenix	Phoenix	Maricopa	AZ	NWS	412.39 m (1351.71 ft)
KIWX	Northern Indiana	North Webster	Kosciusko	IN	NWS	292.3 m (958.01 ft)
KJAX	Jacksonville	Jacksonville	Duval	FL	NWS	10.06 m (32.81 ft)
KJGX	Robins AFB	Jefferson-ville	Twiggs	GA	AFWA	158.8 m (521.65 ft)
KJKL	Jackson, KY	Jackson	Breathitt	KY	NWS	415.75 m (1364.83 ft)
KLBB	Lubbock	Lubbock	Lubbock	TX	NWS	993.34 m (3257.87 ft)
KLCH	Lake Charles	Lake Charles	Calcasieu	LA	NWS	3.96 m (13.12 ft)
KLIX	Slidell	Slidell	St Tammany	LA	NWS	7.32 m (22.97 ft)
KLNX	North Platte	North Platte	Logan	NE	NWS	905.26 m (2969.16 ft)
KLOT	Chicago	Romeoville	Will	IL	NWS	202.08 m (662.73 ft)
KLRX	Elko	Elko	Lander	NV	NWS	2055.57 m (6745.41 ft)
KLSX	St Louis	Weldon Spring	St Charles	MO	NWS	185.32 m (606.96 ft)
KLTX	Wilmington	Shallotte	Brunswick	NC	NWS	19.51 m (65.62 ft)
KLVX	Louisville	Fort Knox	Hardin	KY	NWS	219.15 m (718.5 ft)

ICAO	NEXRAD SITENAME	CITY	COUNTY	STATE	AGENCY	ELEVATION
KLWX	Sterling	Sterling	Loudoun	VA	NWS	82.91 m (272.31 ft)
KLZK	Little Rock	North Little Rock	Pulaski	AR	NWS	173.13 m (567.59 ft)
KMAF	Midland/ Odessa	Midland	Midland	TX	NWS	874.17 m (2867.45 ft)
KMAX	Medford	Medford	Jackson	OR	NWS	2289.96 m (7513.12 ft)
KMBX	Minot AFB	Deering	Mchenry	ND	AFWA	455.07 m (1492.78 ft)
KMHX	Morehead City	Newport	Carteret	NC	NWS	9.45 m (29.53 ft)
KMKX	Milwaukee	Dousman	Waukesha	WI	NWS	292 m (958.01 ft)
KMLB	Melbourne	Melbourne	Brevard	FL	NWS	10.67 m (36.09 ft)
KMOB	Mobile	Mobile	Mobile	AL	NWS	63.4 m (206.69 ft)
KMPX	Minneapolis	Chanhassen	Carver	MN	NWS	288.34 m (944.88 ft)
KMQT	Marquette	Negaunee	Marquette	MI	NWS	430.07 m (1410.76 ft)
KMRX	Knoxville	Morristown	Hamblen	TN	NWS	407.52 m (1338.58 ft)
KMSX	Missoula	Missoula	Missoula	MT	NWS	2394.2 m (7854.33 ft)
KMTX	Salt Lake City	Salt Lake City	Salt Lake	UT	NWS	1969.01 m (6459.97 ft)
KMUX	San Francisco	Los Gatos	Santa Clara	CA	NWS	1057.35 m (3467.85 ft)
KMVX	Fargo/Grand Forks	Grand Forks	Traill	ND	NWS	300.53 m (987.53 ft)
KMXX	Maxwell AFB	Carrville	Tallapoosa	AL	AFWA	121.92 m (400.26 ft)
KNKX	San Diego	San Diego	San Diego	CA	NWS	291.08 m (954.72 ft)
KNQA	Memphis	Millington	Shelby	TN	NWS	85.95 m (282.15 ft)
KOAX	Omaha	Valley	Douglas	NE	NWS	349.91 m (1148.29 ft)
KOHX	Nashville	Old Hickory	Wilson	TN	NWS	176.48 m (577.43 ft)
KOKX	Brookhaven	Upton	Suffolk	NY	NWS	25.91 m (85.3 ft)
KOTX	Spokane	Spokane	Spokane	WA	NWS	726.64 m (2385.17 ft)

ICAO	NEXRAD SITENAME	CITY	COUNTY	STATE	AGENCY	ELEVATION
KPAH	Paducah	Paducah	Mccracken	KY	NWS	119.48 m (390.42 ft)
KPBZ	Pittsburgh	Coraopolis	Allegheny	PA	NWS	361.19 m (1184.38 ft)
KPDT	Pendleton	Pendleton	Umatilla	OR	NWS	461.77 m (1515.75 ft)
KPOE	Ft Polk	Ft Polk	Vernon	LA	AFWA	124.36 m (406.82 ft)
KPUX	Pueblo	Pueblo	Pueblo	CO	NWS	1599.9 m (5249.34 ft)
KRAX	Raleigh/ Durham	Clayton	Wake	NC	NWS	106.07 m (347.77 ft)
KRGX	Reno	Nixon	Washoe	NV	NWS	2529.54 m (8300.52 ft)
KRIW	Riverton/ Lander	Riverton	Fremont	WY	NWS	1697.13 m (5567.59 ft)
KRLX	Charleston, WV	Charleston	Kanawha	WV	NWS	329.18 m (1079.4 ft)
KRTX	Portland, OR	Portland	Washington	OR	NWS	479.15 m (1571.52 ft)
KSFX	Pocatello	Springfield	Bingham	ID	NWS	1363.68 m (4475.07 ft)
KSGF	Springfield	Springfield	Greene	MO	NWS	389.53 m (1279.53 ft)
KSHV	Shreveport	Shreveport	Caddo	LA	NWS	83.21 m (272.31 ft)
KSJT	San Angelo	San Angelo	Tom Green	TX	NWS	576.07 m (1889.76 ft)
KSOX	Santa Ana Mountains	Santa Ana Mountains	Orange	CA	NWS	927 m (3041.34 ft)
KSRX	Western Arkansas	Chaffee Ridge	Sebastian	AR	NWS	195.07 m (639.76 ft)
KTBW	Tampa	Ruskin	Hillsborough	FL	NWS	12.5 m (39.37 ft)
KTFX	Great Falls	Great Falls	Cascade	MT	NWS	1132.03 m (3713.91 ft)
KTLH	Tallahassee	Tallahassee	Leon	FL	NWS	19.2 m (62.34 ft)
KTLX	Norman	Midwest City	Oklahoma	OK	NWS	369.72 m (1213.91 ft)
KTWX	Topeka	Topeka	Wabaunsee	KS	NWS	416.66 m (1368.11 ft)
KTYX	Ft Drum	Montague	Lewis	NY	AFWA	562.66 m (1847.11 ft)
KUDX	Rapid City	New Underwood	Pennington	SD	NWS	919.28 m (3015.09 ft)

ICAO	NEXRAD SITENAME	CITY	COUNTY	STATE	AGENCY	ELEVATION
KUEX	Grand Island	Blue Hill	Webster	NE	NWS	602.28 m (1975.07 ft)
KVAX	Moody AFB	South Stockton	Lanier	GA	AFWA	54.25 m (177.17 ft)
KVBX	Vandenberg AFB	Orcutt	Santa Barbara	CA	AFWA	372.77 m (1223.75 ft)
KVNX	Vance AFB	Cherokee	Alfalfa	OK	AFWA	368.81 m (1210.63 ft)
KVTX	Los Angeles	Los Angeles	Ventura	CA	NWS	830.88 m (2726.38 ft)
KVWX	Evansville, IN (Non-NEXRAD)	Owensville	Gibson	IN	NWS	155.75 m (511.81 ft)
KYUX	Yuma	Yuma	Pima	AZ	NWS	53.04 m (173.88 ft)
LPLA	Lajes AB	Santa Barbara	N/A	AZO RES	AFWA	1016.2 m (3333.33 ft)
PABC	Bethel FAA	Bethel	N/A	AK	FAA	49.07 m (160.76 ft)
PACG	Sitka FAA	Biorka Island	N/A	AK	FAA	63.09 m (206.69 ft)
PAEC	Nome FAA	Nome	N/A	AK	FAA	17.68 m (59.06 ft)
PAHG	Anchorage FAA	Kenai	N/A	AK	FAA	73.76 m (242.78 ft)
PAIH	Middleton Island	Middleton Island	N/A	AK	FAA	20.42 m (65.62 ft)
PAKC	King Salmon FAA	King Salmon	N/A	AK	FAA	19.2 m (62.34 ft)
PAPD	Fairbanks FAA	Fairbanks	N/A	AK	FAA	790.35 m (2591.86 ft)
PGUA	Andersen AFB	Andersen AFB	N/A	GUAM	AFWA	80.47 m (262.47 ft)
PHKI	South Kauai FAA	South Kauai	Kauai	HI	FAA	54.56 m (180.45 ft)
PHKM	Kamuela/Kohala Apt	Kamuela	Hawaii	HI	FAA	1161.9 m (3812.34 ft)
PHMO	Molokai FAA	Molokai	Molokai	HI	FAA	415.44 m (1361.55 ft)
PHWA	South Shore FAA	Naalehu	Hawaii	HI	FAA	420.62 m (1381.23 ft)
RKJK	Kunsan AB	Kunsan Ab	N/A	KOREA	AFWA	23.77 m (78.74 ft)
RKSG	Camp Humphreys	Camp Humphreys	N/A	KOREA	AFWA	15.85 m (52.49 ft)

ICAO	NEXRAD SITENAME	CITY	COUNTY	STATE	AGENCY	ELEVATION
RODN	Kadena AB	Kadena Ab	N/A	JAPAN	AFWA	66.45 m (216.54 ft)
TJUA	San Juan FAA	San Juan	N/A	PR	FAA	851.61 m (2795.28 ft)

17 APPENDIX H: AWC Geographical Area Designator Map

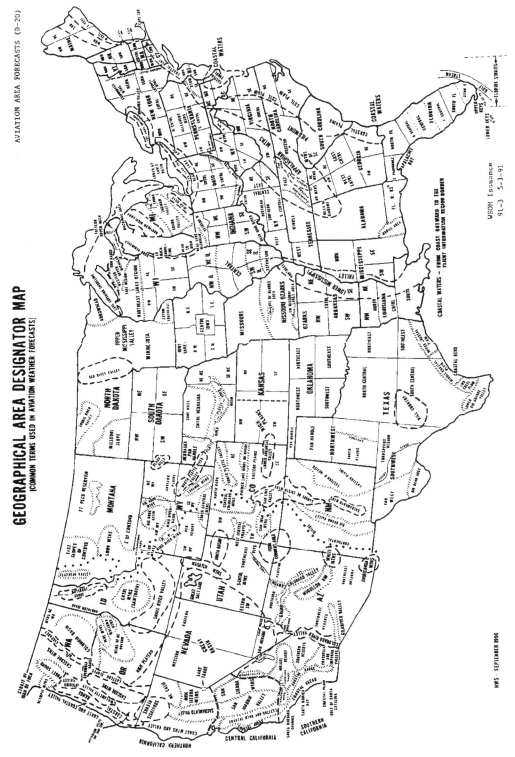

Figure H-1: AWC Geographical Area Designator Map

18 APPENDIX I: Present Weather Symbols

	0	1	2	3	4	5	6	7	8	9
00	Cloud development not observed or not observable during past hour	Clouds generally dissolving or becoming less developed during past hour	State of sky on the whole unchanged during past hour	Clouds generally forming or developing during past hour	Visibility reduced by smoke	Haze	Widespread dust in suspension in the air, not raised by wind, at time of obs	Dust or sand raised by wind, at time of obs	Well developed dust devil(s) within past hour	Duststorm or sandstorm within sight of station or at station during past hour
10	Light fog	Patches of shallow fog at station not deeper than 6 feet on land	More or less continuous shallow fog at station not deeper than 6 feet on land	Lightning visible, no thunder heard	Precipitation within sight, but not reaching the ground	Precipitation within sight, reaching ground, but distant from station	Precipitation within sight, reaching the ground, near to but not at station	Thunder heard but no precipitation at the station	Squall(s) within sight during past hour	Funnel cloud(s) within sight during past hour
20	Drizzle (not freezing, not showers) during past hour, not at time of obs	Rain (not freezing, not showers) during past hour, not at time of obs	Snow (not falling as showers) during past hour, not at time of obs	Rain and snow (not falling as showers) during past hour, not at time of obs	Freezing drizzle or rain (not showers) during past hour, not at time of obs	Showers of rain during past hour, but not at time of obs	Showers of snow, or of rain and snow during past hour, but not at time of obs	Showers of hail, or of hail and rain during past hour, but not at time of obs	Fog during past hour, but not at time of obs	Thunderstorm (with or without precip) during past hour, but not at time of obs
30	Slight or moderate duststorm or sandstorm, has decreased during past hour	Slight or moderate duststorm or sandstorm, no appreciable change during past hour	Slight or moderate duststorm or sandstorm, has increased during past hour	Severe duststorm or sandstorm, has decreased during past hour	Severe duststorm or sandstorm, no appreciable change during past hour	Severe duststorm or sandstorm, has increased during past hour	Slight or moderate drifting snow, generally low	Heavy drifting snow, generally low	Slight or moderate drifting snow, generally high	Heavy drifting snow, generally high
40	Fog at distance at time of obs but not at station during past hour	Fog in patches	Fog, sky discernable, has become thinner during past hour	Fog, sky not discernable, has become thinner during past hour	Fog, sky discernable, no appreciable change during past hour	Fog, sky not discernable, no appreciable change during past hour	Fog, sky discernable, has begun or become thicker during past hour	Fog, sky not discernable, has begun or become thicker during past hour	Fog, depositing rime, sky discernable	Fog, depositing rime, sky not discernable
50	Intermittent drizzle (not freezing), slight at time of obs	Continuous drizzle (not freezing), slight at time of obs	Intermittent drizzle (not freezing), moderate at time of obs	Continuous drizzle (not freezing), moderate at time of obs	Intermittent drizzle (not freezing), thick at time of obs	Continuous drizzle (not freezing), thick at time of obs	Slight freezing drizzle	Moderate or thick freezing drizzle	Drizzle and rain, slight	Drizzle and rain, moderate or heavy
60	Intermittent rain (not freezing), slight at time of obs	Continuous rain (not freezing), slight at time of obs	Intermittent rain (not freezing), moderate at time of obs	Continuous rain (not freezing), moderate at time of obs	Intermittent rain (not freezing), heavy at time of obs	Continuous rain (not freezing), heavy at time of obs	Slight freezing rain	Moderate or heavy freezing rain	Rain or drizzle and snow, slight	Rain or drizzle and snow, moderate or heavy
70	Intermittent fall of snowflakes, slight at time of obs	Continuous fall of snowflakes, slight at time of obs	Intermittent fall of snowflakes, moderate at time of obs	Continuous fall of snowflakes, moderate at time of obs	Intermittent fall of snowflakes, heavy at time of obs	Continuous fall of snowflakes, heavy at time of obs	Ice needles (with or without fog)	Granular snow (with or without fog)	Isolated starlike snow crystals (with or without fog)	Ice pellets (sleet, U.S. definition)
80	Slight rain shower(s)	Moderate or heavy rain shower(s)	Violent rain shower(s)	Slight shower(s) of rain and snow mixed	Moderate or heavy shower(s) of rain and snow mixed	Slight snow shower(s)	Moderate or heavy snow shower(s)	Slight shower(s) of soft or small hail, with or without rain and/or snow	Moderate or heavy shower(s) of soft or small hail, with or without rain and/or snow	Slight shower(s) of hail, with or without rain and/or snow, not assoc with thunder
90	Moderate of heavy shower(s) of hail and/or rain/snow, not associated with thunder	Slight rain at time of obs; thunderstorm during past hour not at time of obs	Moderate or heavy rain at time of obs; TS during past hour not at time of obs	Slight snow and/or rain/hail at time of obs; TS during past hour not at time of obs	Moderate or heavy snow and/or rain/hail at time of obs; TS past hour not at obs time	Slight or moderate thunderstorm without hail but with rain and/or snow at obs time	Slight or moderate thunderstorm with hail at time of obs	Heavy thunderstorm without hail but with rain and/or snow at time of obs	Thunderstorm combined with duststorm or sandstorm at time of obs	Heavy thunderstorm with hail at time of obs

Figure I-1. Present Weather Symbols with Text Explanation

Matching of METAR present weather text to symbol in table below is not necessarily endorsed by the National Weather Service or the World Meteorological Organization. Blue numbers in upper-left corner of white boxes indicate the priority for plotting in event more than one symbol is possible (symbols in gray boxes have no corresponding METAR present weather text). Graphical representation of METARs using this table found at http://adds.aviationweather.gov

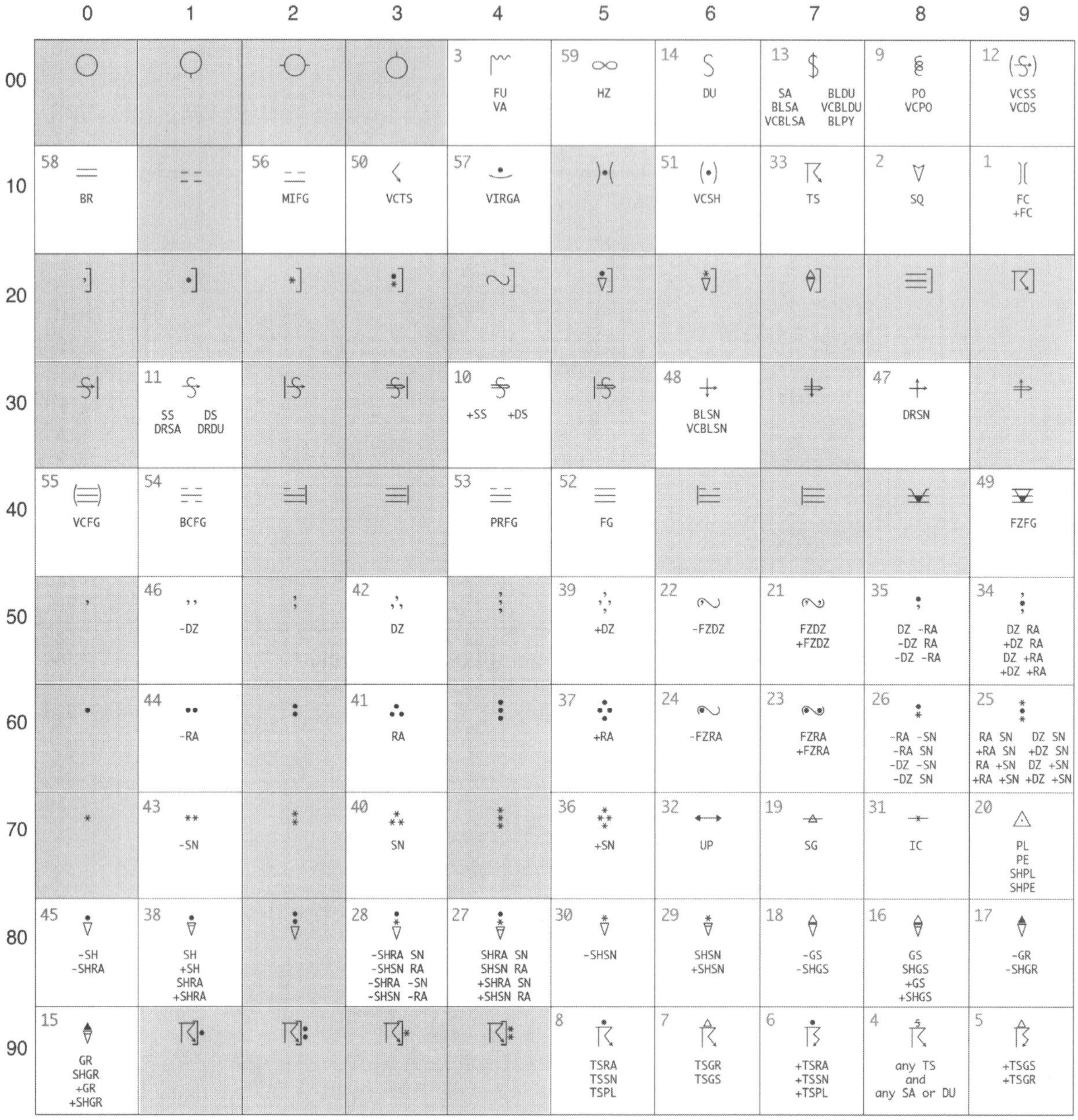

Figure I-2. Present Weather Symbols with Corresponding METAR/SPECI Present Weather Code

19 APPENDIX J: Turbulence and Icing Intensity Depictions

Table J-1. Turbulence Intensity

Intensity	Aircraft Reaction	Symbol
Light	Loose objects in aircraft remain at rest.	∧
Moderate	Unsecured objects are dislodged. Occupants feel definite strains against seat belts and shoulder straps.	⋀
Severe	Occupants thrown violently against seat belts. Momentary loss of aircraft control. Unsecured objects tossed about.	⋀
Extreme	Aircraft is tossed violently about, impossible to control. May cause structural damage.	⋀

Table J-2. Icing Intensity

Intensity	Aircraft Reaction	Symbol
Trace	Ice becomes perceptible. Rate of accumulation slightly greater than sublimation. Deicing/anti-icing equipment is not used unless encountered for an extended period of time (over 1 hour).	∪
Light	The rate of accumulation may create a problem if flight is prolonged in this environment (over 1 hour). Occasional use of deicing/anti-icing equipment removes or prevents accumulation. It does not present a problem if this equipment is used.	ⱡ
Moderate	The rate of accumulation is such that even short encounters become potentially hazardous, and use of deicing/anti-icing equipment or diversion is necessary.	Ѱ
Severe	The rate of accumulation is such that deicing/anti-icing equipment fails to reduce or control the hazard. Immediate diversion is necessary.	Ѱ